Chemistry

for

University Students

Volume 3

Said A. H. Vuai B.Sc.,M.Sc., Ph.D.
Professor
Department of Chemistry,
School of Physical Sciences,
College of Natural &Mathematical Sciences,
The University of Dodoma.

and

H. C. Ananda Murthy B.Sc., M.Sc., M.Phil., Ph.D.
Assistant Professor
Department of Chemistry,
School of Physical Sciences,
College of Natural &Mathematical Sciences,
The University of Dodoma.

TANZANIA EDUCATIONAL PUBLISHERS LTD

Tanzania Educational Publishers Ltd,
TEPU House,
Uganda road, Plot No. 45, Block MDA,
Mob:0685 997583/0758 147871
Email: tepultd@yahoo.com
Website: www.tepu.co.tz
P.O. Box 1222,
Bukoba, Tanzania.

ISBN 978 9987 07 055 8

Dedication

We dedicate this book to our families

CONTENTS

About the Authors

Said A. H. Vuai was born in Pemba, Zanzibar, Tanzania and grew up in Pemba. He received BSc. Degree in Chemical and Process Engineering from University of Dar es Salaam, Tanzania and his MSc. and PhD in Marine and Environmental Sciences majoring in Chemistry from University of the Ryukyus, Okinawa, Japan. After his PhD studies, he joined department of Sciences, of the State University of Zanzibar, where he taught Chemistry from 2004 to 2008. He then moved to the University of Dodoma in the Department of Physical Science where he continued to teach chemistry.In 2013-2014, he was awarded post Doctoral Fellowship to undertake Environmental Sustainability Assessment at the University of Siena, Italy.

Professor Vuai taught wide range of Chemistry courses, including Inorganic chemistry, organometalic chemistry, Analytical Chemistry, Environmental Chemistry and Material Sciences. He also served as external examiner for graduate students from Makerere University, Uganday and Addis Ababa, University, Ethiopia.Professor Vuai has served number of administrative position at university of Dodoma, including Head of Department of Physical Sciences for three years, and Dean, School of Mathematical Sciences for about 5 years. He is enjoying wide collaborations in local regional and international institutions and researchers in Africa, Asia Europe and America. His research interest includes, Kinetic of chemical reactions, organometalic, Environmentalpollution of nutrients and trace metals and Public Health.

H. C. Ananda Murthy has been a sincere, committed and dedicated faculty member at various prestigious universities in India and Tanzania for the last 19 years. He is currently working as Assistant Professor, Department of Chemistry, The University of Dodoma, Tanzania, East Africa.

He completed his B.Sc. from Bangalore University, India in 1995. Later he completed his master's degree in Chemistry from Bangalore University, India in 1997. He also obtained M.Phil. degree in chemistry from Bharathidasan University, India in the year 2006. Then he went on to complete his Ph.D., for his work on corrosion behavior of aluminium composite materials from Visvesvaraya Technological University, Belgaum, India in the year 2012.

He has authored few books, compendia and published several research articles in the journals of international repute. He has taught General chemistry, Coordination chemistry and Organometallic Chemistry, Transition Elements Chemistry, Organometallic Chemistryand Electrochemistry & Corrosion protection, Advanced Coordination Chemistry, Surface Chemistry and Corrosion Science, Environmental Chemistry, Engineering chemistry and Advanced spectroscopyto the university students in India and Tanzania.

His research interest mainly includes Corrosion Science, Composite materials and Materials chemistry. He also presented many papers in national as well as international conferences. He has successfully completed a project titled "Amphibious bicycle" sanctioned by National Innovation Foundation of India (NIFI).

Acknowledgements

We express our profound gratitude and regards to the management of the University of Dodoma for providing us this opportunity to serve at CNMS, UDOM.

It is a great pleasure to express our gratitude to the Principal, CNMS for their constant support and encouragement in bringing out this book. We also place on record our appreciation to fellow colleagues in Department of Chemistry at CNMS, the Dean, School of Physical Sciences and UDOM for their cooperation.

INTRODUCTION

Chemistry for University Students is one book published in three volumes made up of chapters as follows:

Volume 1

Chapter 1: Electronic Structure of Atoms

Chapter 2: Periodic Table

Chapter 3: Chemical Bonding

Chapter 4: Transition Elements Chemistry

Chapter 5: Inner Transition Elements Chemistry

Volume 2

Chapter 6: Coordination Chemistry

Chapter 7: Electronic Spectra of Coordination Compounds

Chapter 8: Organometalic Chemistry

Volume 3

Chapter 9: Electrochemistry

Chapter 10: Environmental Chemistry

Chemistry has great contribution to the world. Chemistry provides thousands of raw materials to get finished products that help keep you safe, warm, cool, on time, in motion, and connected. From heart monitors to satellites to cell phones to microwave ovens to synthetic fibers, packed food items to medicine chemistry makes the things that make modern life possible.

Today's world would be inconceivable without the developments that chemistry and its applications have brought to science, engineering and technology. The achievements of chemistry are omnipresent. Students at universities definitely require chemical knowledge for their better understanding and academic performance. It is understood that a chemistry book comprising many aspects of chemistry is greatly needed by the students of the universities.

This book has been written keeping in mind the requirements of chemistry students at University level comprising of General chemistry, Periodic Table of elements, *d* and *f* block elements, Coordination chemistry, organometallic chemistry, Electrochemistry and Environmental Chemistry (All volumes).

We wish our book will be helpful to the students, as it is prepared to cater to the needs of the students, keeping in mind the relevance of each contents of the U G courses toward better understanding of the subject and with illustration including suitable figures, tables, and examples of structures and configurations of the complexes with simple presentation formats.

Chapter 1

ELECROCHEMISTRY

1.1 INTRODUCTION

All chemical interactions are electrical at the atomic level so that in a sense all chemistry is electrochemistry. Electrochemistry provides an important way of utilizing the free energy available in spontaneous chemical reactions to perform useful work. Electrical work can be derived at the cost of chemical energy. Electrical energy can be applied to bring about the desired chemical change. The interconversion of chemical energy into electrical energy and vice-versa is extensively studied in a separate branch of chemistry known as electrochemistry. Electrochemistry deals with the conversion of chemical energy into electrical energy and vice-versa. An electrochemical cell converts chemical energy into electrical energy. Electrical energy is utilized to bring about chemical energy changes in an electrolytic cell.

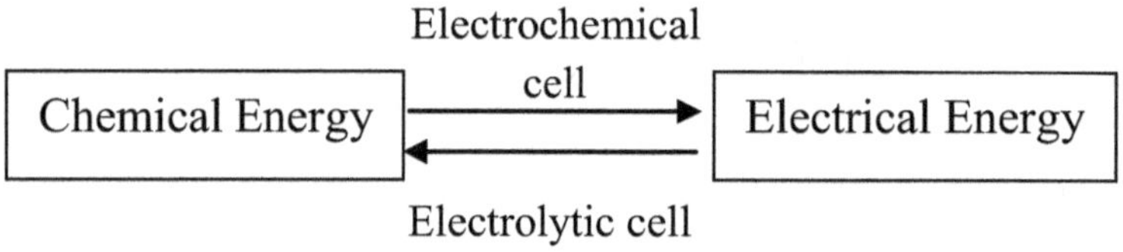

There are many applications of electrochemistry in every-day life. Batteries, fuel cells, super capacitors, control of corrosion, conducting polymers, metallurgy, metal finishing and electroplating are just a few examples of the applications of electrochemistry.

Electrochemistry is a broad subject of covering all areas of charge transfer reactions at interfaces and impacts on daily life in many immediate ways from corrosion and energy storage, through water treatment and purification, clean energy generation and environmental monitoring to medical diagnostics and electrodeposition.

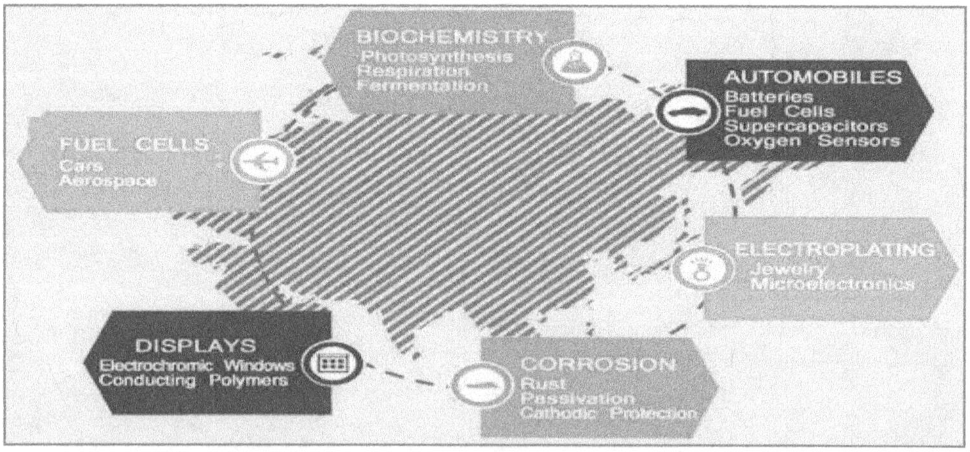

Electrochemistry- Timeline of Events

1800	1791	Galvani discovers a direct link between electricity and muscle contraction
	1800	Volta discovers that contact between dissimilar metals causes a potential difference
	1834	Faraday finds link between the weight of a substance formed and the quantity of electricity passed
	1834	Grove produces electricity by the reaction of hydrogen and oxygen in an electrochemical cell.
	1836	Daniel defines the modern galvanic cell
1850	1867	Leclanche define the modern dry cell
	1891	Stoney proposes that electricity is atomic in nature, names the fundamental unit of charge the "electron"
	1894	Ostwald experimented with catalytic reactions and is known for his contributions in electrochemistry
1900	1905	Tafel discovers a linear relationship between potential and logarithm of current density
	1910	Nernst relates a linear relationship between potential, reactant concentration and temperature with free energy
	1923	Debye-Huckel unite statistical mechanics with dilute ionic solutions
	1930	Frumkin connects current with double layer structure
	1940	Buttler-Volmer relate current density and deposition potential by using transition state theory, confirm Tafel relationship
1950 – 2010	1950-60	The rise of modern electrochemisrty
	1991	Commercial introduction of lithium battery
	2005	Wireless energy transfer- ability to provide electrical energy to remote objects without wires
	2007	Invention of solid state micrometer scale electric double layer capacitors based on advanced superionic conductors
	2007	The nanowire battery, a lithium-ion battery, was invented by a team lead by Dr Yi Cui.

1.2 ELETROLYTES AND NON ELECTROLYTES

Water soluble chemical compounds can be classified into following categories.

1. Electrolytes
2. Non-electrolytes

1. Electrolytes:

Electrolytes are the substances which dissociate into ions in aqueous solution and conduct electric current. Ex. NaCl, KCl etc.

Electrolytes can further be classified into two types:

a) Strong electrolytes

Strong electrolytes are the substances which undergo complete ionization into ions in aqueous solution. Ionisation takes place to an extent of 100%.

Example: NaCl, KCl, HCl, NaOH etc.

$$NaCl \xrightarrow{H_2O} Na^+_{(aq)} + Cl^-_{(aq)}$$

$$KCl \xrightarrow{H_2O} K^+_{(aq)} + Cl^-_{(aq)}$$

$$HCl \xrightarrow{H_2O} H^+_{(aq)} + Cl^-_{(aq)}$$

$$NaOH \xrightarrow{H_2O} Na^+_{(aq)} + OH^-_{(aq)}$$

b) Weak electrolytes

Weak electrolytes are the substances which undergo incomplete [partial] ionization in aqueous solution.

Example: CH_3COOH, NH_4OH etc.

$$CH_3COOH \rightleftharpoons CH_3COO^- + H^+$$

$$NH_4OH \rightleftharpoons NH_4^+ + OH^-$$

Always there exists an equilibrium between the undissociated molecule and dissociated ions of molecule in aqueous solution. Chemical equilibrium is a characteristic of reversible reaction.

2. Non-electrolytes

Non-electrolytes are the substances which dissolve in water but do not dissociate in solution and do not conduct electric current.

Examples: Sugar, urea, glucose, glycerin etc.

Solid electrolytes

- Gel electrolytes – they closely resemble liquid electrolytes. They are essentially liquids in a flexible lattice framework. Various additives are often applied to increase the conductivity of such systems.
- Dry polymer electrolytes – they differ from liquid and gel electrolytes in the sense, that salt is dissolved directly into the solid medium. Usually it's a relatively high dielectric constant polymer

and a salt with low lattice E. In order to increase the mechanical strength and conductivity of such electrolytes, very often composites are used, namely inert ceramic phase is introduced. There 2 major classes of such electrolytes: polymer-in-ceramic, and ceramic-in-polymer.

- Solid ceramic electrolytes (complex nonstoichiometric oxides: i.e. V, Mo and W bronzes of alkali metals A_xMeO_3 – ions migrate through the ceramic phase by means of vacancies and/or interstitials within the lattice. There are also glassy-ceramic electrolytes.

Electrolysis

Electrolysis is the process of decomposition of an electrolyte due to the flow of an electric current through the solution of an electrolyte.

For example, when an electric current is passed through the solution of sodium chloride, sodium chloride decomposes liberating chlorine at the anode and sodium at the cathode (Fig. 92).

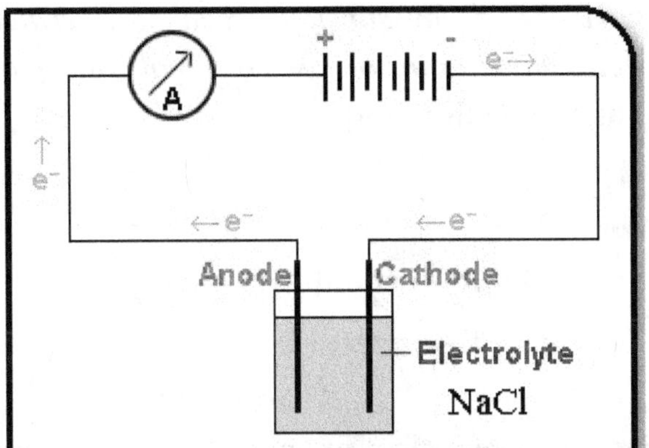

Fig. 92. Electrolytic cell

Chemical reactions that occur during electrolysis are given below:

1. $$NaCl \xrightarrow{H_2O} Na^+_{(aq)} + Cl^-_{(aq)}$$

2. At cathode [Reduction]

$$Na^+ + e \longrightarrow Na$$

3. At anode [oxidation]

$$Cl^- \longrightarrow \frac{1}{2}Cl_2 + e$$

Faraday's Laws of Electrolysis

Michael Faraday (1832) discovered the relationship between the quantity of electricity passing through an electrolyte and the amount of material liberated at the electrodes (Fig. 93).

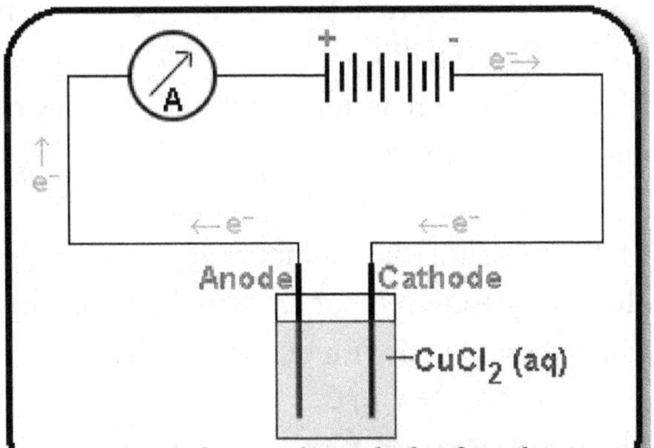

Fig. 93. electrolysis of $CuCl_2$

Faraday's 1st law of electrolysis can be stated as follows:

"The amount of substance liberated or deposited at a particular electrode during electrolysis is directly proportional to the quantity of electricity passed in the solution"

If 'W' g of a substance get deposited at a particular electrode on passing 'Q' coloumbs of electricity , then according to the first law of electrolysis.

$W \propto Q$ but $Q = I \times t$ charge = current x time]

$W \propto I \times t$

$W = Z \times I \times t$

Where Z is the electrochemical equivalent of the substance

If $I = 1$ ampere $t = 1$ second, then $W = Z$

∴ Electrochemical equivalent may be defined as the amount of substance deposited by a current of one ampere passed for one second:

1. One mole of electrons (6.23×10^{23}) carries a charge of 96500 coulomb.

2. One coulomb of current can deposit one electrochemical equivalent weight of the electrolyte. One equivalent of any substance can be liberated at a particular electrode only when the corresponding species either gains or loses one mole of electrons.

At anode [Oxidation]

$$B^-_{(aq)} \longrightarrow B + e^-$$

1 equivalent 1 equivalent 1 mole of electrons

At cathode [Reduction]

$$A^+_{(aq)} + e^- \longrightarrow A$$

1 equivalent 1 equivalent

Since one equivalent of any substance requires one Faraday of electricity to get liberated at a particular electrodes, one Faraday must be the charge carried by one mole of electrons.

1 Faraday = charge carried by one mole of electrons

= charge carried by an electron x Avagadro number

= $1.602 \times 10^{-19} \times 6.023 \times 10^{23}$ C mol^{-1}

1 Faraday = 96488.5 C mol^{-1}

If an electrode reaction involves 'n' moles of electrons, the total quantity of electricity (Q) involved in the reaction is given by

$$Q = nF$$

Thus knowing the number of moles of electrons involved in an electrode reaction, the quantity of electricity required for the reaction can be determined. The mass of substance produced in the reaction can be calculated with the help of Faraday's first law of electrolysis.

Faraday's Second Law of Electrolysis

It can be stated as follows:

"When the same quantity of electricity is passed through the solutions of different electrolytes, the masses of the substances liberated or deposited at the electrodes are directly proportional to their equivalent masses".

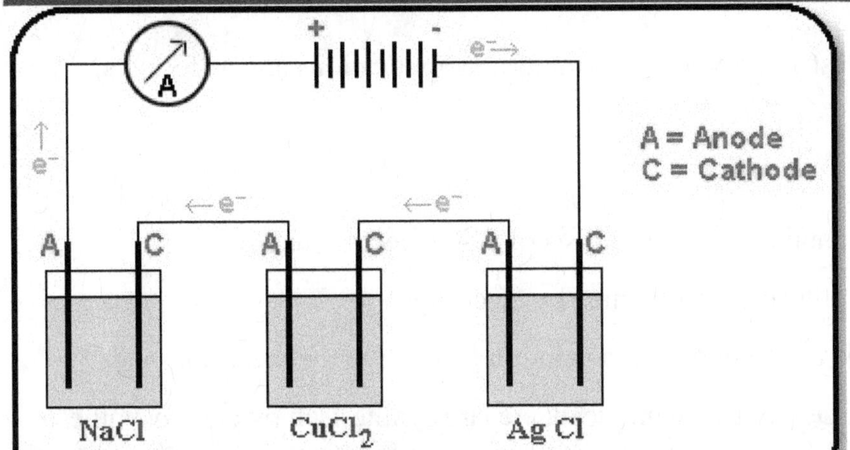

Fig. 94. Electrolysis of chlorides of sodium, copper and silver

On passing electric current, electrolysis takes place in these three cells (Fig. 94) and deposition of Cu and Ag occur.

If the same current is passed for the same time, it is found that

$$\frac{\text{mass of copper deposited}}{\text{mass of silver deposited}} = \frac{\text{Equivalent mass of copper}}{\text{Equivalent mass of silver}}$$

If W_1 and W_2 are the masses of the substance liberated by passing the same quantity of electricity and E_1 and E_2 are their equivalent masses respectively, then according to Faraday's second law of electrolysis

$$\frac{W_1}{W_2} = \frac{E_1}{E_2} \quad ...(1)$$

If Z_1 and Z_2 are the electrochemical equivalents of the two substances and I is the current passed for 't' seconds. We have

$W_1 = Z_1It$ and $W_2 = Z_2It$

Therefore equation (1) can be written as

$$\frac{Z_1 It}{Z_2 It} = \frac{E_1}{E_2}$$

$$\frac{Z_1}{Z_2} = \frac{E_1}{E_2} \quad \text{or} \quad E_1 \propto Z_1$$

$$E_2 \propto Z_2 \quad \text{or} \quad E = F \times Z$$

F → Faraday's constant = 96500 coulombs Electrochemical equivalent $Z = \frac{E}{F}$

$$\therefore \frac{\text{Equivalent mass}}{\text{Faraday's constant}} = \frac{\text{Equivalent mass}}{96500} \text{ g / C}$$

Electrochemical equivalent is expressed in the units of g C^{-1}

Note : *1. Electron reduces and deposits 1M^+ ion at an electrode*

2. n mole of electrons reduce 1 mole of M^{n+} ions

3. 1 mole of electrons reduces1/n mole of M^{n+} ions

Applications of Electrolysis

Electrolysis has wide applications in industries. Some of the important applications are, as follows,

(i) Production of hydrogen by electrolysis of water.

(ii) Manufacture of heavy water (D_2O).

(iii) The metals like K, Mg, Al, etc., are obtained by electrolysis of fused electrolytes.

(iv) Non-metals like hydrogen, fluorine, chlorine are obtained by electrolysis.

(v) In this method pure metal is deposited at cathode from a solution containing the metal ions etc.

(vi) Compounds like NaOH, KOH, Na_2CO_3, $KClO_3$, white lead, etc. are synthesised by electrosynthesis method.

(vii) Electroplating: The process of coating an inferior metal with a superior metal by electrolysis is known as electroplating.

- **Electrorefining**

The anode is the impure metal and any impurities are removed during the process of electrolysis when the metal travels from anode to cathode. During the electrorefining of metals, the cathode has a decomposition of pure metals from a solution containing the metal ion. For example copper is purified through electrolysis in order to be used for applications that require high electrical conductivity. During this process, the cathode is a pure piece of copper, while the anode is an impure piece of copper. The Cu^{2+} from the anode moves through a sulfuric acid-copper(II) sulfate solution into the cathode where it becomes solid copper. While this is occurring, the impurities are left at the bottom of the tank. This leftover residue is called anode mud. The electrolysis is carried out at 0.15 - 0.30V (low voltage) in order to make sure that Ag, Au, and Pt impurities are not oxidized while in the anode and become anode mud. Whereas most of the other components become oxides or hydroxides and form water-soluble species.

- **Electrosynthesis**

a method which uses electrolysis reactions to produce certain products. For example MnO_2 needs to undergo electrolysis in order to be used for alkaline batteries. The solution for the electrosynthesis of MnO_2 is $MnSO_4$ in H_2SO_4. The anode is graphite, where Mn^{2+} is oxidized. While at the cathode, hydrogen is reduced from H^+ to H_2.

Overall Reaction: $Mn^{2+}(aq) + 2H_2O(l) \text{-->} MnO_2(s) + 2H^+(aq) + H_2(g)$

- **Chloro-Alkali Process**

Electrolysis of seawater leads to the production of chlorine and the alkali, sodium hydroxide. There are 3 different methods in which these two components are produced: membrane cell, diaphragm cell, and mercury cell process.

➢ Mercury Cell Process

Electrolysis of seawater in a mercury cell leads to the production of chlorine and sodium hydroxide at the same time. This method involves using mercury as the cathode and graphite as the anode. The mercury attracts either sodium or potassium cations and the mercury forms an amalgam with it. However when the amalgam is introduced to water it forms sodium hydroxide and hydrogen leaving the mercury to be reused later. The chlorine gas is left to form at the anode.

➢ Diaphragm Cell Process

The diaphragm cell has Cl_2 being released from the anode section, while there is H_2 being released from the cathode section. If Cl_2 manages to mix with NaOH, the Cl turns into other products. Therefore the diaphragm cell has a bigger amount of NaCl and a smaller amount of solution in the cathode in order for the NaCl to come in contact with the other solution gradually, while simultaneously preventing backflow of NaOH.

➢ Membrane Cell Process

This process is more efficient than others as it does not use mercury and does not require a significant amount of energy. It contains a cation-exchange membrane which is usually made from flourocarbon polymer. This membrane allows hydrated cations to pass in between the anode and cathode compartments, but does not allow the backflow of the ions, Cl- and OH-. This allows the sodium hydroxide produced to have less contamination by chloride ions.

Electrolysis of water

Hydrogen is the most abundant element in the universe. With the "green-energy" craze and talk of powering our future oil-free economy on hydrogen, it has gotten much attention in the last few years. In pure water at the negatively charged cathode, a reduction reaction takes place, with electrons (e^-) from the cathode being given to hydrogen cations to form hydrogen gas (the half reaction balanced with acid):

Reduction at cathode: $2\ H^+(aq) + 2e^- \rightarrow H_2(g)$

At the positively charged anode, an oxidation reaction occurs, generating oxygen gas and giving electrons to the anode to complete the circuit (Fig. 95);

Oxidation at anode: $2\ H_2O(l) \rightarrow O_2(g) + 4\ H^+(aq) + 4e^-$

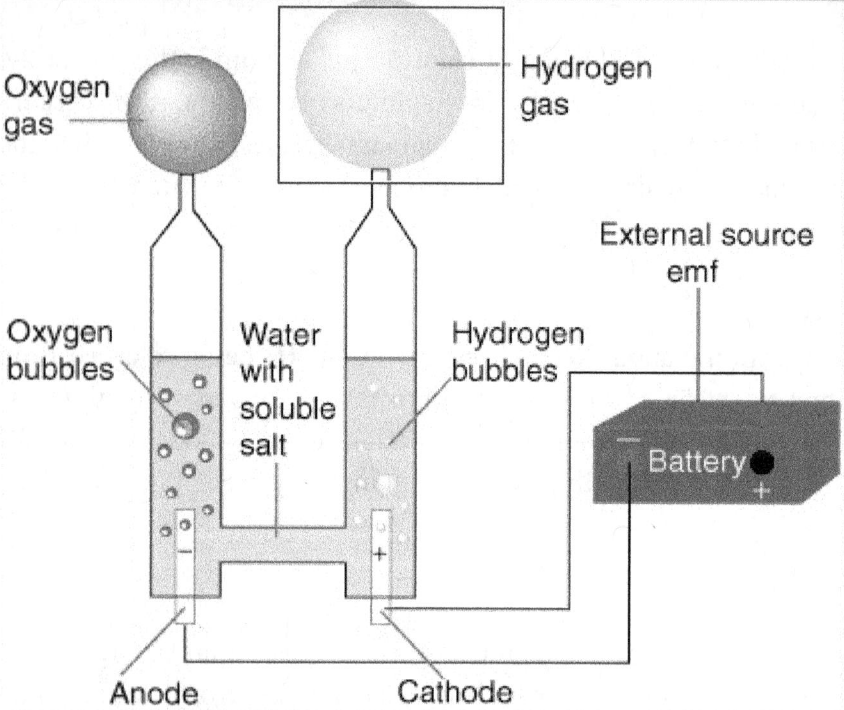

Fig. 95. Electrolysis of water

The same half reactions can also be balanced with base as listed below. Not all half reactions must be balanced with acid or base. Many do, like the oxidation or reduction of water listed here. To add half reactions they must both be balanced with either acid or base.

Cathode (reduction): $2\ H_2O(l) + 2e^- \rightarrow H_2(g) + 2\ OH^-(aq)$

Anode (oxidation): $4\ OH^-(aq) \rightarrow O_2(g) + 2\ H_2O(l) + 4\ e^-$

Combining either half reaction pair yields the same overall decomposition of water into oxygen and hydrogen:

Overall reaction: $2\ H_2O(l) \rightarrow 2\ H_2(g) + O_2(g)$

The number of hydrogen molecules produced is thus twice the number of oxygen molecules. Assuming equal temperature and pressure for both gases, the produced hydrogen gas has therefore twice the volume of the produced oxygen gas. The number of electrons pushed through the water is twice the number of generated hydrogen molecules and four times the number of generated oxygen molecules. Decomposition of pure water into hydrogen and oxygen at standard temperature and pressure is not favorable in thermodynamic terms.

Anode (oxidation): $2\ H_2O(l) \rightarrow O_2(g) + 4\ H^+(aq) + 4e^-$ $E^o ox = -1.23$ V ($E^o red = 1.23$)

Cathode (reduction): $2\ H^+(aq) + 2e^- \rightarrow H_2(g)$ $E^o red = 0.00$ V

Thus, the standard potential of the water electrolysis cell is -1.23 V at 25 °C at pH 0 ($H^+ = 1.0$ M). At 25 °C with pH 7 ($H^+ = 1.0\times10^{-7}$ M).

Manufacture of sodium hydroxide by the electrolysis of brine

Sodium hydroxide is manufactured by the electrolysis of an aqueous solution of sodium chloride [Brine] by using Nelson cell (Fig. 96).

Nelson cell is the most familiar diaphragm cell. It consists.

1. Anode: Graphide rod
2. Cathode: Perforated U-shaped steel tube lined inside with porous asbestos diaphragm.

3. Electrolyte: Saturated NaCl solution.
   Nelson cell consists of perforated U-shaped steel type lined with asbestos diaphragm and is supported in a steel tank. A graphite rod is dipped in sodium chloride solution. The anode is separated from the cathode by the asbestos diaphragm, otherwise the products of electrolysis sodium and chlorine recombine.

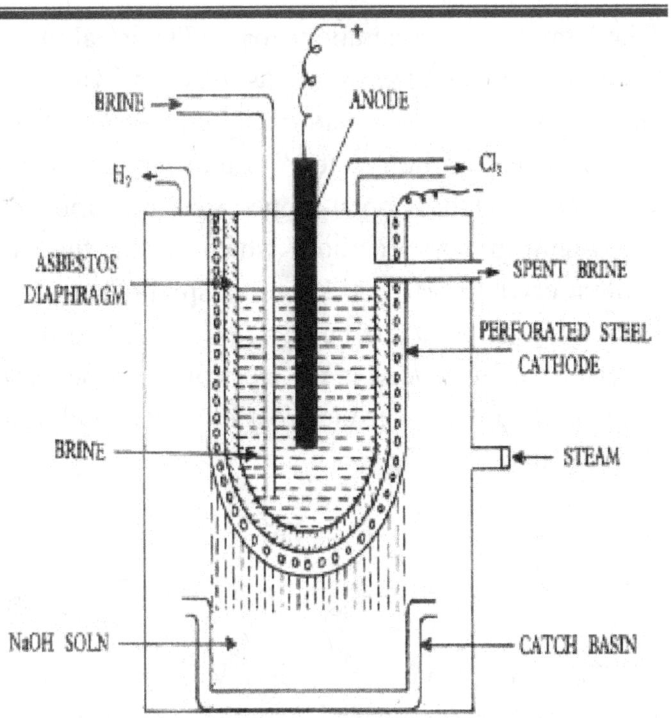

Fig. 96. Nelson cell

On passing electric current through NaCl solution, chloride ions reach the anode and liberate as chlorine gas. The sodium ions pass through Asebstos diaphragm reach the steel tube (cathode) and form sodium. This sodium reacts with steam passed into the steel tank forming sodium hydroxide and hydrogen. The reactions are given below:

Electrolysis: $NaCl \longrightarrow Na^+ + Cl^-$

At Anode: $Cl^- \longrightarrow Cl + e$ $\quad 2Cl \longrightarrow Cl_2 \uparrow$

At Cathode: $Na^+ + e \longrightarrow Na$

$$2Na + 2H_2O \longrightarrow 2NaOH + H_2$$

The sodium hydroxide thus formed trickles out of the perforated cathode and collects in the catch basin. The sodium hydroxide solution is drawn out evaporated to dryness at 160°C and the molten sodium hydroxide is cast into stics or made into pellets.

Sodium hydroxide is used in the manufacture of soap, paper, rayon, mercirised cloth, dyes, drugs and a number of chemicals. It is used in the refining of petroleum and vegetable oils. It is used in the preparation of sodium metal and several sodium salts It is used in the purification of bauxite It is used as a reagent in the laboratory.

Electrolytic reduction of aluminium oxide to aluminium

Primary manufacture of aluminium from bauxite involves four processes:

a) Extraction of the ore, bauxite

b) Purification of bauxite to pure aluminium oxide (alumina)

c) Synthesis of cryolite, Na_3AIF_6 and aluminium fluoride, to be used in the electrolytic reduction process

d) Electrolytic reduction of aluminium oxide to aluminium

The commercial operation for reducing alumina to aluminium is an electrolytic process invented independently but simultaneously by Paul Heroult in France and Charles Hall in the US in 1886. Huge quantities of electricity are required (10-13 kWh of electricity per 1 kg of aluminium) so aluminium smelters need to be located close to sources of cheap electric power. The process uses a reduction cell (Fig. 97), a shallow open-topped steel box approximately 8 m by 4 m by 1 m deep, the floor of which is lined with graphite (carbon) which acts as the cathode connection. In operation, the box is filled with molten cryolite (at 1200 K) and a small amount (2-5 %) of alumina. The side walls of the box are also lined with carbon. The boxes are designed so that cryolite solidifies in a thin layer on the surface of the electrolyte. Suspended in the electrolyte are several blocks of graphite which form the anode.

The exact nature of the electrolyte is not fully known but the overall reactions at the electrodes can be represented as:

⊖ Cathode: $Al^{3+} + 3e^- \longrightarrow Al$

⊕ Anode: $2O^{2-} \longrightarrow O_2 + 4e^-$

Molten aluminium is denser than cryolite and collects on the cell bottom where it forms the operating cathode. It is normal to operate a cell with a pool of aluminium 10 cm deep, the metal being siphoned off daily to maintain this level. The metal is generally cast as ingots, which are at least 99 % pure with small amounts of iron and silicon being the main impurities. Carbon is consumed in the anode process, much of it reacting with the liberated oxygen:

$C(s) + O_2(g) \longrightarrow CO_2(g)$ The anode blocks must thus be renewed regularly.

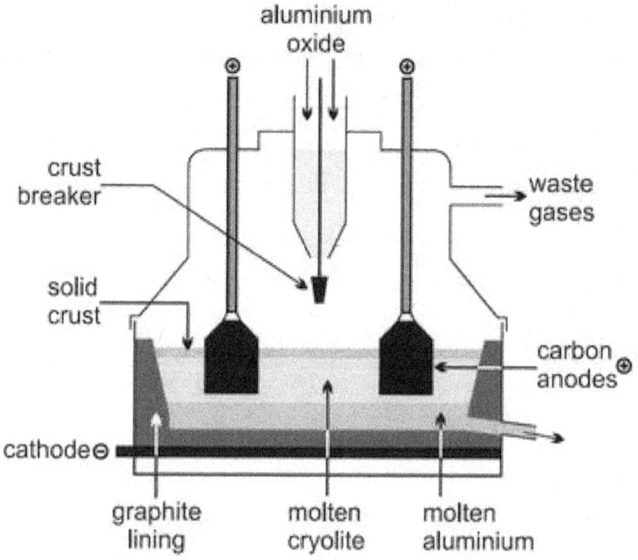

Fig. 97. *An electrolytic cell for the manufacture of aluminium.*

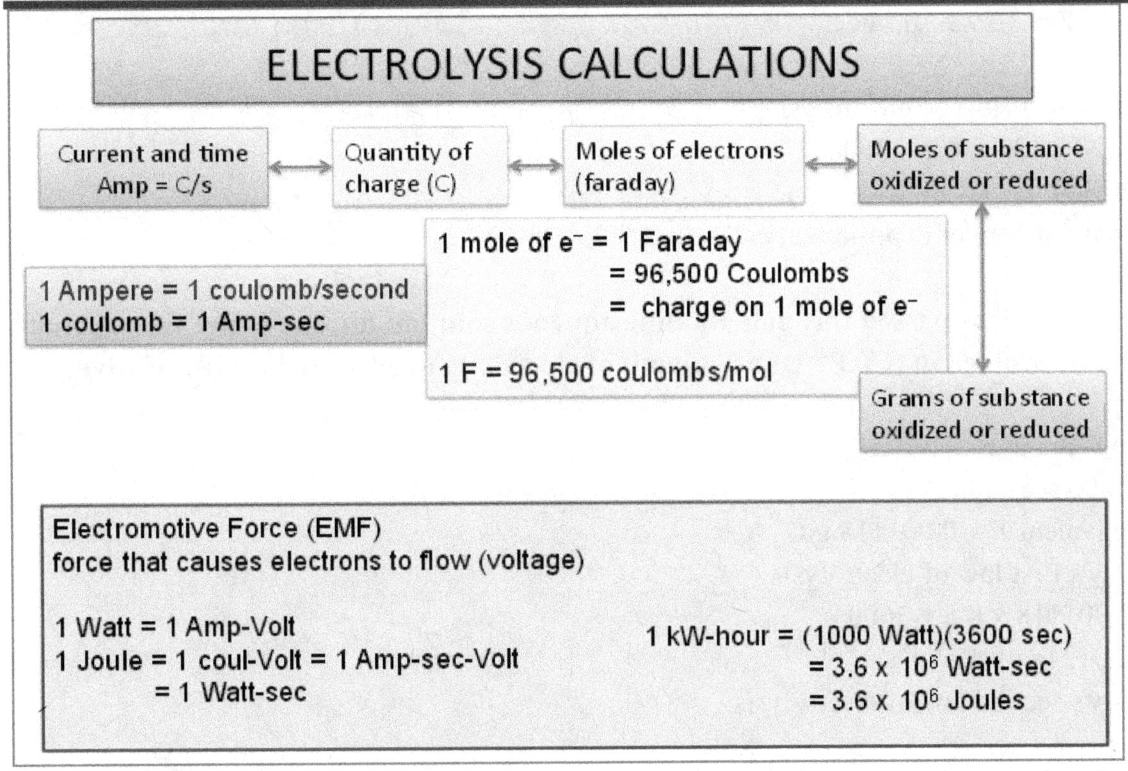

1.3 NUMERICAL PROBLEMS ON FARADAYS LAWs OF ELECTROLYSIS

A metal wire carries a current of 1 ampere. How many electrons pass a point in the wire in one second?

Solution:

Charge in coulomb = Current in Ampere x Time

= 1 x1 = 1

1 F (96500 C) of electricity is carried out by 1 mole of electrons i.e., 6.023×10^{23} electrons

1 coulomb shall involve $\frac{6.023 \times 10^{23}}{96500}$

$= 6.24 \times 10^{18}\ e^-_{ns}$

Thus 6.24×10^{18} electrons pass a point in the wire in 1 second.

1. **A current of 0.2 ampere is passed in aqueous solution of copper sulphate for one hour. Calculate the mass of copper deposited at the cathode. [Given: equivalent mass of Cu = 31.75]**

Solution:

Given I = 0.2 ampere

t = 1 hour = 60 x 60 = 3600 sec

Electrochemical equivalent (Z) of copper is given by

$$Z = \frac{\text{Equivalent mass}}{96500}$$

$$= \frac{31.75}{96500}$$

$Z = 3.29 \times 10^{-4}\ gC^{-1}$

According to Faraday's first law of electrolysis:

$W = ZIt = 3.29 \times 10^{-4} \times 0.2 \times 3600$

$= 0.2369$ g

Hence the mass of copper deposited at cathode is 0.2369g

2. **A current of 0.5 ampere is passed through alkaline aqueous solution for one hour. What volume of hydrogen will be evolved at S.T.P? One coulomb of electricity deposits 0.001118g of silver.**

Solution:

Given $I = 0.5$ ampere

$t = 3600$ sec

Electrochemical equivalent $Z = 0.001118\ g\ C^{-1}$

According to Faraday's first law of electrolysis

$W = ZIt = 0.001118 \times 0.5 \times 3600$

$= 2.0124$g of Ag

According to Faraday's second law of electrolysis

$$\frac{W_{H_2}}{W_{Ag}} = \frac{E_{H_2}}{E_{Ag}}$$

$$W_{H_2} = \frac{1.00}{108} \times 2.0124 = 0.01863g$$

1 mole of H_2 [2g] occupies 22,400 cm^3

∴ 0.01863g occupies.... ?

$$\frac{0.01863 \times 22,400}{2}$$

Volume of H_2 liberated at S.T.P = 208.7 cm^3

3. **In an electrolysis experiment, electric current was passed for 5 hours through two cells connected in series. The first cell contains a gold salt and the second contains $CuSO_4$ solution. 9.83g of gold were deposited in the first cell. If the oxidation number of Gold is +3, find the amount of copper deposited in the second cell. Also calculate the magnitude of current in amperes. At. Wt of Au = 197, At. Wt of Cu = 63.5**

Solution:

$$\text{Equivalent mass of Au}[E_{Au}] = \frac{\text{Atomic mass}}{\text{Oxidation number}} = \frac{197}{3} = 65.67$$

$$\text{Equivalent mass of Cu}\ [E_{Cu}] = \frac{\text{Atomic mass}}{\text{Oxidation number}} = \frac{63.5}{2} = 31.75$$

According to Faraday's second law of electrolysis

$$\frac{W_{Au}}{W_{Cu}} = \frac{E_{Au}}{E_{Cu}}$$

$$W_{Cu} = \frac{W_{Au} \times E_{Cu}}{E_{Au}} = \frac{9.83 \times 31.75}{65.67} = 4.753 \text{ g}$$

Hence amount of copper deposited in the second cell is 4.753 g

According to first law of electrolysis

$W_{Cu} = ZIt$

$$Z \text{ for Cu} = \frac{\text{Equivalent mass}}{96500} = \frac{31.75}{96500} = 3.29 \times 10^{-4} \text{ g C}^{-1}$$

t = 5 hours = 5 x 60 x 60 = 18000s

$$I = \frac{W_{Cu}}{Zt} = \frac{4.753 \text{ g}}{3.29 \times 10^{-4} \times 18000} = 0.8026 \text{ amp}$$

4. Calculate the number of electrons lost or gained during electrolysis of (a) 2g of Cl^- ions (b) 1g of Zn^{2+} ions

Solution:

(a) Number of equivalents of Cl^- = $\frac{\text{Weight}}{\text{Equivalent wt}} = \frac{2}{35.5} = 0.0563$

1 gram equivalents of Cl^- = 1 Faraday charge = 6.023×10^{23} electrons

'0.0563' gram of Cl^- = ?

$$= \frac{6.023 \times 10^{23} \times 0.0563}{1} = 3.39 \times 10^{22} \text{ electrons}$$

(b) Number of gram equivalents of Zn^{2+} = $\frac{\text{Weight}}{\text{Equivalent weight}} = \frac{1}{32.6} = 0.0306$

1 gram equivalent of Zn^{2+} = 1 Faraday = 6.023×10^{23} electrons

'0.0306' gram equivalent of Zn^{2+} = $6.023 \times 10^{23} \times 0.0306$

= 1.85×10^{22} electrons

5. 30 ml of 0.13M of $NiSO_4$ is electrolyzed using a current of 360 milliamperes for 35.3 minutes. How much of the metal would have been plated out if current efficiency was only 60%.

Solution:

Charge in coulomb = Current (ampere) x Time (sec)

$$\text{Charge in Faraday} = \frac{0.36 \times 35.3 \times 60}{96500}$$

= 0.0079 F

1 mole of electricity (1F) produces 1 equivalent of substance (28.5)

∴ 0.079F of electricity produces?

= 0.079 x 28.5 = 0.2251 g

Since current efficiency is 60%. The amount of Zinc deposited

$$= \frac{60}{100} \times 0.2251 \text{ g} = 0.135 \text{ g}$$

6. How long a current of 3 amps has to be passed through a solution of $AgNO_3$ to coat a metal surface of 80 cm^2 with 0.005 mm thick layer? [Density of Ag is 10.5 g/cm^3]

Solution:

$$\text{Number of equivalents of Ag deposited} = \frac{\text{Weight of Ag deposited [g]}}{\text{Equivalent wt of Ag}}$$

$$= \frac{\text{Volume} \times \text{density}}{\text{Equivalent weight}}$$

$$= \frac{\text{Surface are} \times \text{thickness} \times \text{Density}}{\text{Equivalent weight}}$$

$$= \frac{80 \times 0.0005 \times 10.5}{108}$$

$$= \frac{7}{1800} = 0.0039$$

Quantity of electricity required to deposit 0.0039 Equivalent of Ag = 0.0039 x 96500 C

= 376.35 coulomb

∴ Duration of electrolysis

$$= \frac{\text{Ch arg e in coulomb}}{\text{Current in amp}}$$

$$= \frac{376.35}{3}$$

=125.5 minutes

1.4 DEBYE-HUCKEL THEORY

The Debye- Huckel model is the basic model of the electrolytic solutions, proposed in the Year 1923. Arrhenius suggested that "electrolytes, when dissolved in water, are dissociated to varying degrees into electrically opposite positive and negative ions. The degree to which this dissociation occurred depended above all on the nature of the substance and its concentration in the solution". In 1923, Debye- Huckel starting from Arrhenius theory of dissociation and using the concept of chemical potential and excess Gibbs energy developed the first complete model of a solution containing the electrolytic ions in the dissolved state.

Ionic solutions are so important that they are often ignored. Chemical reactions have been studied for more than a century in ionic solutions. Life occurs in electrolyte solutions made of mixtures of 'bio-ions' (sodium Na+, potassium K^+, calcium Ca^{2+}, and chloride Cl^-), along with many other charged components. Ionic solutions are neutral systems formed by a solute of positive and negative ions immersed in a neutral polar solvent.

This category includes systems of very different degree of complexity, ranging from electrolyte solutions with cations and anions of comparable size and charge, to highly asymmetric macromolecular ionic liquids in which macro ions like polymers, micelles, proteins coexist with microscopic counter ions.

The basic idea of Debye-Huckel Theory is that ions in solution are not uniformly distributed. Overall, the solution is electrically neutral, but near any given ion there is a higher concentration of counter ions, ions with the opposite charge. Over time, more counter ions than like ions pass a given ion, and their directions are uncorrelated.

Averaged over time, this appears as a spherical haze with the same net charge as the ion, but with the opposite charge. This is called the ionic atmosphere. The average, spherical charge distribution of these anions is the ionic atmosphere.

Ionic atmosphere

The principle of electro- neutrality requires the presence of anions. Consider a positively charged ion in solution. Due to the Columbic force of attraction between oppositely charged ions, the positive ion will be surrounded by an ionic atmosphere consisting of an excess of negatively charge ions as illustrated in Figure. Similarly, a negatively charged ion will be surrounded by its own ionic atmosphere of excess positively charged ions.

Thus, surrounding an ion of a certain charge Z, it will always find a higher than average density of ions of the opposite charge. While enhancing the local density of the opposite charges, this screening layer will greatly reduce the Columbic force of this ion Z on charges far away.

The thickness of the screening layer, which is given the symbol κ^{-1} in Debye- Huckel theory, effectively cuts off the long- ranged nature of the Coulomb interactions. Beyond the screening length κ^{-1}, ions are effectively non- interacting while within κ^{-1}, Debye- Huckel theory provides an estimate for the charge density, and the interaction between charge density and the charge Z can be used to determine the chemical potential of each ion.

The inter- space between the spherical ionic layer, and the central ion is called Co- sphere. Solutions of electrolytes are non-ideal at relatively low concentrations.

The activity of ions in solution is relatively large compared to neutral compounds. Ions interact through a Coulombic potential that varies as $1/r$ (r is the distance between ions). Neutral solutes interact through London dispersion forces that vary as $1/r^6$.

The greater, the charge on the ions the larger the deviations from ideality. For example, per mole of $CaCl_2$ dissolved the deviation from ideal behavior is larger than for NaCl due to the 2+ charge of calcium ion.

The thickness of the ion's atmosphere (Fig. 98) decreases with the increasing of the charge value and ion's concentration and the ionic strength of the solution. The general interaction of the ions increasing with the increasing of solution's concentration according to the reducing of the average distance between the ions. Increase of the ion's interaction coursing the reducing of the ionic activity.

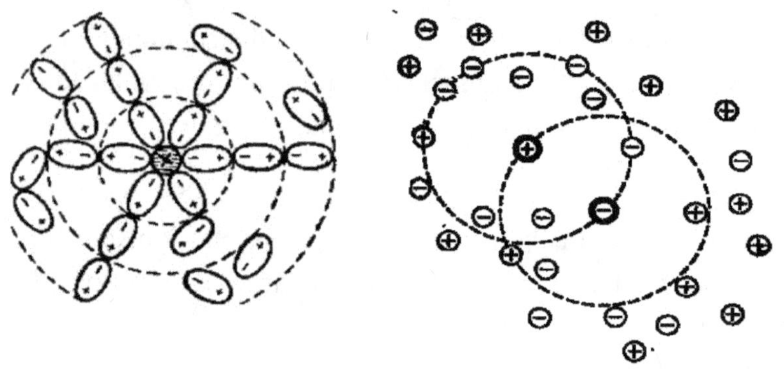

Fig. 98. Ionic atmosphere

The Debye length

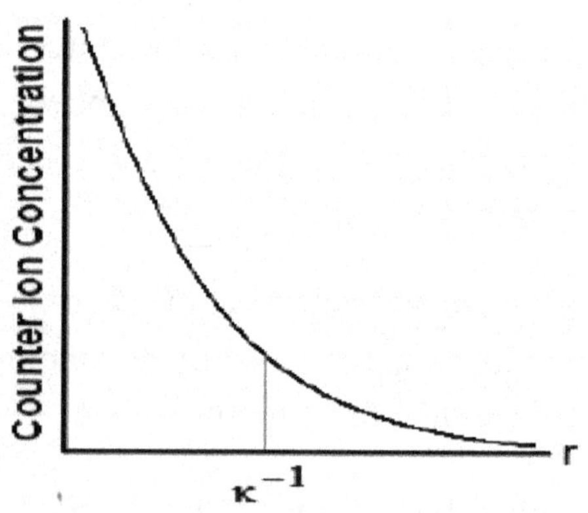

If we consider the ionic atmosphere around an isolated ion in solution, we find that the concentration of counter ions decreases exponentially as we move further from the ion.

$C \sim e^{-\kappa r}$

where the inverse of the decay rate κ^{-1} is called the Debye Length. The thickness of the ionic atmosphere is equal to the Debye Length. Physically, it is the distance from the isolated ion we have to go before the ions do not feel each other.

Trends of κ^{-1}

The Debye length:

1. Decreases as the concentration of ions increases
2. Decreases as the valency of the ions increases
3. Increases with temperature

Debye-Huckel Theory- Postulates

The Debye- Huckel model treated ions as point charges, allows one to describe the main features of structural and dynamical properties of very dilute electrolyte solutions. A quantitative treatment of ionic interactions requires (a) the spatial distribution of the ions, (b) The forces which act on the ions as a result of their ionic atmospheres. These two aspects of the problem are inter- dependent. The inter-ionic forces depend on the distribution of the ions.

In order to develop the necessary expressions for ionic distribution and inter-ionic forces, Debye- Huckel made the following assumptions in their model:

(a) Strong electrolytes are completely dissociated into ions.

(b) Inter- ionic interactions obey Coulomb's Law.

(c) Ionic distribution is controlled by Coulomb's forces and thermal motion.

(d) There are no external forces.

(e) Solvent is a continuous dielectric medium with dielectric constant 'ε'.

When an electrolyte is free of all external forces, the ionic atmosphere around each ion must be the same in all directions, i.e., it is spherically symmetrical. Thus the ionic distribution functions as well as the electric field are functions of distance r, but not of direction. All the equations presented use standard notation of symbols.

The magnitude of the Columbic force between the two charged ions can be represented as,

$$F = (|z_1 z_2| e^2 / \varepsilon_0 \varepsilon_r r^2)$$

where z_1 and z_2 : valencies of the two ions under consideration, located at a distance r,
e : electronic charge ,
ε_0: permittivity of free space,
ε_r: relative permittivity (dielectric constant) of the solvent medium.

The force F acts inwards into the central ion under consideration, if it is attractive or outwards away from the central ion if it is repulsive.

The charge density at the central ion and the divergence of the gradient of potential are given by Poisson Boltzmann equation (in Cartesian coordinate system) as:

$$[\text{Div}\ (\text{grad}\ \psi_0)] = \left[-\frac{4\pi\rho_c}{\varepsilon}\right]$$

where ϕ_0 : the potential ,
ρ_c: the average charge density at the central ion .
The chemical potential or partial molar free energy of an ionic species Aj is given by

$$\mu_j = \mu_j^o + RT \ln \{A_j\}$$

Where μ^o_j: the standard chemical potential and is constant under specified conditions of temperature and pressure:

{Aj}: the activity of species A.
The activity may be considered as an idealized or effective concentration.
Another way of expressing the chemical potential is,

$$\mu_j = \mu_j^o + RT \ln\gamma_j m_j$$

where m_j : molal concentration of species Aj (i.e., moles of A per kg H_2O),
γ_j : corresponding activity coefficient.

The activity coefficient is a measure for the deviation from ideal solubility.
The ionic strength is calculated using:

$$I_c = \frac{1}{2}[z_+^2 c_+ + z_-^2 c_-]$$

where z+ and z- are the charge of the positive and negative ions, respectively and c+ and c- are their concentrations. It is important to note that z+ and z- ions under consideration that make up the solution are not necessarily the same charges as those involved in the chemical reaction. In other words, excess ions in solution contribute to the ionic atmosphere.

Debye-Huckel calculation of potential and charge density for the electrolytic system
Consider an electrolyte solution consisting of different ionic species represented by the subscripts 1, 2...s., the bulk concentrations n^1, n^2,. . . ns ions/cm^3, and the values of the charges z^1, z^2,... z^s respectively. Further choose a coordinate system with one of the j ions at the origin. The electrical potential ψ_jI at a point r from the origin is due to the interaction between the central ion j and its ionic atmosphere. Let n_1^1, n_2^1 . . . n_s^1 be the local concentrations of ions 1 to s in the ionic atmosphere. The local ionic concentrations can be related to the bulk concentrations through
Boltzmann's distribution law as:

$$n_i^1(r) = n_i e^{-[U_i/kT]}$$

Further it can write Ui as the potential energy of the ith ion at point r. If we assume that this potential energy is due solely to the electrical field then,

$$U_i = z_i e \psi_j I$$

Leading to

$$n_i'(r) = n_i e^{-[\frac{z_i e \psi_j}{kT}]}$$

The charge density ρ (charge per cm^3) at a distance r from the j ion is given by

$$\rho_j(r) = \sum_{i=1}^{s} n_i'(r) z_i e$$

Deviations from ideality:
The D-H theory deals with departures from ideality in electrolyte solutions. The main experimental evidence for this non- ideality is that:
1. Concentration equilibrium constants are variables.
2. Rate constants depend on concentration.

3. Molar conductivities for strong electrolytes vary with concentration.
4. Freezing points of electrolytes are different from what would be expected for ideal behaviour.

These departures from ideality reduce as the concentration decreases and they were attributed to the Columbic law of electrostatic interaction.

The activities of ions in solution are relatively large compared to neutral compounds. Ions interact through a Columbic potential which varies as reciprocal of r, where r is the distance between ions. Neutral solutes interact through London dispersion forces that vary as r^{-6}. The greater the charge on the ions, the larger the deviation from the ideality.

Debye-Huckel Limiting Law

The thermodynamic consequences of electrostatic interactions between ions in solutions of strong electrolytes can be dealt with through using activities and activity coefficients.

We previously defined the chemical potential for a non ideal solution in terms of the activity a_i:

$$\mu_i = \mu_i^* + RT\ln a_i$$

We can now define the activity in terms of the molar concentration:

$$a_i = \gamma_i c_i$$

where γ_i is the activity coefficient.

The activity coefficient of a particular ion in the solution is given (without derivation) by the expression,

$$\log \gamma_i = -z_i^2 B I^{1/2}$$

where $B = 0.509\ (l/mol)^{1/2}$ and I is the **ionic strength** of the solution:

$$I = \frac{1}{2}\sum_{ions} c_i z_i^2$$

where c_i is the molar concentration of ions of type i , z_i is the valency of ion i and the sum is taken over all ions in solution. [This value of B assumes that the density of the solution is 1.00 kg / l . If it is not, then all concentrations are given in terms of molality. For purposes of this course, we will accept this assumption.]

In practice, it is impossible to measure the activity coefficient of a particular ion in solution because the solution thermodynamics are directly related to the presence of the counter ions. As a result, we introduce the mean activity coefficient:

$$\gamma_\pm = \left(\gamma_+^{\nu_+}\gamma_-^{\nu_-}\right)^{1/(\nu_+ + \nu_-)}$$

where ν_+ and ν_- are the numbers of ions produced by the electrolyte.

For example, for $ZnCl_2$, $\nu_+ = 1$ and $\nu_- = 2$. Since ν_i is related to z_i, we can take the original expression for $\log \gamma_i$ and apply it to each ion, then find the mean activity coefficient. In doing this, we obtain:

$$\log \gamma_\pm = -Bz_+|z_-|I^{1/2}$$

This is called the Debye-Huckel Limiting Law. It works well for dilute solutions of strong electrolytes.

Example 1:

We can more accurately represent the solubility product for the dissociation of $ZnCl_2$:

$$ZnCl_2(s) \rightarrow Zn^{2+}(aq) + 2Cl^-(aq)$$
$$K_{sp} = a_{Zn^{2+}} a_{Cl^-}^2 = ([Zn^{2+}]\gamma_+)([Cl^-]\gamma_-)^2$$

The individual activity coefficients can be replaced by the mean activity coefficient:

$$K_{sp} = [Zn^{2+}][Cl^-]\gamma_\pm^3$$

Example 2:

Calculate the mean activity coefficient for 0.001 M solution of $CaCl_2$.

First, we calculate the ionic strength:

$$I = \frac{1}{2}\sum_{ions} c_i z_i^2 = \frac{1}{2}\left\{\left[(0.001M)(+2)^2\right] + \left[(0.002M)(-1)^2\right]\right\} = 0.003M$$

Now, we calculate the log of the mean activity coefficient:

$$\log\gamma_\pm = -Bz_+|z_-|I^{1/2} = -(0.509M^{-1/2})(+2)|-1|(0.003M)^{1/2} = -0.0558$$

so: $\boxed{\gamma_\pm = 0.880}$

1.5 ELECTRONIC AND IONIC CONDUCTORS

Conductors are generally substances which have the property to pass different types of energy. In the following, the conductivity of electricity is the value of interest.

Metals: The conductivity of metals is based on the free electrons (so-called Fermi gas) due to the metal bonding as shown in Fig. 99. Already with low energy electrons become sufficiently detached from the atoms and conductivity is achieved.

Metallic bonding: fixed ions and free valence electrons (Fermi gas)

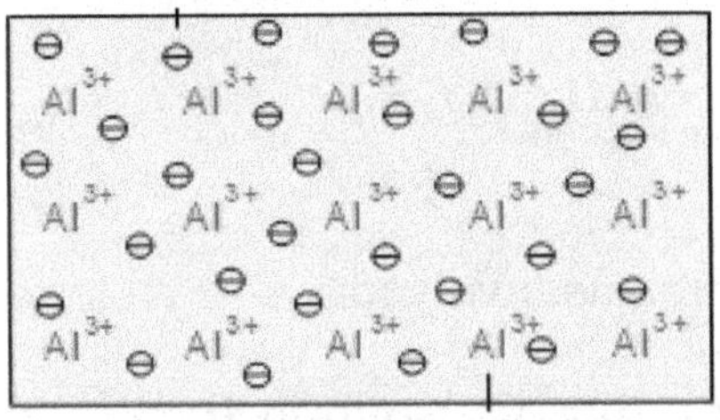

Fig. 99. Aluminium lattice

The conductivity depends, inter alia, on the temperature. If the temperature rises, the metal atoms swing ever stronger, so that the electrons are constrained in their movements. Consequence, the resistance increases. The best conductors, gold and silver, are used relatively rare because of the high costs (gold e.g. for the contacting of the finished chips). The alternatives in the semiconductor technology for the wiring of the individual components of microchips are aluminum and copper.

Salts: In addition to metals, salts can also conduct electricity. There are no free electrons, so the conductivity depends on ions which can be solved when a salt is melting or dissolving, so that the ions are free to move.

Insulators

Insulators possess no free charge carriers and thus are non-conductive.

The atomic bond: The atomic bond is based on shared electron pairs of nonmetals. The elements which behave like nonmetals have the desire to catch electrons, thus there are no free electrons which might serve as charge carriers.

The ionic bond: In the solid state, ions are arranged in a grid network. By electrical forces, the particles are held together. There are no free charge carriers to enable a current flow. Thus substances composed of ions can be both conductor and insulator.

Semiconductors

Semiconductors are solids whose conductivity lies between the conductivity of conductors and insulators. Due to exchange of electrons - to achieve the noble gas configuration - semiconductors arrange as lattice structure. Unlike metals, the conductivity increases with increasing temperature.

Increasing temperatures leads to broken bonds and free electrons are generated. At the location at which the electron was placed, a so-called defect electron ("hole") remains.

Cut-out of a silicon lattice

The electron flow is based on the conductivity properties of semiconductors. The electronic band structure illustrates why semiconductors behave like this.

The band theory of solids

Band theory is a quantitative model of bonding in solids.

– The wave functions of the valence electrons interact with each other.

– Bonding molecular orbitals result from constructive interference.

– Antibonding molecular orbitals result from destructive interference.

• The number of molecular orbitals formed equals the number of atomic orbitals involved.

– For every bonding molecular orbital formed, one antibonding molecular orbital is also formed.

– For odd number of bonding atoms, a nonbonding molecular orbital is formed.

– The molecular orbitals formed belong to the entire set of atoms.

Bonding in Metals –Molecular Orbitals of Solids

Band Theory

- Extension of MO theory
  - Valence orbitals on N atoms combine to form N molecular orbitals
  - for large N, the energies of the MO's are closely spaced
  - Each valence orbital will generate a band with different energy range

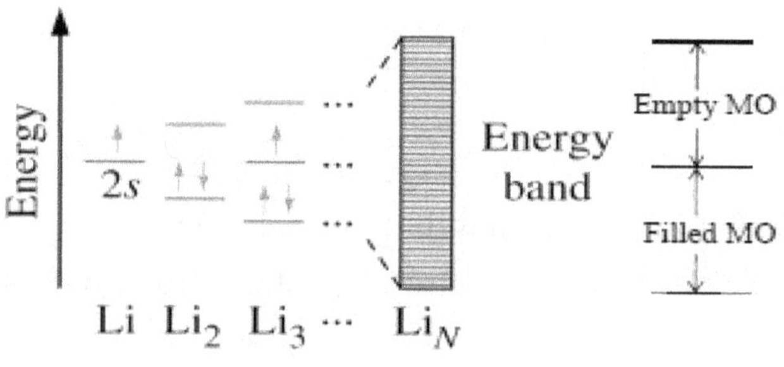

•N/2 are filled.
•The valence band.
•N/2 are empty.
•The conduction band.

The electronic band structure is an energy schema to describe the conductivity of conductors, insulators, and semiconductors. The schema consists of two energy bands (valence and conduction band) and the band gap. The valence electrons - which serve as charge carriers - are located in the valence band, in the ground state the conduction band is occupied with no electrons. Between the two energy bands there is the band gap, its width affects the conductivity of materials.

The energy bands If we consider a single atom, there are according to the Bohr model of atoms sharply distinct energy levels, which may be occupied by electrons. If there are multiple atoms side by side they are interdependent, the discrete energy levels are fanned out. In a silicon crystal, there are approximately 10^{23} atoms per cubic centimeter, so that the individual energy levels are no longer distinguishable from each other and thus form broad energy ranges.

Energy levels of atoms which are in interdependency with other atoms

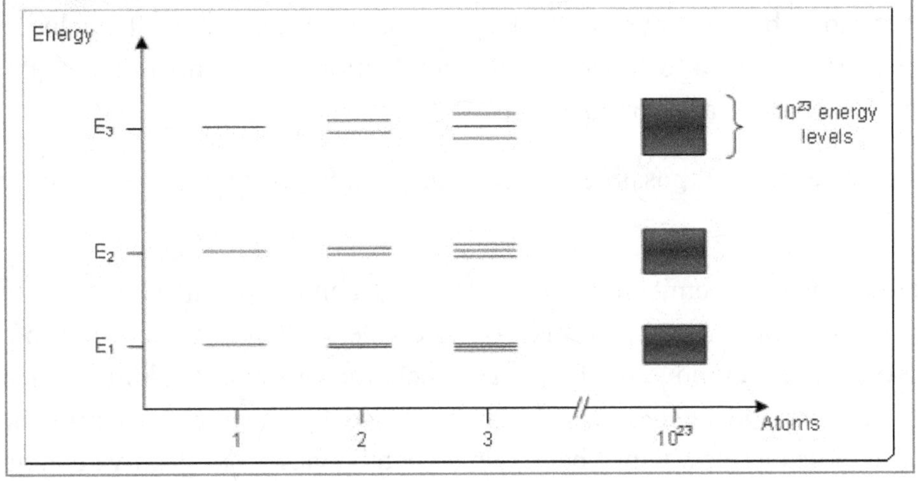

The width of the energy bands depends on how strongly the electrons are bound to the atom. The valence electrons in the highest energy level interact strongly with those of neighboring atoms and can be solved relatively easily from an atom; with a very large number of atoms, a single electron can no longer be assigned to one single atom. As a result, the energy bands of the individual atoms merge to a continuous band, the valence band (Fig. 100).

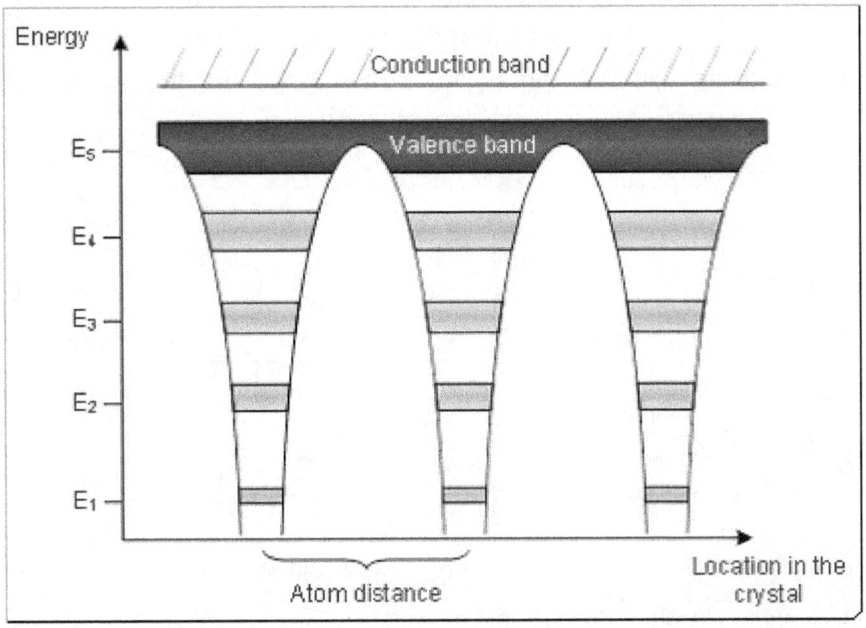

Fig. 100. Enenrgy bands of atoms which are in interdependency with other atoms

The band model of conductors: In conductors, the valence band is either not fully occupied with electrons, or the filled valence band overlaps with the empty conduction band. In general, both states occure at the same time, the electrons can therefore move inside the partially filled valence band or inside the two overlapping bands. In conductors there is no band gap between the valence band and conduction band.

The band model of insulators: In insulators the valence band is fully occupied with electrons due to the covalent bonds. The electrons cannot move because they're "locked up" between the atoms. To achieve conductivity, electrons from the valence band have to move into the conduction band. This prevents the band gap, which lies in-between the valence band and conduction band.

Only with considerable energy expenditure (if at all possible) the band gap can be overcome; thus leading to a negligible conductivity.

The band model of semiconductors: Even in semiconductors, there is a band gap, but compared to insulators it is so small that even at room temperature electrons from the valence band can be lifted into the conduction band (Fig. 101). The electrons can move freely and act as charge carriers. In addition, each electron also leaves a hole in the valence band behind, which can be filled by other electrons in the valence band. Thus one gets wandering holes in the valence band, which can be viewed as positive charge carriers.

There are always pairs of electrons and holes, so that there are as many negative as positive charges, the semiconductor crystal as a whole is neutral. A pure undoped semiconductor is known as intrinsic semiconductor. Per cubic centimeter there are about 10^{10} free electrons and holes (at room temperature).

Since the electrons always assume the energetically lowest state, they fall back into the valence band and recombine with the holes if there is no energy supply. At a certain temperature equilibrium is arranged between the electrons elevated to the conduction band and the electrons falling back. With increasing temperature the number of electrons that can leap the band gap is increased, and thus increasing the conductivity of semiconductors.

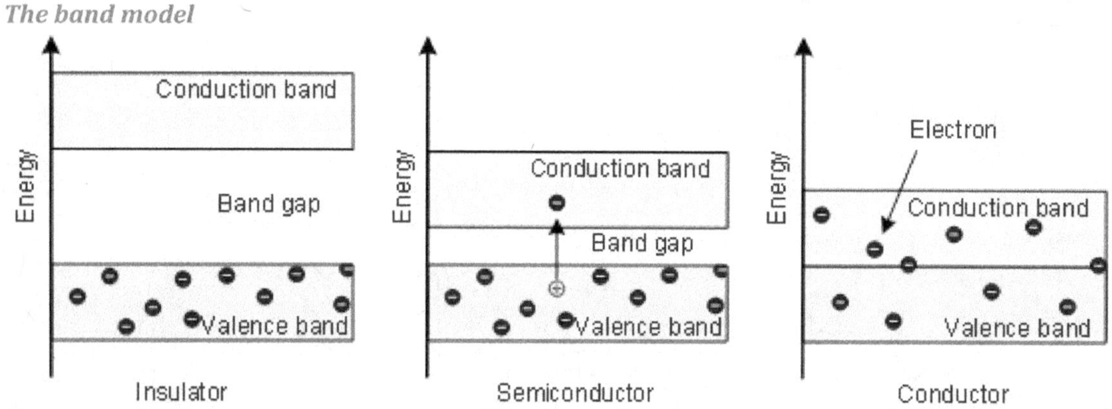

Fig. 101. Band structures of Insulator, semiconductor and conductor

Since the width of the band gap represents a certain energy corresponding to a particular wavelength, one tries to alter the width selective in order to obtain certain colors of light emitting diodes (LED). This may be achieved by combining different materials. Gallium arsenide (GaAs) has a band gap of 1.4 eV (electron volts, at room temperature) and thus emits red light.

The intrinsic conductivity of silicon is of no interest for the functioning of components, since it depends, inter alia, on the supplied energy. Which means that it changes with the temperature; in addition a conductivity comparable to metals is only possible at very high temperatures (several hundred degrees Celsius). In order to deliberately influence the conductivity of semiconductors, impurity atoms can be introduced into the regular silicon lattice to alter the number of free electrons and holes.

Material	Conductivity	Type
Copper	10^{6} S/cm	Conductor
Silicon	10^{-6} S/cm	Semiconductor
Silicon dioxide	10^{-12} S/cm	Insulator

Electronic band structure in doped semiconductors

Through the introduction of a dopant with five outer electrons, in n-doped semiconductors there is an electron in the crystal which is not bound and therefore can be moved with relatively little energy into the

conduction band. Thus in n-doped semiconductor one finds a donator energy level (Fig. 102) near the conduction band edge, the band gap to overcome is very small.

Analog, through introduction of a 3-valent dopant in a semiconductor, a hole is available, which may be already occupied at low-energy by an electron from the valence band of the silicon. For p-doped semiconductors one finds an acceptor energy level near the valence band.

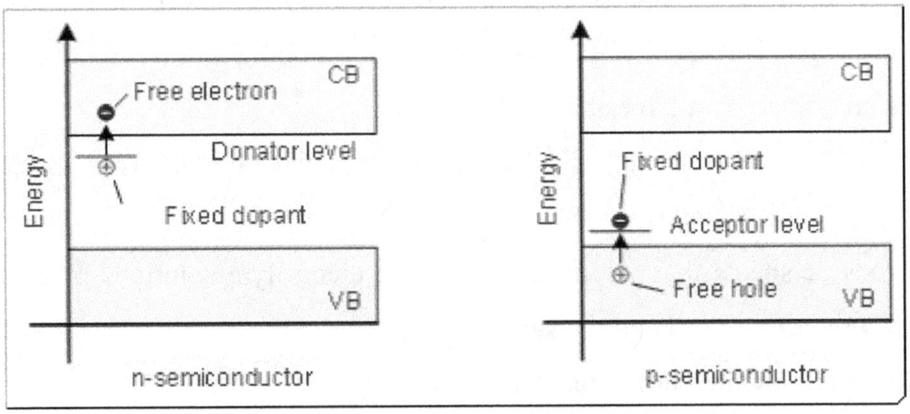

Fig. 102. Band structure of doped semiconductors

1.6 ELECTROCHEMICAL CELLS

Electrochemistry always involves an oxidation-reduction process. In a chemical reaction process, you can't have one process without the other. Hence they are called redox reactions. When the oxidation and reduction processes are *separated* from each other in two separate beakers (**half cells)** and the electrons from the reducing agent are made to pass around an external circuit to the oxidising agent, electrical energy can be generated. **This set-up is known as an electrochemical cell or a galvanic cell.**

For example, i.e. when a rod of zinc metal (the *anode*) is immersed in a Zn^{2+} solution and connected by a wire to a rod of copper metal (the *cathode*) in a Cu^{2+} solution, the zinc is oxidized, releasing electrons which travel through the wire and reduce Cu^{2+} in the other half-cell, producing an electrical current.

A *salt bridge* between the half-cells maintains electrical neutrality and completes the circuit. The chemical energy in the reactants can then be used to light a bulb or run an electrical device. This particular type of galvanic cell is called a **Daniell cell.**

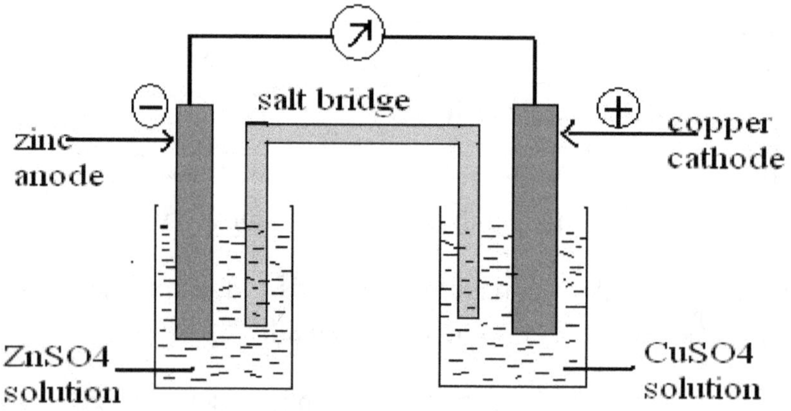

Fig. 103. Daniel cell

In general, in a **galvanic (voltaic) cell**, the components of each half-reaction are placed in separate containers called **half-cells**, consisting of an **electrode (electronic conductor)** dipping into an **electrolyte (ionic conductor)** solution; the cells are joined by a **wire** and a **salt bridge**.

The **electrolyte** solutions are solutions of ions (usually dissolved in water) that are involved in the reaction or that carry the charge between the electrodes.

The **electrodes** are usually metal rods or sheets that are immersed in the electrolyte solutions of each half-cell; they conduct the electricity between the cell and the surroundings.

An electrode is classified by the half-reaction which takes place there:

- The **anode** is where the **oxidation** half-reaction takes place. Electrons are *given up* by the substance being oxidized (reducing agent) and *leave the cell* at the anode.
- The **cathode** is where the **reduction** half-reaction takes place. Electrons are *taken up* by the substance being reduced (oxidizing agent) and *enter the cell* at the cathode.

Electrochemical cells are the devices which involve inter conversion of electrical energy and chemical energy. They are of two types: Galvanic and Electrolytic cell.

Galvanic also called as Voltaic cells generate electricity from chemical reactions that occur spontaneously at the electrodes in the cell.

A **galvanic cell** (or **voltaic cell**) uses a **spontaneous reaction** ($\Delta G° < 0$) to generate electrical energy. – The difference in potential energy between higher energy reactants and lower energy products is converted into electrical energy. – *The reactants do work on the surroundings.* However, when a greater electromotive force is applied from the external circuit using a battery or power source, it forces the species within the cell to perform the reverse reaction to that which they would normally tend to do – such a cell is called electrolytic cell. i.e an **electrolytic cell** uses electrical energy to drive a **non spontaneous reaction** ($\Delta G° > 0$). – Lower energy reactants are converted into higher energy products by electrical energy. – *The surroundings do work on the reactants.*

	Galvanic cell	**Electrolytic cell**
1	It does not require a source of external energy	It requires a source of electrical energy
2	It converts chemical energy to electrical energy.	It converts electrical energy into chemical energy.
3	The redox reaction is spontaneous	The redox reaction is non-spontaneous
4	Each electrode is dipped in its own ionic solution in two separate compartments.	Both the electrodes are immersed in the same electrolyte solution (single compartment).
5	Salt bridge is required.	No salt bridge is required.
6	Cathode is positive electrode and anode is negative electrode.	Cathode is negative electrode and anode is positive electrode.

Origin of electrode potential

Nature seems to strongly discourage any process that would lead to an excess of positive or negative charge in matter. Suppose, for example, that we immerse a piece of zinc metal in pure water. A small number of zinc atoms go into solution as Zn ions, leaving their electrons behind in the metal: $Zn(s) \rightarrow Zn^{2+} + 2e^-$

As this process goes on, the electrons which remain in the zinc cause a negative charge to build up within the metal which makes it increasingly difficult for additional positive ions to leave the metallic phase. A similar buildup of positive charge in the liquid phase adds to this inhibition. Very soon, therefore, the process comes to a halt, resulting in a solution in which the concentration of Zn^{2+} is still too low (around 10^{-10} M) to be detected by ordinary chemical means.

There would be no build-up of this opposing charge in the two phases if the excess electrons could be removed from the metal or the positive ions consumed as the electrode reaction proceeds. For example, we could drain off the electrons left behind in the zinc through an external circuit that forms part of a complete electrochemical cell; this we will describe later. Another way to remove these same electrons is to bring a good electron acceptor (that is, an oxidizing agent) into contact with the electrode. A suitable acceptor would be hydrogen ions; this is why acids attack many metals. The degree of charge unbalance that is allowed produces differences in electric potential of no more than a few volts, and corresponds to unbalances in the concentrations of oppositely charged particles that are not chemically significant. There is nothing mysterious about this prohibition, known as the *electroneutrality principle*; it is a simple consequence of the thermodynamic work required to separate opposite charges, or to bring like charges into closer contact. The additional work raises the free energy change of the process, making it less spontaneous.

At the electrode, when a metal is in contact with a solution containing its own ions, one of the following two reactions can take place,

(i) Metal ions ***in the solution*** can gain electrons and undergo reduction reaction $M^{n+} + ne^- \rightarrow M$

or

(ii) Metal atoms ***on the electrode*** can lose electrons, undergo oxidation to give metal ions which go into solution $M \rightarrow M^{n+} + ne^-$

After sometime, a dynamic equilibrium exists between the two reactions resulting in charge separation.

$M^{n+} + ne^- \leftrightarrow M$

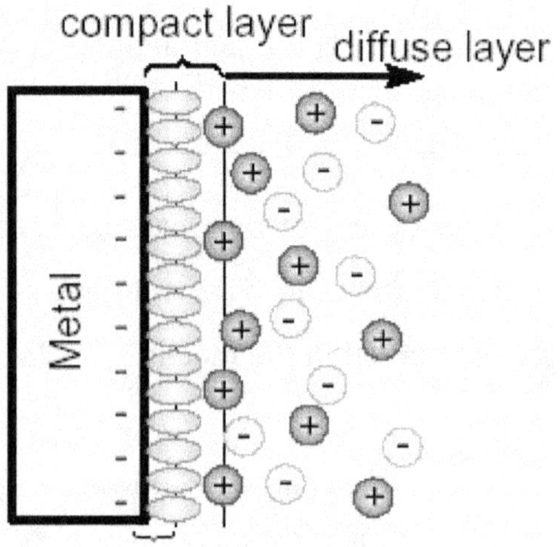

Fig. 104. Helmholtz double layer

Negative charge developed on the Zinc rod attracts the positively charged M^{n+} ions present in the solution. An electrical double layer consisting of electrons on the zinc rod and a layer of M^{n+} ions close to the negatively charged zinc rod is formed. The transition region between two phases consists of a region of charge unbalance known as the *electric double layer (Fig. 104).* This is called **"Helmholtz electrical double layer"** (HDL) and is similar to a parallel plate capacitor. Consequently, a potential difference is established between the metal and the solution. **This is the origin of a single electrode potential.**

Given any metal electrode in a solution that contains ions of that metal, equilibrium between the metal at the electrode and its ions in solution is established as $M \rightarrow M^{n+} + ne^-$ hence a potential difference between the metal and the solution will develop. The magnitude of this potential is different for different metals depending on the ease with which the metals form the ions. ***"Single electrode potential is thus defined as the potential established at the interface between the metal (M) and its ionic solution (M^{n+}) when it is in contact with a solution of its own ions".***

The interfacial potential differences which develop in electrode-solution systems are limited to only a few volts at most. This may not seem like very much until you consider that this potential difference spans a very small distance. In the case of an electrode immersed in a solution, this distance corresponds to the thin layer of water molecules and ions that attach themselves to the electrode surface, normally only a few atomic diameters. Thus a very small voltage can produce a very large potential gradient. Interfacial potentials are not confined to metallic electrodes immersed in solutions; they can in fact exist between any two phases in contact, even in the absence of chemical reactions. In many forms of matter, they are the result of adsorption or ordered alignment of molecules caused by non-uniform forces in the interfacial region. Thus colloidal particles in aqueous suspensions selectively adsorb a given kind of ion, positive for some colloids, and negative for others. However, the absolute value of this potential difference between the metal and the solution cannot practically be measured directly. Hence, the difference in potential between this electrode and another standardised electrode called a ***reference electrode*** is measured in comparison. This involves two potential differences, one at each interface between electrode and solution. The standard reference electrode used is called a standard hydrogen electrode (SHE), whose potential is defined as zero.

Standard electrode potential

The emf measured when a metal / metal ion electrode is coupled to a standard hydrogen electrode under standard conditions of 1M (solution) or 1atm (gas) at 25^0C is known as the **standard electrode potential** of that metal / metal ion combination. Standard electrode potential is given the symbol **E°**.

Redox potentials - Electrochemical Series

The standard reduction potentials of various metals can be determined by using S.H.E. Different electrodes have different standard electrode potentials. The standard reduction potentials (E^o) for some electrodes are negative while for some others E^o values are positive. The arrangement of various electrodes in the order of increasing standard reduction potentials is called electrochemical series.

Standard electrode	Electrode reaction	Standard reduction potential
Li^+/Li	$Li^+ + e \rightleftharpoons Li$	-3.05V
K^+/K	$K^+ + e \rightleftharpoons K$	-2.93 V
Ba^{2+}/Ba	$Ba^{2+} + 2e \rightleftharpoons Ba$	-2.90 V
Sr^{2+}/Sr	$Sr^{2+} + 2e \rightleftharpoons Sr$	-2.890V
Ca^{2+}/Ca	$Ca^{2+} + 2e \rightleftharpoons Ca$	-2.87V
Na^+/Na	$Na^+ + e \rightleftharpoons Na$	-2.71V
Mg^{2+}/Mg	$Mg^{2+} + 2e \rightleftharpoons Mg$	-2.37
Al^{3+}/Al	$Al^{3+} + 3e \rightleftharpoons Al$	-1.66V
Zn^{2+}/Zn	$Zn^{2+} + 2e \rightleftharpoons Zn$	-0.76V
Fe^{2+}/Fe	$Fe^{2+} + 2e \rightleftharpoons Fe$	-0.44V
Ni^{2+}/Ni	$Ni^{2+} + 2e \rightleftharpoons Ni$	-0.25V
$H^+/\frac{1}{2}\ H_2.Pt$	$H^+ + e \rightleftharpoons H$	0.0V
Cu^{2+}/Cu	$Cu^{2+} + 2e \rightleftharpoons Cu$	+0.34V
Ag^+/Ag	$Ag^+ + e \rightleftharpoons Ag$	+0.80V
Au^{3+}/Au	$Au^{3+} + 3e \rightleftharpoons Au$	+1.50V

Standard Hydrogen Electrode (SHE)

Hydrogen electrode is the primary standard electrode. It consists of a small platinum strip coated with platinum black (thin film of fine powder of Pt) as to adsorb hydrogen gas. A platinum wire is welded to the platinum strip and sealed in a glass tube as to make contact with the outer circuit; glass tube has an inlet for hydrogen gas at the top (Fig. 105). The platinum strip is placed in an acid solution which has H^+ ion concentration 1 M. Pure hydrogen gas is circulated at one atm. pressure. Catalysed by the platinum, equilibrium is established between the adsorbed hydrogen and hydrogen ions in the solution.

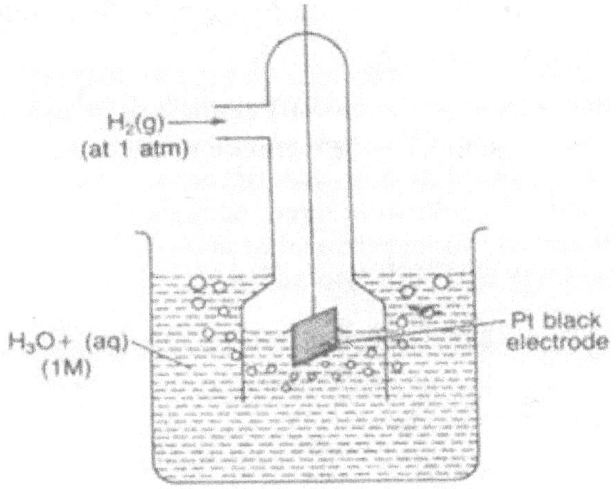

Fig. 105. Standard hydrogen electrode

$$2H^+_{(aq)} + 2e^- \rightleftharpoons H_{2(g)}$$

The hydrogen electrode can act as cathode or anode with respect to other electrode.

SHE half reaction	Electrode potential
$H_2 \rightarrow 2H^+ + 2e^-$	0.0 V (Anode)
$2H^+ + 2e^- \rightarrow H_2$	0.0 V (Cathode)

when it acts as a cathode, the electrode is represented as: $H^+(aq) \mid H_2(g) \mid Pt(s)$ and

when it acts as an anode, it is represented as: $Pt(s) \mid H_2(g) \mid H^+(aq)$

MEASUREMENT OF ELECTRODE POTENTIAL

The electrode potential of a given electrode is measured by forming a galvanic cell by connecting it with a standard hydrogen electrode (SHE) through a salt bridge. i.e the hydrogen electrode forms one half-cell and electrode whose potential is to be measured forms the other half-cell of the voltaic cell.

E° values give a way of comparing the positions of equilibrium when these elements lose electrons to form ions in solution. $M^{n+} + ne \Leftrightarrow M$

- Large positive E° (or less negative) value, means the more the element or compound gets reduced, the further the equilibrium lies to the right.
- A large negative E° means the element or compound is hard to reduce. The more negative the E° value, the further the equilibrium lies to the left - the more readily the element loses electrons, gets oxidized forming the ions.

i) Measurement of standard electrode potential of Zn^{2+}/Zn (Anode):

a) Construction of the cell

The cell is constructed by coupling Zn electrode with a SHE through a salt bridge (Fig. 106) and EMF of the cell assembly is determined using an electronic voltmeter.

Cell representation

$Zn \mid Zn^{2+}(1M) \parallel H^{+}(1M) \mid H_2(1\ atm) \mid Pt$

The electrode reactions are:

At anode: (Oxidation) $Zn \rightarrow Zn^{2+} + 2e^{-}$

At cathode: (Reduction) $2H^{+} + 2e^{-} \rightarrow H_2$

Net cell reaction: $Zn + 2H^{+} \rightarrow Zn^{2+} + H_2$

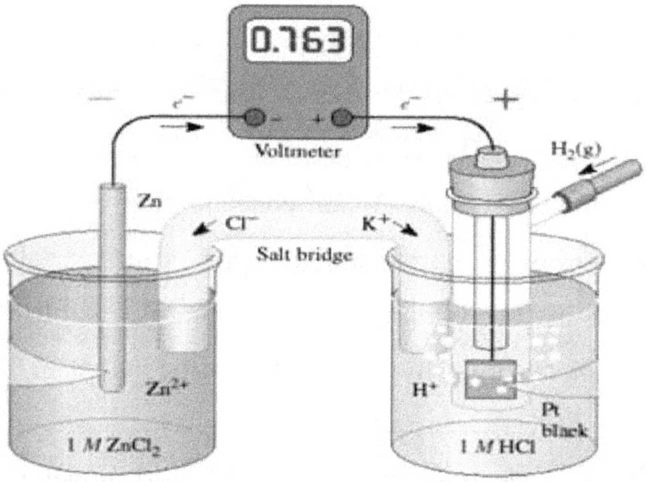

Fig. 106. Zn half cell coupled with SHE

b) Assignment of sign on electrode:

When the two electrodes are connected to voltmeter direction of the current indicates sign of the electrode. In case of Zn electrode it takes negative sign and is assigned as anode. SHE takes positive sign and assigned as cathode.

c) EMF Calculation:

$$E^\circ_{cell} = E^\circ_{cathode} - E^\circ_{anode}$$

$$E^\circ_{cell} = E^\circ_{SHE} - E^\circ_{Zn^{2+}|Zn}$$

The potential of the cell is found to be 0.76V. E°_{SHE} is considered as zero.

$$0.76 = 0 - E^\circ_{Zn^{2+}|Zn}$$

$$\mathbf{E^\circ_{Zn^{2+}|Zn} = -0.76V}$$

ii) Measurement of standard electrode potential of Cu^{2+}/Cu (Cathode):

a) Construction of the cell assembly:

The cell is constructed by coupling Cu electrode with a SHE through a salt bridge (Fig. 107) and EMF of the cell assembly is determined using an electronic voltmeter.

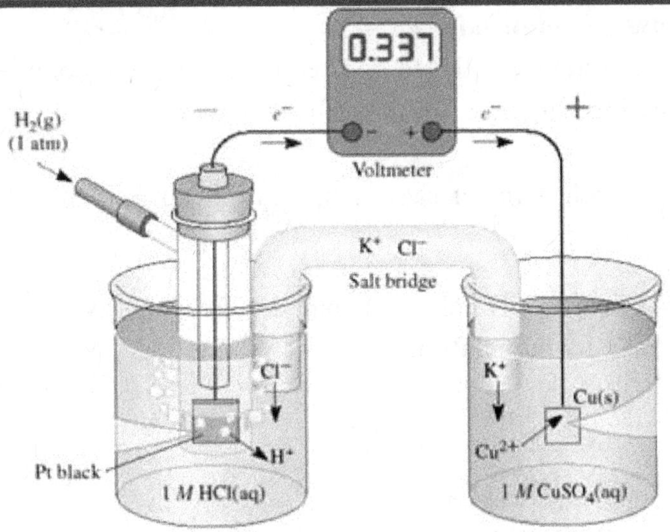

Fig. 107. Copper half cell coupled with SHE

Pt | H_2 (1 atm) | H^+ (1M) || Cu^{2+} (1M) |Cu

The electrode reactions are

At anode: $H_2 \rightarrow 2H^+ + 2e^-$ (Oxidation)

At cathode: $Cu^{2+} + 2e^- \rightarrow Cu$ (Reduction)

Net cell reaction: $H_2 + Cu^{2+} \rightarrow 2H^+ + Cu$

b) Assignment of sign on electrode:

In this assembly copper electrode is cathode (+ve) and SHE is anode (-ve).

c) Calculation:

$$E°_{cell} = E°_{cathode} - E°_{anode}$$

$$E°_{cell} = E°_{Cu^{2+}|Cu} - E°_{SHE}$$

The potential of the cell is found to be 0.34V. $E°_{SHE}$ is considered as zero.

$$0.34V = E°_{Cu^{2+}|Cu} - 0$$

$$\mathbf{E°_{cu^{2+}|Cu} = 0.34V}$$

Nernst Equation (for single electrode potential)

The Nernst Equation allows the electrode potential to be predicted when the conditions are not standard. Walter Nernst developed the following equation in the late 1800's while studying the thermodynamics of electrolyte solutions:

Consider an electrode process in which a metal (M) is in equilibrium with its ions (M^{n+}) in solution

$$\mathbf{M^{n+} + ne^- \leftrightarrow M}$$

We know from thermodynamics (Van't Hoff relation) that the free energy change of this process is related to the equilibrium constant 'K' by the relation:

$$\Delta G = \Delta G^0 + RT\ \ln K$$

Where K = [product] / [reactant] = [M] / [M^{n+}]

$$\Delta G = \Delta G^0 + RT\ \ln \frac{[M]}{[M^{n+}]}$$

Also we know that, the free energy change accompanying this reversible spontaneous process in the cell produces the current in the external circuit; and from second law of thermodynamics it is equal to the electrical work done.

The electric work *W* is the product of electricity passing, 'nF' and the potential *E* in Volt i.e nFE volt-coulombs or joules

(where 'n' is the number of moles of electrons and 'F' is one Faraday of charge in Coulomb (C) 'nF' is no. of coulombs of charge flowing).

As the process is spontaneous, there is decrease in free energy. i.e $-\Delta G = nFE$

Under standard conditions, $\Delta G^0 = -nFE^0$

On substituting these, we get $-nFE = -nFE^0 + RT\ln \frac{[M]}{[M^{n+}]}$

Divide equation by -nF, on both sides, and convert natural log into base 10, we get

$$E = E^0 - \frac{2.303RT}{nF} \log_{10} \frac{[M]}{[M^{n+}]}$$

Under standard conditions [M] = 1, hence the above equation becomes

$$E = E^0 + \frac{2.303\ RT}{n\ F} \log_{10} [M^{n+}]$$

On substituting the values for R = 8.314 J/K/mol; T = 298 K; F = 96500 C/mol, the above equation reduces to

$$E = E^0 + \frac{0.0591}{n} \log_{10} [M^{n+}]$$

This equation is known as the **Nernst equation for single electrode potential.**

Nernst equation for Galvanic cell

Let us consider a cell,

$M_1 \mid M_1^{n+} \parallel M_2^{n+} \mid M_2$

Which is obtained by connecting two half cells

The net cell reaction is given as

$M_1 + M_2^{n+} \rightarrow M_1^{n+} + M_2$

The emf of the cell is

$$E_{cell} = E^o_{cell} + \frac{2.303\ RT}{n\ F} \log 10 \frac{[\text{Metal ion concentrat ion at cathode}]}{[\text{Metal ion concentrat ion at anode}]}$$

$$E_{cell} = E^o_{cell} + \frac{0.0591}{n} \log 10 \frac{[\text{reduced species}]}{[\text{oxidised species}]}$$

$$\mathbf{E_{cell} = E^o_{cell}} + \frac{0.0591}{n} \mathbf{\log_{10}} \frac{[M_2^{n+}]}{[M_1^{n+}]}$$

$E^o_{cell} = E^o_{cathode} – E^o_{anode} = E^o M_2^{n+}/M_2 – E^o M_1^{n+}/M_1$

where, $[M_1^{n+}]$ = The metal ion concentration at Anode

$[M_2^{n+}]$ = The metal ion concentration at Cathode

Numerical problems

Example 1: Calculate the electrode potential at a copper electrode dipped in a 0.1 M solution of copper sulphate at 25° C. The standard electrode potential of Cu^{2+}/Cu system is 0.34 volt at 298 K.

Solution: We know that $E_{red} = E^o_{red} + 0.0591/n\ \log_{10}[ion]$

Putting the values of E^o_{red} =0.34 V, n = 2 and $[Cu^{2+}]$= 0.1 M

$E^o_{red} = 0.34+0.0591/2\ \log_{10}[0.1]$

$= 0.34 + 0.02955 \times (-1)$

= 0.34 - 0.02955 = 0.31045 volt

Example 2: Calculate the standard electrode potential of Cu^{2+}/Cu if its electrode potential at 25^0C is 0.296V when $[Cu^{2+}] = 0.015$

$$E = E^{o} + \frac{0.0591}{n} \log [M^{n+}]$$

$$E^{o} = E - \frac{0.0591}{n} \log [Cu^{2+}]$$

$$E^{0} = 0.296 - \frac{0.0591}{2} \log 0.015$$

$\mathbf{E^0 = 0.349V}$

Example 3: What is the single electrode potential of a half-cell for zinc electrode dipping in 0.01 M $ZnSO_4$ solution at 25° C? The standard electrode potential of Zn/Zn^{2+} system is 0.763 volt at 25° C.

Solution: We know that $E = E^o_{red} + 0.0591/n \log_{10}[ion]$

substituting $E^o ox = 0.763$ V, ($E^o_{red} = -0.763V$), n=2 and $[Zn^{2+}]=0.01$ M

$E = -0.763 + 0.0591/2 \log [0.01]$

$= -0.763 + 0.02955 \times (-2) = (-0.763 - 0.0591)$ volt $= -0.8221$ volt

EMF of a cell, Sign convention and cell notation,

Galvanic cell; Cell notation & conventions

Drawing a full diagram to represent a galvanic cell takes too long. Instead, the cell is represented in the form of notation as:

Anode | Anode's Electrolyte ‖ Cathode's Electrolyte | Cathode

Cell notation

- By convention, the electrode written to the *left* of the salt bridge in this cell notation is always taken to be the *anode*, and the associated half-reaction is always written as an *oxidation*. The right-hand electrode is therefore always the *cathode*, and the half-equation is always written as a *reduction*.
- A single vertical line ' | ' indicates a phase boundary such as that between solid and aqueous solution or gas and aqueous solution.
- Concentrations of aqueous solutions are written in parentheses after the symbol for the ion or molecule.
  - wo or more components in the same phase are separated by commas

- A double vertical line '||' is used to indicate the junction between the half-cells or a salt bridge.
- The electron flow is left to right in this notation
- The term electrode potential always refers to reduction potential and is represented as $E_{M^{n+}/M}$
- The reduction and oxidation electrode potentials are conventionally denoted as $E_{M^{n+}/M}$ (or $E_{M/M^{n+}}$).

The voltage indicates the tendency for current to flow in the external circuit, that is, it shows how strongly the anode reaction can push electrons into the circuit and how strongly the cathode reaction can pull them out.

Cell Reaction:

Given a short-hand representation of a cell, the half-cell reactions are written at the two electrodes (oxidation at anode and reduction reaction at cathode) and the cell reaction corresponding to the given shorthand description is obtained by summing the half-equations after multiplying by any factors necessary to equalize the number of electrons lost at the anode with the number gained at the cathode. First, write out the half-cell reactions. One half cell reaction will show the oxidation (at the anode) and the other half cell reaction will show the reduction (at the cathode). An overall redox reaction is the sum of the half reactions associated with each half-cell process:

Anode $\quad Zn(s) \rightarrow Zn^{2+}(aq) + \cancel{2e^-}$ (oxidation)

Cathode $\quad \underline{Cu^{2+}(aq) + \cancel{2e^-} \rightarrow Cu(s)}$ (reduction)

Cell reaction $\quad Zn(s) + Cu^{2+}(aq) \rightarrow Zn^{2+}(aq) + Cu(s)$

EMF of a standard cell: Calculating the cell potential, E_{cell} (under standard conditions: 1M/ 1atm at 25^0C):

$$E^\circ_{cell} = E^\circ_{cathode} - E^\circ_{anode}$$

The maximum potential difference which can be measured for a given cell is called the **electromotive force**, abbreviated emf and represented by the symbol *E*.

The reduction potential for $Zn^{2+}(aq) + 2e^- \rightarrow Zn(s)$ is –0.76 V and

the reduction potential for $Cu^{2+}(aq) + 2e^- \rightarrow Cu(s)$ is 0.34 V.

Using these reduction potentials, calculate the cell potential:

$$E^\circ_{cell} = E^\circ_{cathode} - E^\circ_{anode} = 0.34\ V - (-0.76\ V) = +1.10V$$

An $E^\circ > 0$ means the cell will be spontaneous (produce a voltage).

An E° < 0 means the cell must be supplied that voltage from an external power supply in order for the cell reaction to run in the direction it is written. If anything, that cell would want to run in the reverse direction. Often these non-spontaneous cells are called electrolytic cells

$E°cell = E°cathode - E°anode$

EMF of a non- standard cell:

EMF of the cell is the difference in the electrode potentials,

i.e $E_{cell} = E_{cathode} - E_{anode}$

$$E_{cell} = E^0_{cell} + \frac{0.0591}{n} \log_{10} Q$$

$$E^0_{cell} = (E^0_{cathode} - E^0_{anode})$$

where Q is the constant of the overall redox reaction $\frac{[M_C^{n+}]}{[M_A^{n+}]}$

and 'n' is the number of moles of electrons after balancing the redox reactions.

Numerical problems on electrode potential and EMF of a cell,

1. Calculate the voltage of a cell which consists of a rod iron immersed in a 1M solution of $FeSO_4$ and rod manganese immersed in a 0.1M solution of $MnSO_4$ at 25^0C.Write the cell reaction. Given E^oFe^{2+}/Fe = – 0.44 V and E^oMn^{2+}/Mn = – 1.18V.

Solution: (i)Cell representation: $Mn|Mn^{2+}$ (0.1 M) || Fe^{2+} (1.0 M) | Fe

(ii) Electrode reactions:

At anode: $Mn \rightarrow Mn^{2+} + 2e^-$

At cathode: $Fe^{2+} + 2e^- \rightarrow Fe$

Net cell reaction: $Mn + Fe^{2+} \rightarrow Mn^{2+} + Fe$

(iii)The emf of the cell is

$$E_{cell} = E^0_{cell} + \frac{0.0591}{n} \log \frac{[Fe^{2+}]}{[Mn^{2+}]}$$

$$E_{cell} = (E^o_{Fe2+/Fe} - E^o_{Mn2+/Mn}) + \frac{0.0591}{2} \log \frac{1.0}{0.1}$$

$= [-0.44-(-1.18)] + 0.02955 = 0.74+0.02955$ $E_{cell} = 0.7696$ V

2. An electrochemical cell consists of an iron electrode, dipped in 0.1M $FeSO_4$ and silver electrode dipped in 0.05M $AgNO_3$ solution. Write cell representation, cell reaction and calculate the emf of the cell at 298K. Given SRP of Fe and Ag are -0.44 and +0.8V respectively.

Solution: (i) Cell representation: $Fe|FeSO_4$ (0.1 M) || $AgNO_3$ (0.05 M) | Ag

(ii) Electrode reactions At anode: $Fe \rightarrow Fe^{2+} + 2e^-$

At cathode: $[Ag^+ + e^- \rightarrow Ag] \times 2$

Net cell reaction: $Fe + 2Ag^+ \rightarrow 2Ag + Fe^{2+}$

(iii)The emf of the cell is

$$E_{cell} = E^0_{cell} + \frac{0.0591}{n} \log \frac{[Ag^+]^2}{[Fe^{2+}]}$$

$$E_{cell} = (E^o_{Ag+/Ag} - E^o_{Fe2+/Fe}) + \frac{0.0591}{2} \log \frac{[0.05]^2}{[0.1]}$$

$= [0.8-(-0.44)] + 0.02955 \log 0.025 = 1.24-0.0473$

E_{cell} = 1.1927 V.

Concentration cells

Concentration cell is an electrochemical cell in which two identical electrodes are immersed in the same ionic solution but of different concentration (Fig. 108). As a result, the potential difference arises due to transfer of substance from a solution of higher concentration to a solution of lower concentration.

Derivation of expression for the cell potential of a concentration cell:

Concentration cell can be represented as

$Zn| Zn^{2+} (C_1) || Zn^{2+} (C_2)| Zn$

The half cell reactions are

At anode: $Zn \rightarrow Zn^{2+} (C_1) + 2e^-$

At cathode: $Zn^{2+} (C_2) + 2e^- \rightarrow Zn$

Net cell reaction: $Zn^{2+} (C_2) \rightarrow Zn^{2+} (C_1)$

C_1 and C_2 are concentrations of Zn^{2+} in the two half cells respectively and $C_2 > C_1$.The left hand electrode is the anode and the right hand electrode is the cathode.

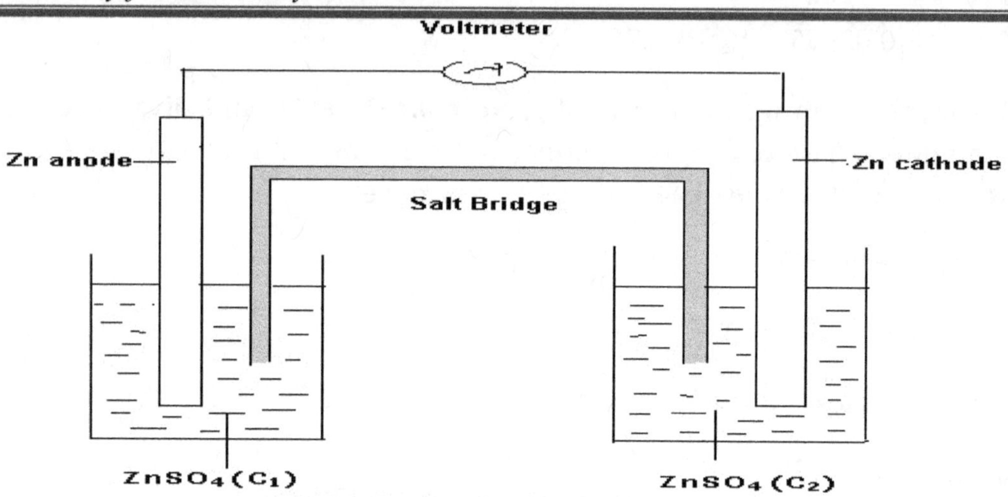

Fig. 108. Concentration cell

Emf of a concentration cell can be calculated using Nernst Equation as follows:

$$E_{anode} = E^0_{electrode} + \frac{2.303RT}{nF} \log C_1$$

$$E_{cathode} = E^0_{electrode} + \frac{2.303RT}{nF} \log C_2$$

$$E_{cell} = E_{cathode} - E_{anode}$$

$$= \left(E^0 + \frac{2.303RT}{nF} \log C_2\right) - \left(E^0 + \frac{2.303RT}{nF} \log C_1\right)$$

$$\mathbf{E_{cell} = \frac{0.0591}{n} \log \frac{C_2}{C_1}}$$

Numerical problems

1. The emf of the cell $Cd/CdSO_4(0.0093\ M)\ //\ CdSO_4(x\ M)\ /Cd$ is 0.086V at 25^{0}C. Find the value of x.

$$E_{cell} = \frac{0.0591}{n} \log \frac{C_2}{C_1}$$

$$0.086 = \frac{0.0591}{2} \log \frac{x}{0.0093}$$

$$\log \frac{x}{0.0093} = 2.910$$

log x- log0.093=2.910

logx + 2.0315=2.910

logx =2.910-1.0315 = 0.8785

x = antilog (0.8785) = **7.55 M**

2. Represent the cell formed by the coupling of two Cu electrodes immersed in CuSO4 solutions concentration of cupric ions in one electrode system in 100 times more concentrated than other. Write the cell reaction and calculate the potential at 300K.

(i)Cell representation:

$Cu \mid CuSO_4$ (x M || $CuSO_4$ (100x M $\mid Cu$

(ii) Electrode reactions:

At anode: $Cu(s) \rightarrow Cu^{2+}$ (x M) + $2e^-$

At cathode: Cu^{2+} (100x M) + $2e^- \rightarrow Cu(s)$

Net cell reaction: Cu^{2+} (100x M) $\rightarrow Cu^{2+}$ (x M)

(iii)The emf of the concentration cell,

$$E_{cell} = \frac{2.303RT}{nF} \log \frac{C_2}{C_1}$$

$$E_{cell} = \frac{2.303 \times 8.314 \times 300}{2 \times 96500} \log \frac{100x}{x}$$

$$= 0.0297 \log 100$$

$$= 0.0297 \times 2$$

$$E_{cell} = 0.0595V$$

1.7 INERT AND SOLUBLE ELECTRODES

Types of Electrodes: Two general types - metallic and membrane electrodes An electrochemical cell can be conveniently constructed by selecting suitably any of the following types of electrodes or half cells, each of which has its own developed potential values.

They are:

Metallic Indicator Electrodes:

Electrodes of the first kind - respond directly to changing activity of electrode ion

Example: Copper indicator electrode

*1) **Metal-metal ion electrode**:* A half cell of this kind consists of a metal dipped in a solution containing its own ions.

The half cell is represented as: **M/M^{n+} (Oxidation)**

The reaction at the electrodes: **$M \rightarrow M^{n+} + ne^-$**

$E_{elect} = E^0_{elect} + (0.0591/n) \log [M^{n+}]$

Example : Zn/Zn^{2+}, Cu/Cu^{2+}

2) ***Metal-metal salt-ion electrode:*** In this type of half cell a metal is in contact with a sparingly soluble salt of the same metal dipped in a solution containing anion of the salt.

The electrode is represented as: **$M \mid MX(s) \mid X^-$**

The electrode reaction is: $MX(s) + ne^- \rightarrow M(s) + X^-$

$E_{elect} = E^0_{elect} + (0.0591/n) \log [X^-]$

Calomel and silver-silver chloride electrodes are the common electrodes generally used in cells

(i) Calomel electrode, $Hg/Hg_2Cl_2/Cl^-$

(ii) Silver-silver chloride electrode, $Ag/AgCl_{(s)}/Cl^-$.

3) ***Gas electrode:*** A gas electrode consists of a particular gas flushed around an inert electrode (Pt) which is dipped in a solution containing ions to which it is reversible. At the metal interface, there is equilibrium between the gas and its ions. The standard hydrogen electrode, (SHE)

The half cell is represent as : **$Pt / H_2(g)$ (1 atm) / H^+ (1M) or $Pt / X_2(g) / X^-$**

The electrode reaction is $X_2 + 2e^- \rightarrow 2X^-$

$E_{elect} = E^0_{elect} + (0.0591/n) \log [H^+]$ or $E_{elect} = E^0_{elect} + (0.0591/n) \log [X^-]$

The electrode reaction is reduction, if the half cell is made as the cathode.

4) ***Oxidation-reduction electrode:*** In this type of half-cell, often, **inactive** (or **inert**) **electrodes** are used, usually made of graphite or platinum:

The half cell is represent as: **$Pt / M_{oxid}, M_{red}$**

The electrode reaction is: $M_{oxid} + ne^- \rightarrow M_{red}$

$E_{elect} = E^0_{elect} + (0.0591/n) \log [M_{oxid}] / [M_{red}]$

For Example: $2I^-(aq) \rightarrow I_2(s) + 2e^-$ [anode]

$MnO_4^-(aq) + 8H^+(aq) + 5e^- \rightarrow Mn^{2+}(aq) + 4H_2O(l)$ [cathode]

In this case, the species involved can't act as electrodes, and so something else must be used. The electrode potential at an inert electrode like Pt arises due to the presence of both oxidized and reduced species of the same metal in the solution It is also called indicator electrode. The electrolyte solutions must contain all of the species involved in the half-reaction.

The developed potential could be described due to the tendency of one form to change into other more stable form. The potential is sensed by the inert electrode, Pt.

Examples: Pt / Fe^{2+}, Fe^{3+}; Pt / Sn^{2+}, Sn^{4+}

5) ***Ion selective electrode:*** Ion Selective Electrodes (ISE) are membrane electrodes that respond selectively to specific ions in the presence of others. The most commonly used ISE is the pH probe. Different types of ion-selective electrodes or chemical sensors are available; that are sensitive selectively to a particular ionic species. ISE are also available for measuring other ions like fluoride, bromide, cadmium, and gases in solution such as ammonia, carbon dioxide, and nitrogen oxide. The potential developed at the ion sensitive electrode sensor is a measure of the concentration of the ionic species of interest.

These electrodes are frequently used as potentiometric electrodes during laboratory analysis as well as in industry, process control, physiological measurements, and environmental monitoring.

An *ion-selective membrane* is the key component of all potentiometric ion sensors. The voltage developed between the sensing and the reference electrodes is a measure of the concentration of the reactive ion being measured.

Reference electrodes

Primary and Secondary Reference Electrodes:

A reference electrode is an electrode which is reversible, has a known and a constant reproducible electrode potential.

Limitations of Primary Reference Electrode: (SHE)

1) It is rather difficult to regulate the pressure of the H_2 gas to be at exactly 1atm throughout the experiment.
2) Excess of H_2 bubbling out carries little HCl with it and hence the H^+ concentration decreases. In such a system, it is difficult to maintain the concentration of HCl at 1M.

3) Platinum foil gets easily poisoned by the impurities present in the gas and HCl. In fact, the attainment of equilibrium is ensured by trial and error.

4) If the solution contains any oxidizing agent, the H_2 electrode cannot be used.

Secondary Reference electrodes: Construction and working of calomel electrode and Ag -AgCl electrode Calomel **Electrode:**

Since a standard hydrogen electrode is difficult to prepare and maintain, it is usually replaced by other reference electrodes, which are known as secondary reference electrodes. These are convenient to handle and are prepared easily. Two important secondary reference electrodes are described here.
It consists of mercury at the bottom over which a paste of mercury-mercurous chloride is placed. A solution of potassium chloride is then placed over the paste. A platinum wire sealed in a glass tube helps in making the electrical contact. The electrode is connected with the help of the side tube on the left through a salt bridge with the other electrode to make a complete cell (Fig. 109).

The potential of the calomel electrode depends upon the concentration of the potassium chloride solution inside. If potassium chloride solution is saturated, the electrode is known as saturated calomel electrode (SCE) and if the potassium chloride solution is 1 N, the electrode is known as normal calomel electrode (NCE) while for 0.1 N potassium chloride solution, the electrode is referred to as decinormal calomel electrode (DNCE).

Half-cell for Calomel Electrode: Pt, Hg, Hg_2Cl_2 / KCl(Satd.)

The **electrode reaction if SCE acts as anode** is: $2Hg(l) + 2Cl^- \rightarrow Hg_2Cl_2\,(s) + 2e^-$

The **electrode reaction if SCE acts as cathode** is

$$Hg_2Cl_2\,(s) + 2e^- \rightarrow 2Hg(l) + 2Cl^-$$

Position of equilibrium affected by $[Cl^-]$ from KCl so E^0 depends on $[Cl^-]$. Most common saturated calomel electrode SCE ($[Cl^-]$~4.5 M)

The potential of the electrode is given by $E_{cal} = E^0 - 0.0591 \log [Cl^-]$

The reduction potentials of the calomel electrodes on hydrogen scale at 298K are as follows:

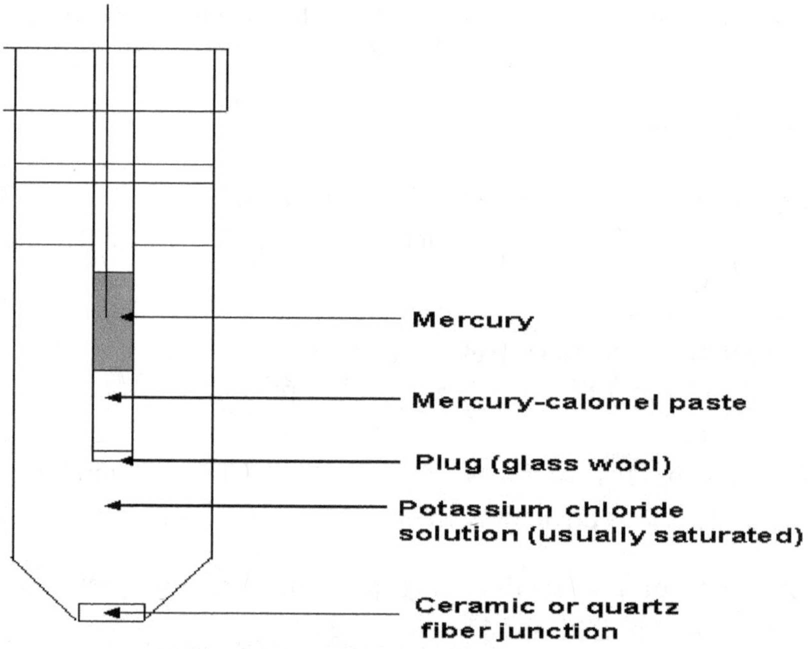

Fig. 109. Calomel electrode

Saturated KC1	0.2415 V
1.0NKC1	0.2800 V
0.1NKC1	0.3338 V

The electrode potential of any other electrode on hydrogen scale can be measured when it is combined with calomel electrode. The emf of such a cell is measured. From the value of electrode potential of calomel electrode, the electrode potential of the other electrode can be evaluated.

(ii) Silver-silver chloride electrode: This is another widely used reference electrode. It is reversible and stable and can be combined with cells containing chlorides without inserting liquid junctions. These are three phase electrodes reversible to anions in solution – i.e the potential is determined by the activity of anion in solution. The three phases in equilibrium with each other are silver, solid silver chloride and a solution containing chloride.

Silver chloride is deposited electrolytically on a silver or platinum wire and it is then immersed in a solution containing chloride ions (Fig. 110). Its standard electrode potential with respect to the standard hydrogen electrode is 0.2224 V at 298 K. **The half-cell of** electrode is represented as:

Ag | AgCl | Cl^-

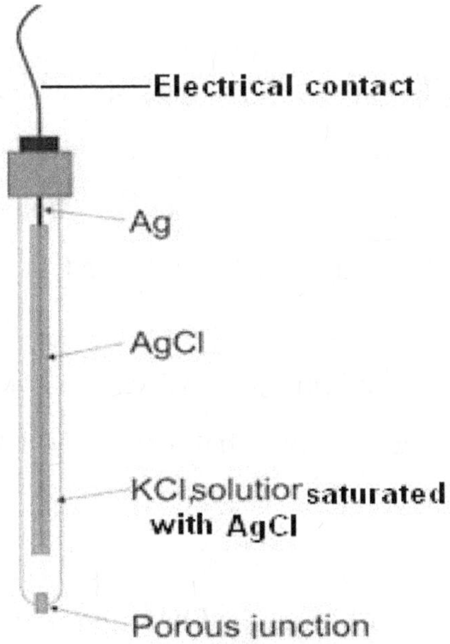

Fig. 110. Ag-AgCl electrode

The electrode functions as a redox electrode and the electrode reaction is between the silver metal (Ag) and its salt - silver chloride, AgCl.

$$AgCl(s) + e^- \leftrightharpoons Ag(s) + Cl^-$$

The potential of the reference electrode is related to the concentration of the Cl^- ion in the filling solution as:

$$E = E^0_{Ag/Ag^+} + \frac{RT}{F} ln \frac{K_{sp}}{[Cl^-]}$$

$$E = E^0_{Ag/Ag^+} + \frac{RT}{F} \ln[K_{sp}] - \frac{RT}{F} ln[Cl^-]$$

$$E = E^0_{Ag/Ag^+} - \frac{RT}{F} ln[Cl^-]$$

the dependence of the potential of the silver-silver(I) chloride electrode on the activity or effective concentration of chloride-ions:

$$E = E^0 - \frac{RT}{F} \ln a_{Cl^-}$$

The silver-silver chloride reference electrode develops a potential proportional to the chloride concentration, whether it is sodium chloride, potassium chloride, mmonium chloride or some other chloride salt and remains constant as long as the chloride concentration remains constant. Most of reference electrodes use a saturated KCl solution with an excess of KCl crystals.

Ion selective electrode (ISE): Glass electrode – Construction

Ion Selective Electrodes (ISE) are electrodes with ***membrane*** which respond selectively to a specific ion in solution, in the presence of other ions. An Ion-Selective Electrode (ISE) produces a potential that is proportional to the concentration of the ion in analyte.

The most commonly used ISE is the probe to measure H^+ ion (pH) concentration and it is called the ***'Glass electrode'***. Ion selective electrodes are also available to measure other ions like fluoride, bromide, cadmium, and gases in solution such as ammonia, carbon dioxide, and nitrogen oxide.

An ***ion-selective membrane*** is the key component of all potentiometric ion sensors. The nature of the membrane determines the selectivity of the electrode. It is across this membrane that the charge develops.

Construction of Glass electrode:

A typical glass electrode is made of glass tube ending with small glass bulb. Active part of the electrode is the membrane in the form of *glass bulb* of wall thickness 0.05 – 0.2mm. The bulb is made of special variety glass called Corning-015, made up of SiO_2 72 %, Na_2O 22 % and CaO 6 % (SiO_4^{4-} framework with charge balancing cations). The bulb has relatively higher electrical conductivity and is highly hygroscopic in nature.

Inside of the glass electrode bulb is usually filled with buffered solution of chlorides, in which silver wire covered with silver chloride is immersed. pH of internal solution is generally either 1.0 (0.1M HCl) or 7.0. The glass electrode can be represented as:

Ag| AgCl(s)| HCl (0.1M)| Glass membrane‖…

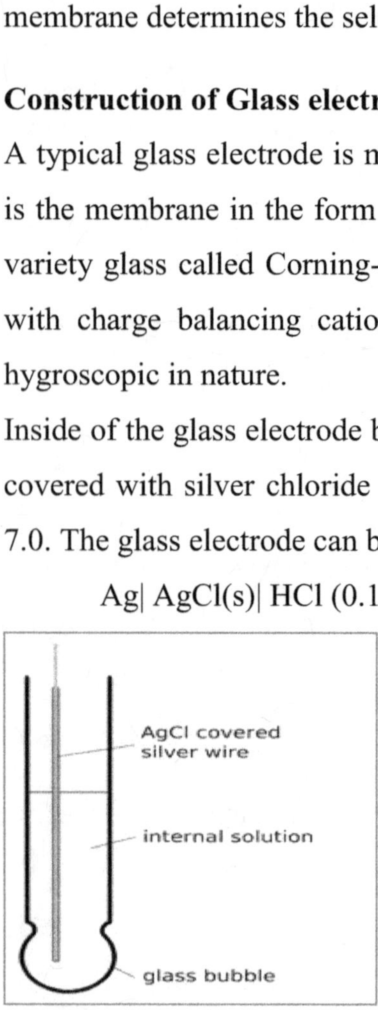

Working: When the bulb is immersed in dilute acid solution, the inside and the outside of the bulb form a hydrated layer across which mobile protons from the solutions penetrate and exchange with the alkali ions in the glass surface and attain an equilibrium. Due to the difference in concentrations of H^+ on either side of the bulb and hence a difference in the rate of exchange of H^+, a potential called the boundary potential develops across the membrane of the glass bulb. This boundary potential can be expressed in the form of Nernst equation viz.

$E_b = \frac{2.303\,RT}{nF}\log\frac{C_x}{C_1}$ where c_x and c_1 are the concn.[H^+] ions in the outside test solution and inside the glass bulb respectively.

$$\text{or } E_b = \frac{2.303\,RT}{nF}\log C_x - \frac{2.303\,RT}{nF}\log C_1$$

$E_b = L - 0.0591pH$ ....(1)

where $L = (-\frac{2.303\,RT}{nF} \log C_1)$

$-\log C_x = pH$ and $(2.303\ RT/nF) = 0.0591$ when $n = 1$ and $T = 298K$

Ideally when the concentration of H^+ ions are equal on either side of the glass membrane (i.e $c_x = c_1$), the boundary potential E_b should equal 0. However, practically it is found that it is not zero but there is a residual potential called as asymmetry potential due to unevenness of the glass membrane surface and is represented as E_{asymm}.

Further, the silver (Ag) wire, on to which is coated a layer of silver chloride (AgCl), which is used to electrically connect the internal buffer in the glass bulb electrode to the pH meter (voltmeter) contributes a constant potential. [Silver is the best electrical conductor of all metals. A common name for the Ag/AgCl wire is reference element].

Thus, the potential of the glass electrode E_G has three components

$$\mathbf{E_G = E_b + E_{assy} + E_{Ag/AgCl}}$$

Substituting for E_b from (1) we have

$$\mathbf{E_G = E_G^0 - 0.0591pH}$$

(where $E_G^0 = L + E_{assy} + E_{Ag/AgCl}$ sum of three constant terms)

Measuring pH of a solution using Glass Electrode:

In practice, the glass electrode is connected to an external reference electrode like saturated calomel electrode and the potential difference measured.

The total cell with the two electrodes can be represented as:

Ref 1 Analyte Glass Electrode

$SCE \,||\, H_3O^+ (C_x) \,|\, Membrane \,|\, H_3O^+ (C_1), Cl^- (a = 1\ M), AgCl\ (sat'd) \,|\, Ag$

Ref 2

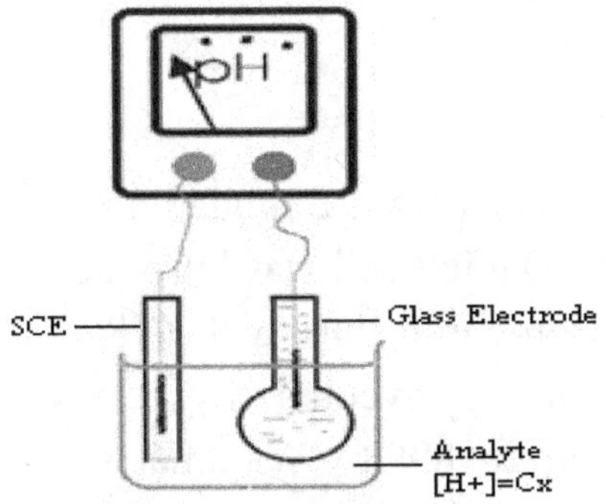

The emf of the complete cell is given by

$E_{cell} = E_G - E_{SCE}$

$E_{cell} = (E^0_G - 0.0591\ pH) - E_{SCE}$

(Where $E_G = E^0_G - 0.0591\ pH$)

$E_{cell} = K - 0.0591\ pH$ (Where $K = E^0_G - E_{SCE}$)

$$pH = \frac{K - E_{cell}}{0.0591}$$

A pH meter is a precise voltmeter that measures the potential difference, in thousandths of a volt (mV), between the reference electrode and the measuring pH electrode. It's scaled in such a way that it displays not the measured potential, but converts it to a display of pH. According to Nernst equation a standard pH electrode generates a voltage of about 59 mV per pH, and at pH 7 (neutral pH) the electrode produces 0 volts. Acids produce negative and bases positive voltages.

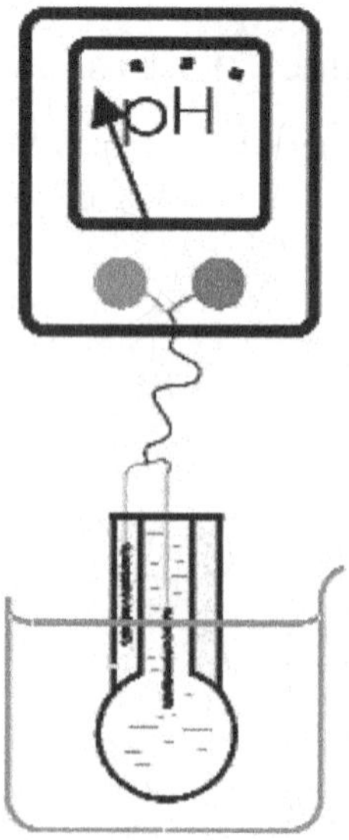

A typical pH meter measures the cell potential and electronically converts the potential to the pH. To evaluate K value, a solution of known pH (standard buffer solution) is used and cell potential is determined.

However, these days, the external calomel reference electrode is replaced with Ag/ AgCl electrode which is built in the glass electrode in a concentric double barrel body of the device and the electrode assembly is called a "combined glass electrode".

Alkaline Error:

At high pH (>10), glass electrode indicates pH less than true value. This is because, at low $[H^+]$, alkali metal ions in solution also exchange with membrane.

$H^+ + Gl^- \rightarrow H^+Gl^-$ (small) $Na^+ + Gl^- \rightarrow Na^+Gl^-$

1.8 BATTERY TECHNOLOGY

Science & technology has helped in the invention of devices which can convert energy available in one form into forms that can perform various tasks. One of the major breakthroughs is the invention of batteries or galvanic cells that convert chemical energy directly into electrical energy. Batteries serve as portable source of electricity. Even though the batteries give limited power out put, they have not lost the importance since the portability is the prime requisite for the efficient use of energy. The development in the field of battery technology continues to meet today's energy requirement.

History of batteries

The history of batteries goes a way back to as early as 250 BC where Parthians from Baghdad had used copper based batteries. The voltaic cell was developed in 1798 – 1800. The Daniell cell was introduced in 1836 by Daniell and the lead – acid cell in 1859 by Gaston Plante. Sometime during the 1860's a Frenchman, Call and invented the gravity cell. In 1866, Georges Leclanche invented dry cell. Carl Gassner invented the modified dry cell in 1887. Ni – Cd cell was introduced in 1899 by Waldmar Junger.

In the year1989, the important invention has been the commercialization of the Ni - Metal hydride cell in which Cd is replaced by a metal hydride. The most significant recent change has been the rapid advance of Lithium battery technology. The excellent electrical characteristics & outstanding performance of such cells now make them the normal choice for primary applications where cost is not the overriding consideration.

Importance of batteries

The number of chemical reactions, which can be successfully harnessed to provide electrical energy, is limited; the range of energy requirements for particular applications extends over many orders of magnitude from 100mWh for a button cell to 20 KWh for an electric vehicle battery. The same is true for power from 25-100 μW for a cardiac pacemaker to 50 kW for a military thermal reserve battery. The diversity of use also requires many specific electrical characteristics which intern calls for different battery chemistry, materials & cell. Consumer electrical equipments know the importance of batteries in regular life. There is high demand for packed electricity in the defense sector, for communication systems, radiosensor, emergency signals, detonators, and submarines. Operating public transport on electricity minimizes pollution. Utility items like flashlights, wrist watches, tape recorders, photo flash devices, heart pacers etc are operated by batteries.

Classification of batteries

Batteries are classified as

(1) primary cells
(2) secondary cells
(3) reserve batteries

Primary cells: primary cells are designed for single use. The electrical energy available depends on the quantity of active materials put in. They have limited shelf life & are not rechargeable.

Ex; Li-MnO_2 cell, Zn-MnO_2 cell.

Secondary cells: Secondary cells are also known as rechargeable batteries or accumulators or storage cells. Secondary cells can be recharged with the help of external source of electricity. The cell reaction is reversible. The charge – discharge cycle can be repeated several hundred times.

Ex: Pb-acid storage cell, Ni – Cd battery.

Reserve batteries: Reserve batteries are primary cells designed for emergency use particularly in military applications. In reserve batteries, one vital component is separated from rest of the components prior to activation. The battery needs to be activated by the addition of this component. The battery is capable of good shelf – life as the chemical reactions during storage is prevented by removing the vital component. The key component is usually electrolyte that is separated. In some of the reserve batteries, water is separated or even heat may be the activator. Heating results in melting of the electrolyte to cause conduction & reactions.

Ex: Zn-Ag_2O, Mg – water activated batteries

Super Capacitors

This is an electrochemical capacitor that has high energy density when compared to common capacitors. These are also called electrical double layer capacitors (EDLCs).

EDLCs store electrical energy using electrochemical cells. These capacitors use two large surface area electrodes (layers) of the same substrate and do not have a conventional dielectric. Super capacitors make use of a nonporous material, typically activated charcoal or metal oxide or conducting polymers in place of conventional insulating barrier.

The EDLC effect was first noticed in 1957 by general electric engineers experimenting with porous carbon – electrodes. Modern version was eventually developed by researches at standard oil of Ohio in 1916.To operate at higher voltages, super capacitors are connected in series.

These are energy storage devices used in vehicles, home solar systems, PC cards, digital cameras, flash lights, Media players, MP3, AM/FM radiators, cell phones and emergency kits.

Commercial cells

In principle, several hundreds of combinations of electrodes of inorganic & organic materials are possible to from a battery. However, in practice, only a few combinations are feasible. Some of the fundamental requirements of commercial cells are

1. Portability
2. Large emf
3. Electrode materials must be stable
4. Good shelf life
5. Long & useful life
6. Compact & agreeable power - weight ratio.
7. Rechargeable
8. High energy efficiency
9. Economically priced

1.9 BASIC CONCEPTS OF BATTERIES

The battery has two half cells, and is represented as

M1|electrolyte: Electrolyte |M2

Oxidation reaction occurs at anodic half cell & is given as

$$M \rightarrow M^{n+} + ne^-$$

& reduction reaction occurs at the cathodic half cell.

$$M^{n+} + ne^- \rightarrow M$$

The electrons from the anodic half cell are transferred to the cathodic half cell. This electron flow occurs along an external conductor and constitutes the flow of electric current. The cell behaves as a source of electrical energy. This process is called **'discharging'**. During discharging, galvanic cells convert chemical energy into electrical energy. Electrochemical oxidation & reduction processes must be fast in electrochemical cells and the equilibrium is established rapidly and steady potential difference is developed.

During Charging, electrical energy is supplied from the external source of power supply. The reactions are reversed i.e., reduction takes place at anode & oxidation at cathode. The electroactive species are restored and the battery can be discharged again.

The reactions during charging are

At anode, $M^{n+} + ne^- \rightarrow M$

At cathode, $M \rightarrow M^{n+} + ne$

Principal components of a battery

A battery consists of

1. Anode
2. Cathode

The electrode materials must be stable over a long period of time in contact with the electrolyte in the prevailing environment. Commercially four anode materials are used and these are Pb, Zn, Fe and Cd. The cathode is invariably oxides or salts of Mn, Hg and Cu or O_2. Some of the combinations like Li-Ag even though yield high emf, are often not preferred because Li is more active and Ag makes the system expensive.

3. Electrolyte

The electrolyte must be highly concentrated with cent percent ionic conduction, non corrosive and cost effective.

Anode and cathode are separated by a separator like cellulose, cellophane, and nafion membranes. This separator prevents the internal short circuiting & allows the electrolyte to pass through it.

The cells are available in various sizes & shapes like cylindrical, disc, and button & prismatic.

Cell characteristics / performance characteristics.

Performance of a battery depends on the cell characteristics. These characteristics are important to suit a battery to particular or specific applications. Some of the important cell characteristics are

1. Voltage:

i) The cell voltage depends on the change in free energy of the system. Free energy change is given by

$$\Delta G = -EnF$$

Emf of the cell is higher when free energy change for the overall reaction is a higher negative value.

ii) emf of a cell is given by the Nernst Equation, $E_{Cell} = E^0_{Cell} - \frac{2.303RTlogK}{nF}$

emf of a cell depends on E^0_{cell} ,

$$E^0_{cell} = E^0_{Cathode} - E^0_{anode}$$

E^0_{cell} is higher when the difference in SRPs of the electrodes is greater.

iii) E_{cell} depends on reaction quotient K,

$$K = \frac{\text{Product of molar conc.of products}}{\text{Product of molar conc. of reactants}}$$

As K increases, emf of the cell decreases.

iv) E_{cell} decreases with increase in temperature and vice versa.

v) Emf of cell depends on internal resistance and over potential. If the electron transfer reactions are rapid, internal resistance of the cell decreases and emf increases.

2. Capacity

Capacity of a battery is the total charge in A-h, it is given by

$$C = \frac{WnF}{M}$$

W is the mass of active material,
M is the molar mass of the active material,
1 A-h is the ability to supply 1A current for 1 hr. or equivalent.

The capacity of a battery depends on its size, which is determined by Faraday's relation. According to Faraday's I law of electrolysis, the amount of a substrate deposited, liberated or dissolved at an electrode is directly proportional to the quantity of electricity passing through the electrolyte.

$$W \propto c \times t$$

$$W = ect \qquad (1)$$

'W' is the amount substrate deposited by Q Coulombs of electricity
& Q = ct
'c ' is the current in amperes.
'e' is the proportionality constant called electrochemical equivalent.

To deposit one mole of atoms of any metal of valiancy 'n' we need n moles of electrons. Thus,

$$M^{n+} + ne^- \rightarrow M$$

$$\therefore M = enF \qquad (2)$$

F is the charge carried by 1 mole of e^-s.
M is the molar mass of the active species.
form equation (1) & (2) we get,

$$C = \frac{WnF}{M}$$

Capacity of a battery is determined discharging a battery at a fixed current. The emf is measured over a definite period of time till the emf reaches a minimum. The variation of cell voltage verses time at constant current discharge is given by the flatness of the curve. Capacity of the battery is higher when the flatter region of the curve is longer. This is one of the most important properties of the battery.

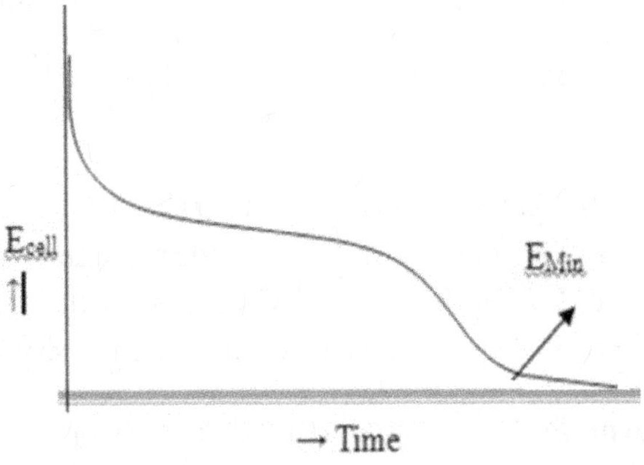

3. **Current:** Current of a battery depends on the internal resistance of the battery. Rapid electron transfer reactions decrease the internal resistance and increases the current. Current of a battery is the rate at which the battery is discharging.

4. **Power density**

   Power density is expressed in W Kg^{-1}.

   Given by iE_{cell}/Weight in kg.

   It is the power per unit weight of a battery. This is a critical characteristic in assessing a battery for its application.

5. **Energy Density**

   Energy density is the energy per unit weight of a battery given by,

   iE_{Cell} t/ W in Kg.

   Expressed in $WhKg^{-1}$,Where t is the time to reach $^{av}E_{Cell}$ at a fixed current discharge.

6. **Specific Capacity** or **energy storage density**.

   It is defined as the capacity per unit weight of the cell and has a unit $AhKg^{-1}$.

7. **Energy Efficiency**

   A good battery possesses high energy efficiency. Energy efficiency depends on the discharge conditions and internal resistance.

   $$\% \text{ Energy Efficiency} = \frac{\text{Energy released on discharge}}{\text{Energy needed for charging}} \times 100$$

This characteristic is applicable to secondary cells.

8. **Cycle life**

   Number of times a battery can be discharged or charged before the failure of a battery occurs is the cycle life. Failure of a battery can occur due to short circuit, corrosion of contacts or falling of active materials from the plate causing unavailability of them for charging.

9. **Shelf life:** Some of the batteries can be stored for many years. The duration of storage of a cell without self – discharge or loss of performance is called shelf life. After the storage period battery should be able to perform the specified characteristics.

10. Batteries should possess tolerance to service conditions like variation in temperature, vibration & shock.

1.10 CLASSICAL BATTERIES

1. Leclanche's dry cell Or Zn-MnO_2 Cell

The dry cell is a modified version of the Leclanche's Cell. The cell was introduced by Gassner in 1888. The contents of the cell are not literally dry but they are pasty.

Anode: Zinc Container (99.9 % purity)

Cathode: Graphite rod

The electrolyte is in the form of jelly. Cathode is surrounded by a paste of MnO_2 & powdered carbon (20%). The zinc container is in contact with a paste of $ZnCl_2$ (91%) NH_4Cl (26%) gelled with starch. This is separated from MnO_2 & Carbon paste by a muslin bag (Fig. 111).

Representation

$Zn \mid ZnCl_2, NH_4Cl \mid MnO_2, C$

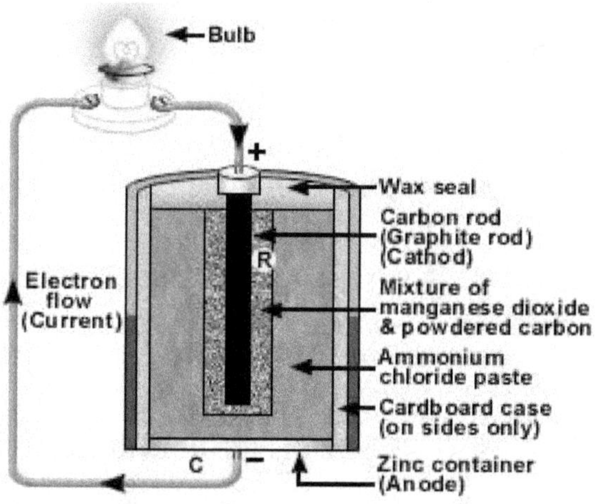

Fig. 111. Zn- MnO_2 cell

Reactions

At Anode, oxidation of zinc takes place

$$Zn \rightarrow Zn^{2+} + 2e^- \quad (1)$$

At cathode, MnO_2 is reduced from +4 to Mn_2O_3 i.e. +3.

$$2MnO_2 + H_2O + 2e^- \rightarrow Mn_2O_3 + 2OH^- \quad (2)$$

Mangannic oxide

The net cell reaction is

$$Zn + 2MnO_2 \rightarrow Zn^{2+} + Mn_2O_3 + 2OH^-$$

$$E_{cell} = 1.5 \text{ V}$$

Secondary reactions

Acidic NH_4Cl reacts with OH^- ions liberating NH_3.

$$NH_4Cl + OH^- \rightarrow NH_3 + Cl^- + H_2O$$

NH_3 evolved under high current drain is absorbed by $ZnCl_2$.

$$Zn^{2+} + 2NH_3 + 2Cl^- \rightarrow ZnCl_2 \,.\, 2NH_3$$

(Unstable complex)

Electrolyte is mainly NH_4Cl. $ZnCl_2$ is added to NH_4Cl to prevent the drying of the electrolyte and to maintain the electrolyte acidic. $ZnCl_2$ absorbs NH_3, which disrupts the current flow.

Advantages

i) Economically priced.

ii) Emf remains constant for small discharge current from 1.5 to 1.45 V for most of the useful battery life.

Disadvantages

i) Low power output when current is drawn rapidly.

ii) Has indefinite shelf – life NH_4Cl being acidic corrodes the zinc container & the cell starts leaking.

Uses:

Used in radios, torch, & flash lights.

Secondary Cells / Storage cells

(i) Lead acid storage cell

The lead – acid storage battery is a secondary cell; it is rechargeable. The cell reaction can be reversed to its original condition.

Construction

Anode: Anode consists of lead alloyed with antimony grids coated with red lead.

Cathode: lead alloyed with antimony grid coated with lead dioxide.

$BaSO_4$, lampblack and graphite are also incorporated as expanders for heavy current discharge batteries. Additives prevent contraction and solidification of spongy lead.

Electrolyte: 27-39% by weight sulphuric acid specific gravity 1.2 g/cm^3 at 15 ^{0}C.

Battery consists of number of positive plates connected in parallel & number of negative plates connected in parallel of the battery. The positive and negative electrodes are interleaved with the separators (Fig. 112). The porous separators used are alkali treated wood, porous rubber, ceramic, microporuos polyvinyl chloride. The electrodes are dipped in H_2SO_4 (5M). Acid corrosion resistant materials are used for the containers, covers and vents.

Representation

$$Pb|\ H_2SO_4|\ PbO_2,\ Pb$$

Or

$$Pb|\ PbSO_4,\ H_2SO_4\ |\ PbSO_4\ |PbO_2,\ Pb$$

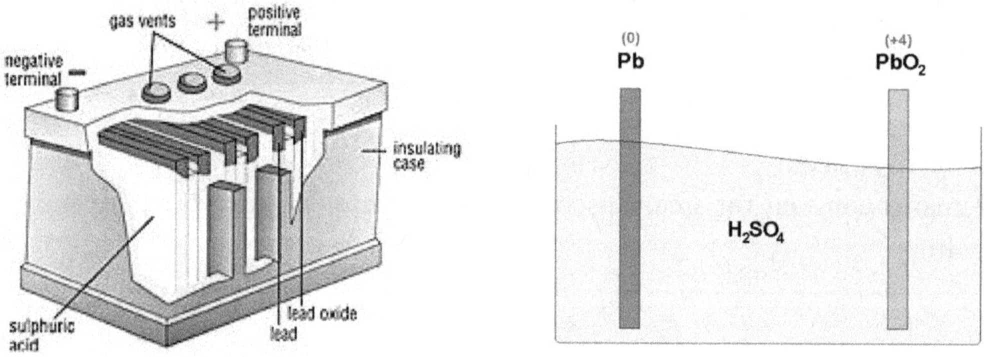

Fig. 112. Lead acid battery

Reactions

During Discharging

At anode, lead is oxidized to lead sulphate

$$Pb \rightarrow Pb^{2+} + 2e^-$$

$$Pb^{2+} + SO_4^{2-} \rightarrow PbSO_4$$

$$Pb + SO_4^{2-} \rightarrow PbSO_4 + 2e^- \qquad (1)$$

The electrons liberated at anode flow towards cathode in the external circuit.

At Cathode, PbO_2 undergoes reduction from +4 state to +2 state in presence of H^+ ions. Pb^{2+} ions combine with SO_4^{2-} ions to form $PbSO_4$

$$PbO_2 + 4H^+ + 2e^- \rightarrow Pb^{2+} + 2\ H_20$$
$$Pb^{2+} + SO_4^{2-} \rightarrow PbSO_4$$

$$PbO_2 + 4H^+ + SO_4^{2-} + 2e^- \rightarrow PbSO_4 + 2\ H_20 \qquad (2)$$

Net cell reaction during discharging is (1) + (2)

$$Pb + PbO_2 + 4H^+ + SO_4^{2-} \rightarrow 2PbSO_4 + 2\ H_20 \qquad (3)$$

The voltage of each cell is 2.0 V

During discharging, $PbSO_4$ is formed at both the electrodes. H_2SO_4 is consumed and H_2O is one of the products. Water formed dilutes the electrolyte. In this condition, the cell is discharged.

Charging

By connecting the cell to an external source of electricity the electrons are forced to flow in the opposite direction. The reactions are reversed.

At anode $PbSO_4 + 2e^- \rightarrow 2Pb + SO_4^{2-}$ (4)

At cathode $PbSO_4 + 2H_2O \rightarrow 2PbO_2 + 4H^+ + SO_4^{2-} + 2e^-$ (5)

The net cell reaction during charging, (4) + (5)

$$2PbSO_4 + 2H_2O \rightarrow Pb + PbO_2 + 4H^+ + 2S0_4^{2-} \quad (6)$$

Uses: Used in automobiles for ignition, in labs, hospitals, telephone exchanges, electric vehicles UPS, invertors.

Disadvantages:

1. Efficiency of the cell decreases with decrease in temperature.
2. $PbSO_4$ clogs the pores in the electrode plates & obstructs penetration of the acid preventing it from interacting with the active materials.
3. PbO_2 & $PbSO_4$ shedding.
4. Excessive charging results in electrolysis of H_2O liberating H_2 and O_2. This builds up the pressure & there are chances of explosion.
5. When $PbSO_4$ grows into large crystals at the electrodes and fall from the surface, battery can no longer be recharged. This is called sulphating & it is an important cause of battery failure.

(ii) Ni-Cd battery

Ni-Cd battery is an alkaline storage cell. Alkaline storage cells were introduced in 1919.

Construction

The electrodes are made up of Nickel steel strips.

Anode: Spongy Cd with 78% Cd $(OH)_2$, Iron (18%), graphite (3%), Nickel.

Cathode: Ni $(OH)_3$ (80%), $Co(OH)_2$ (2%), graphite (18%) & barium compound.

Electrolyte: 6M KOH with few drops of LiOH.

The additives improve the life of the electrode and efficiency of the cell. LiOH increases the capacity of the cell and found to improve cell performance.

For proper action of the cell, positive active material must occupy more space than negative active material. The electrodes are separated by a separator containing an absorbent to hold the electrolyte.

Representation: Cd, $Cd(OH)_2$ | KOH (6M)| $Ni(OH)_3$, $Ni(OH)_2$

In a button type cell (Fig. 113), the anode is the top cover with nickel plated steel cap & cathode forms the bottom can. The electrodes are separated by a thin ebonite separator with absorbent & sealed using a plastic washer.

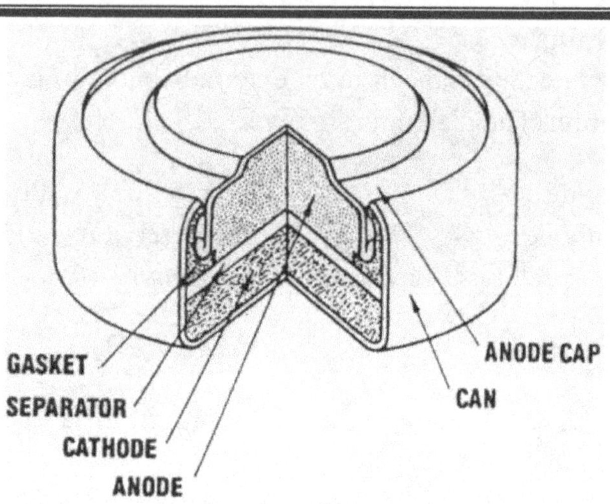

Fig. 113. Ni-Cd battery (Button Type)

Reactions:
During discharge,

At Anode,

$$Cd \rightarrow Cd^{2+} + 2e^{2-}$$
$$Cd^{2+} + 2OH^{-} \rightarrow Cd(OH)_2$$

$$Cd + 2OH^{-} \rightarrow Cd(OH)_2 + 2e^{-} \quad (1)$$

At cathode, $$Ni(OH)_3 + e^{-} \rightarrow Ni(OH)_2 + OH^{-} \quad (2)$$

Net cell reactions is

$$Cd + 2Ni(OH)_3 \rightarrow Cd(OH)_2 + 2Ni(OH)_2 \quad (3)$$

Emf of the cell is 1.3- 1.4V

During charging,

At Anode, $$Cd(OH)_2 + 2e^{-} \rightarrow Cd + 2OH^{-} \quad (4)$$

At Cathode $$Ni(OH)_2 + OH^{-} \rightarrow Ni(OH)_3 + e^{-} \quad (5)$$

Net cell reaction, $$Cd(OH)_2 + 2Ni(OH)_2 \rightarrow Cd + 2Ni(OH)_3 \quad (6)$$

Uses:
Used in pocket calculators, alarm systems, propulsion of industrial trucks, locomotives, mine locomotives, photoflash units & hearing aids.

Advantages:
1. Good ruggedness.
2. Good performance even at low temperature.
3. Low internal resistance.
4. Life span is 5 years.
5. Products formed stick to the electrodes & concentration of KOH is invariant during discharging.

1.11 MODERN BATTERIES

1. Zinc – Air cell

In metal – air cells, one of the active materials is air, which does not contribute to the mass of the battery. High capacity batteries can be made with low overall weight. Oxygen of air is the cathodic depolarizer.

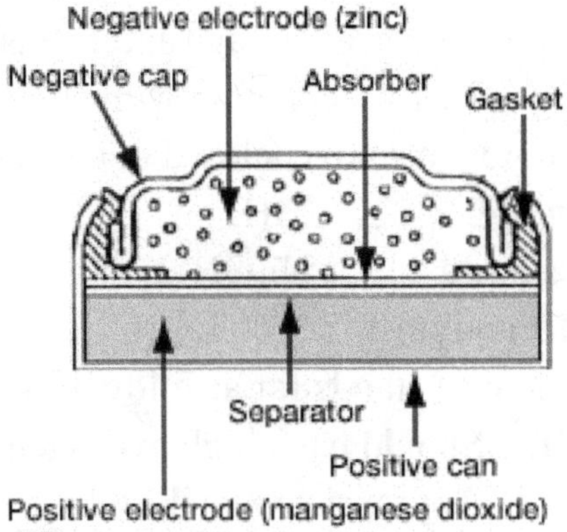

Fig. 114. Zinc-air cell

Construction

The outer container is made of ebonite casing. Rectangular flat plates of zinc are placed on either side of non-graphitic porous carbon cathode. 20% NaOH is used as electrolyte (Fig. 114).

Anode : Zinc

Cathode **: Porous carbon**

Electrolyte : 20% NaOH

Representation: Zn | NaOH(20%) | O_2, C

Reactions:

At anode, zinc is oxidized to Zn^{2+} ions which react with OH^- ions to form zinc oxide.

$$Zn + 2OH^- \rightarrow ZnO + H_2O + 2e^- \quad (1)$$

At Cathode,

$$O_2 + 2H_2O + 4e^- \rightarrow 4OH^- \quad (2)$$

The net cell reaction is, $2Zn + O_2 \rightarrow 2ZnO$

The solution adjacent to cathode becomes progressing alkaline during discharge. Open circuit voltage is 1.45 V.

Advantages:

1. Cell voltage remains constant within 0.1V.
2. High capacity.
3. High energy density and power density.
4. Operating temperature -10^0 to 55^0 C.
5. Low cost.

Disadvantages:

1. Shelf life is limited.
2. H_2 evolution occurs whenever O_2 supply is reduced.
3. Pure O_2 should be used for high current. Presence of CO_2 clogs the pores by forming carbonates.
4. Nitrogen in air clogs the pores when high current is drawn

Uses:

Used in hearing aids, in pagers, remote communication, railways & military applications.

2. Nickel – metal hydride battery (Ni-MH battery)

Ni-MH battery has been developed on a commercial basis by the Ovonic battery company in USA, essentially for electric vehicles.

Construction

The positive electrode is the conventionally sintered chemically impregnated $Ni(OH)_3$ electrode. The negative electrode is a metal hydride and stores hydrogen. Anodic material allows the reversible electrochemical storage and release of hydrogen during charging & discharging of the cell. Polypropylene is used as a separator with an absorbent to hold the electrolyte. Electrolyte used is KOH. All these components are enclosed in a steel case (Fig. 115).

Anode : MH,

Cathode : $Ni(OH)_3$,

Electrolyte : 30% KOH,

Where M stands for an alloy, this alloy

(i) Should have high hydrogen storage capacity.

(ii) Appropriate metal-hydrogen bond strength.

(iii) Posses' fast H_2 recombination kinetics.

The alloys developed as anodic material contain V, Ti, Zr, Ni & Cr with transition element modifies such as Co, Mn or Fe

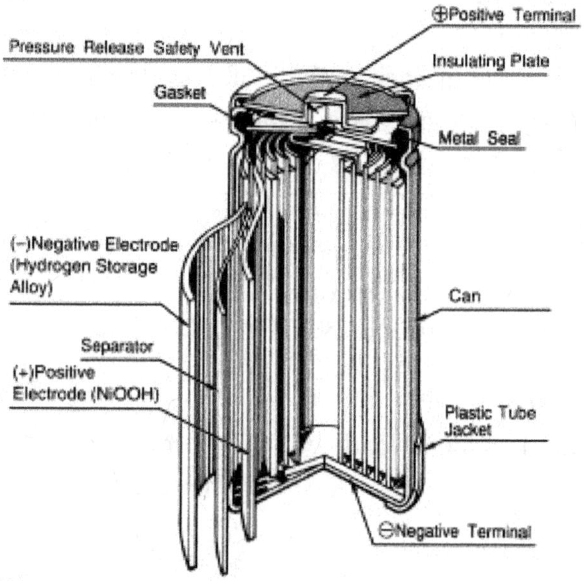

Fig. 115. Ni-MH battery

Representation

MH | KOH (30%) | $Ni(OH)_2$, $Ni(OH)_3$

Reactions:

During discharge,

At anode,

$$MH + OH^- \rightarrow M + H_2O + e^- \quad (1)$$

At cathode,

$$Ni(OH)_3 + e^- \rightarrow Ni(OH)_2 + OH^- \quad (2)$$

The net cell reaction (1) + (2)

$$MH + Ni(OH)_3 \rightarrow M + H_2O + Ni(OH)_2 \quad (3)$$

$E_{cell} = 1.35$ V

During charging,

At anode,

$$M + H_2O + e^- \rightarrow MH + OH^- \quad (4)$$

At cathode,

$$Ni(OH)_2 + OH \rightarrow Ni(OH)_3 + e^- \quad (5)$$

The net cell reaction (4) + (5)

$$M + Ni(OH)_2 + H_2O \rightarrow MH + Ni(OH)_3$$

Advantages:

1. Replaces cadmium, minimum environmental hazards.
2. High capacity
3. Maintenance free.
4. Long cycle life & high energy density
5. Tolerance to overcharge & over discharge

Disadvantages:

1. High cost
2. Safety during operations

Uses: Used in electric vehicles, cell phones, laptops and in space crafts.

3. Li –MnO_2 cell

Lithium batteries are classified as primary and secondary cells. They are also grouped as solid cathode, soluble cathode & liquid cathode or solid state, molten state and high temperature systems. Li-MnO_2 is a primary battery (Fig. 116) and not chargeable.

Construction

Anode : Lithium

Cathode : MnO_2, heat treated and other additives.

Electrolyte : Lithium halide like $LiAlCl_4$, $LiBF_4$ & $LiAsF_6$ in organic solvent like propylene carbonate or 1, 2 dimethoxy ethane

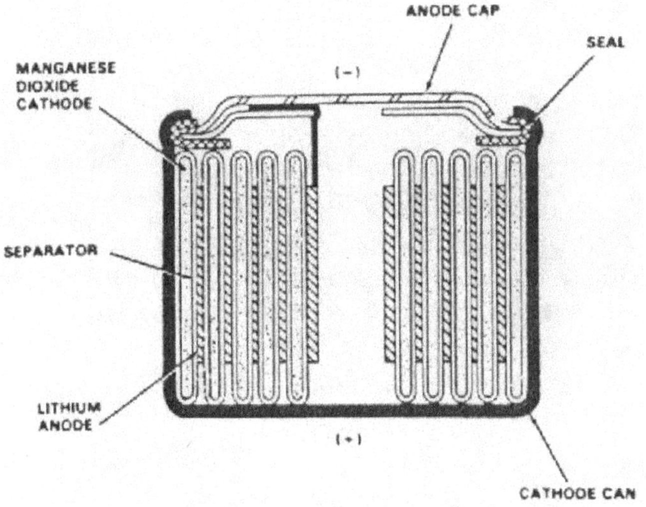

Fig. 116. Li- MnO_2 Cell

Water is not used as a solvent as lithium dissolves in water and also rate of corrosion of lithium is slow in organic solvents. It is constructed using lithium as anode. Most commercial lithium cells use insoluble solid as cathode. Heat treated MnO_2 is used as cathode. Most commercially used solvents are ethylene carbonate, propylene carbonate, esters & sulphoxides. The solutes are $LiClO_4$, $LiACl_4$, $LiBF_4$, & $LiAsF_6$.

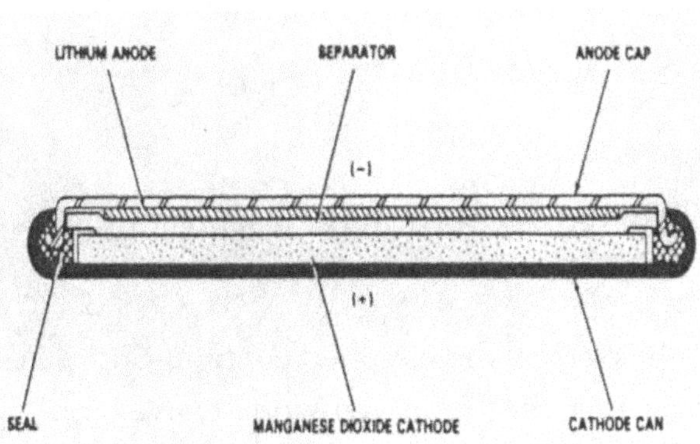

The cell is constructed by using lithium material as the cap of the battery serves as anodic terminal. The bottom can is made up of MnO_2, which serves as the cathodic terminal. Polypropylene acts as a separator and as absorbent to hold the electrolyte.

At anode, oxidation of Li takes place producing Li^+ ions. Li^+ ions move into the electrolyte and electrons move in the external circuit to reach the cathode. At cathode, MnO_2 is reduced from +4 oxidation state to +3 to give MnO_2Li^+.

Reactions:

At Anode, $Li \rightarrow Li^+ + e^-$ (1)

At cathode, $MnO_2 + Li^{+} + e^{-} \rightarrow MnO_2.\ Li^{+}$ (2)

Net cell reaction, (1) +(2)

$$Li + MnO_2 \rightarrow MnO_2Li^{+}$$

Interstitial lithium ion in host MnO_2 lattice

$E_{cell} = 3.0\ V$

Advantages:

1. Lithium is the lightest metal & Li batteries have high power density & energy density.
2. Lithium has lowest SRP.
3. It has high electro chemical equivalence.

Applications:

Li cells find applications in electric watches, cardiac pacemakers, hearing aids, safety & security devices, and memory back up. Secondary Li cells find application in space.

"Baghdad Battery" - 1000 BCE?

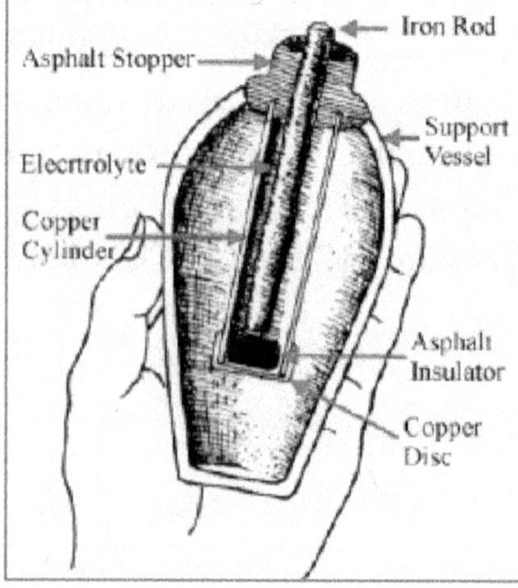

Earthenware jars containing an iron rod surrounded by a copper cylinder were discovered near Baghdad in 1938. They are believed to have been used by the Parthian civilization that occupied the region about 2000 years ago as a source of electricity to plate gold onto silver.

1.12 FUEL CELLS

Fuel cells are the galvanic cells, which convert the free energy of the electrochemical reaction of fuel - oxidant system directly into electrical energy.

Fuel cells consist of electrodes and electrolyte. The fuel and the oxidant are supplied continuously. These fuel cells give more efficient way to generate electricity as it avoids heat –to-work conversion and directly convert chemical energy into electrical energy through electrochemical reactions. The products formed in the fuel cells are non objectionable and government agencies classify the vehicles run by fuel cells as zero –emission vehicles (zev). The first company to adopt fuel cells vehicles is Honda. The cars run by fuel cells are called Honda FCCs.

Fuel cells differ from batteries. The major differences are

* Batteries are energy storage devices; fuel cells are energy conversion devices
* Secondary batteries are rechargeable; Fuel cells are not chargeable in conventional manner. However, they give continuous direct current as long as the reactants are fed at the electrodes
* The reactants and products form integral parts of batteries; in fuel cells there is continuous movement of fuel, oxidant & reaction products in and out of cells.

The fuel cell was first introduced by Sir William Grove in 1839. This cell was planned for a central power station. And today they are exclusively used in space research.

Fuel cells are represented as

Fuel | Anode | Electrolyte | cathode | oxidant

History of Fuel cells

1939 William Grove	H_2-O_2 fuel cell
1889 Mond and Longe	H_2-O_2 fuel cell with Pt electrodes with Pt black
1913 Seigel	Improved the Mond and Longe cell & several molten slat systems were introduced.

The first modern fuel cell is the Bacon cell introduced in 1954 and later commercialized.

Classification of Fuel cells

Fuel cells are classifieds in many ways

(i) Based on the temperature

Low temperature fuel cells (<100^0C)

Intermediate temperature fuel cells (100-250^0 C)

High temperature fuel cells (>500^0 C)

(ii) **Based on the electrolyte**

(a) Alkaline fuel cells

(b) Molten carbonate fuel cells

(c) Solid oxide fuel cells

(d) Phosphoric acid fuel cells

(e) Solid polymer electrolyte fuel cells

(f) Biochemical fuel cells

The reaction taking places in a fuel cell are,

at anode, fuel undergoes oxidation to give electrons

$$\text{Fuel} \rightarrow \text{oxidized product} + ne^-$$

Electrons flow in the external circuit towards cathode, and

at the cathode, reduction of the oxidant takes place.

$$\text{Oxidant} + ne^- \rightarrow \text{reduced products}$$

Basic components of fuel cell

Electrolyte should have high conductivity. It requires the use of highly concentrated solutions. Acids and alkalis are preferred. Many a times, alkalis are less corrosive than acids and electrochemical reduction of O_2 is reversible in alkaline media.

Electrodes should allow the redox reactions to occur at the interface and act on electrolyte. Most commonly used electrodes are Graphite, Platinum or Nickel.

Catalysts used are embedded in the electrodes. Gaskets prevent the leakage of gases between the electrodes.

To generate high current and desired voltage, fuel cells are stacked in series.

Advantages of fuel cells

1. High power efficiency, approximately, 75%.
2. Non objectionable emission & eco friendly.
3. No moving parts and there is no wear and tear, high reliability in electricity generated.
4. Space required for fuel cell is less.
5. Produces direct current for a long time.

Disadvantages

1. Storage of fuel and oxidant.
2. Electrolytes and electrodes are costly.
3. Gives DC output and should be converted to AC.

H_2-O_2 fuel cell

Alkaline H_2-O_2 fuel cell (Fig. 117) has found extensive applications in space flights and automobiles. Pure H_2 gas is used as a fuel and CO_2 free O_2 is used as an oxidant. The electrodes are made up of porous carbon. Electrocatalysts are Pt and Ag_2O impregnated in the pores of the carbon at anode and cathode respectively. Electrolyte is 30-45% aqueous KOH solution.

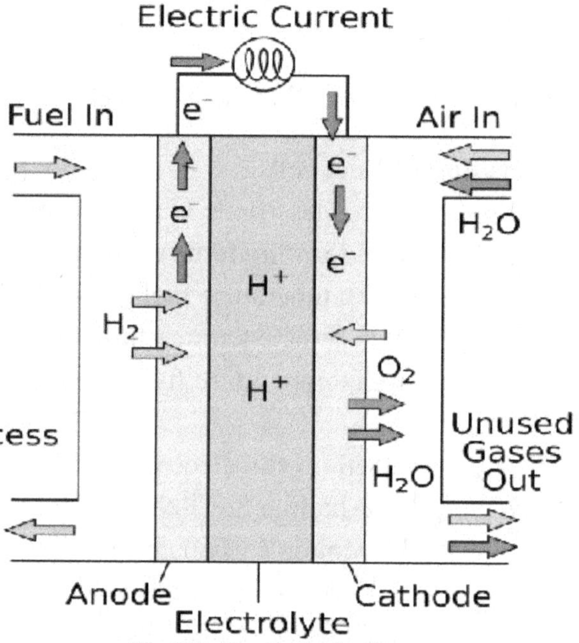

Fig. 117. H_2 - O_2 Fuel cell

Electrodes : Porous Carbon
Fuel : H_2
Oxidant : O_2
Electrolyte : KOH
Electrocatalyst: powdered Pt at anode, Ag_2O at cathode

Working

At anode, H_2 gas at a pressure of 50 atm is continuously fed. H_2 gas diffuses through the pores of the electrode and reacts with the electrolyte, undergoes oxidation to form H_2O.

$$H_2 \rightarrow 2H$$
$$2H + 2OH^- \rightarrow 2H_2O + 2e^-$$

$$H_2 + 2OH^- \rightarrow 2H_2O + 2e^- \quad (1)$$

Electrons produced at the anode flow through the external circuit towards the cathode.
At cathode, O_2 is diffused through porous carbon electrode impregnated with Ag_2O catalyst, where it is reduced to OH^- ions

$$O_2 + 2H_2O + 4e^- \rightarrow 4OH^- \quad (2)$$

The overall reaction of H_2-O_2 fuel cell is

$$2H_2 + O_2 \rightarrow 2H_2O$$

The standard emf H_2-O_2 fuel cell is 1.23 V. The operating voltage is 0.8 to 1.0 V. The electrical output is a result of flow of electrons from anode to cathode through the external circuit. In addition to the electricity, the water formed is also a valuable product.

The temperature is maintained around 80^0C to evaporate the water formed. Otherwise, water dilutes the electrolyte and efficiency of the fuel cell decreases.

Applications

Used in space vehicles, submarines and automobiles.

Methanol-O_2 fuel cell

Methanol is preferred as a fuel in fuel cells because of the following reasons.

1. It has low carbon content.
2. It possesses a readily oxidisable alcoholic group and
3. It has high solubility in aqueous electrolytes.

Construction

Electrodes are made up of porous nickel sheets impregnated with electro catalysts Pt/Pd catalyst is deposited on anode and silver on cathode (Fig. 118).

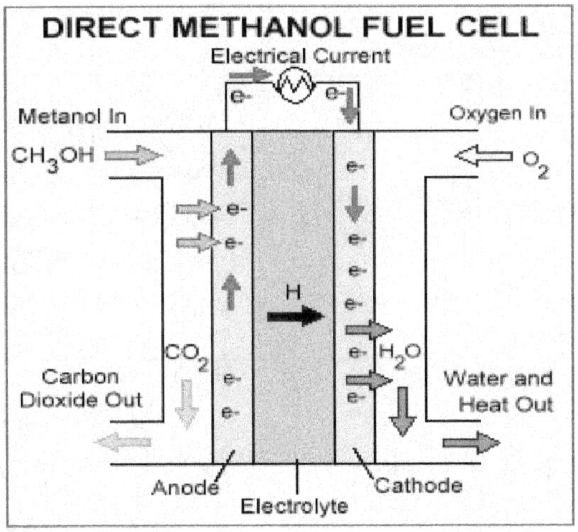

Fig. 118. Methanol-oxygen fuel cell

Fuel : Methanol
Oxidant: O_2
Electrolyte: H_2SO_4

KOH is not used as electrolyte as it reacts with CO_2 and gets converted to carbonate. Thus the conductivity of the electrolyte decreases as well as the efficiency.

At anode, CH_3OH undergoes oxidation to CO_2.

$$CH_3OH + H_2O \rightarrow CO_2 + 6H^+ + 6e^- \quad (1)$$

Electrons flow towards the cathode and are responsible for the electrical output.

At cathode, O_2 gets reduced in presence of H^+ ions.

$$\frac{3}{2}O_2 + 6H^+ + 6e^- \rightarrow 3H_2O \qquad (2)$$

The net cell reaction of CH_3OH –O_2 fuel cell is given by

$$CH_3OH + \frac{3}{2}O_2 \rightarrow CO_2 + 2H_2O$$

The maximum theoretical voltage is 1.186 V at 25^{0}C. The potential developed is a mixed potential, involving reactions of HCHO & HCOOH species. The catalyst is easily poisoned by C-OH species. The cell has military applications.

Considerable research work is going on since the invention of fuel cell in 1839 to meet the power requirements of day-to-day life.

Different types of fuel cells, classified on the basis of type of electrolyte present

	Alkaline FC	Phosphoric acid FC	Molten Carbonate FC	Solid Oxide FC	Solid Polymer electrolyte FC	Biochem FC
Fuel	H_2	H_2	H_2	H_2 + CO	H_2	H_2 or Glucose urea
Oxidant	O_2	O_2	O_2	O_2	O_2	O_2,Carbonate / sulfate
Electrodes	Carbon or Pt Porous	Carbon coated with Pt or Pt -Co-Cr alloy.	Anode: Ni-Cr alloy Cathode : NiO doped with Li	Anode: Co or Ni Cathode : In_2O_3 +SnO_2	Graphite coated with Pt black	Bio organism
Electrolyte	KOH	Phosphoric acid	$LiAlO_2$, Li_2CO_3, K_2CO_3	ZnO_2- CaO doped with Y_2O_3	Ion exchange membrane (Nafion R1170)	Water/ sea water pH=7 or IEM
Catalyst	Pt/Pd, Ag_2O	Pt-Co-Cr alloy	Nickel Oxide	Not required	Pt	Enzymes
Operating temperatur e	80^{0}C	150-200^{0}C	650^{0}C	1000^{0}C	80^{0}C	0-40^{0}C
Power density	1	0.8-1.9	1.5-2.6	0.1-1.5	4.8-6.5	Low power o/p

(kW/m^3)	–					
Applications	Space shuttles & Submarines	Stationary power in hotels hospitals, office buildings & vehicles	Used as a DC device in chloroalkali & aluminum industries	Electric power plants, locomotives	Space shuttle	Space research, as low power source for detonators.

1.13 CORROSION: Introduction

The corrosion of metals is one of the most significant problems faced by advanced industrial societies. It is estimated that about 30-40 % of iron and Steel produced annually is used just to replace the rusted Iron objects. Corrosion costs of automobiles-fuel systems, radiators, exhaust systems, and bodies- are in the billions. Corrosion touches all- inside and outside the home, on the road, on the sea, in the plant, and in aerospace vehicles. Total annual costs of floods, hurricanes, fires, lightning and earthquakes are less than costs of corrosion.

Metals are thermodynamically unstable and found in nature that require greater amounts of energy to convert them to useful engineering materials that show a greater tendency to corrode while those that require less energy show a lower tendency to corrode. Examples of the former include magnesium, zinc, aluminum, and steel, while examples of the latter include gold, silver and platinum. Corrosion is the spontaneous slow chemical interaction of metal or alloy with it's environment resulting in the formation of one of its compounds like oxide, hydrated oxide, carbonate, sulphide, sulphate etc. Extraction of metals from their ores is an endothermic process. Pure metals, being highly energetic, have natural tendency to revert back to their combined states.

$$\text{Metal ores} \underset{\text{Corrosion}}{\overset{\text{Metallurgical extraction}}{\rightleftarrows}} \textit{Metal}$$

Corrosion is defined as the destruction of metals or alloys by the surrounding environment through chemical or electrochemical changes.

Corrosion is the deterioration of materials by chemical interaction with their environment. The term corrosion is sometimes also applied to the degradation of plastics, concrete and wood, but generally refers to metals.

The most familiar example of corrosion is rusting of iron exposed to the atmospheric conditions.

$$2\,Fe\,(s) + \frac{3}{2}O_2(g) + x\,H_2O\,(l) \longrightarrow \underset{\text{Chemical Composition of rust}}{Fe_2O_3 \cdot x\,H_2O\,(s)}$$

In general any type of corrosion cell (Fig. 119) consists of an anode, a cathode, an metallic (electronic) path between them, and an electrolyte providing ionic path. A voltage difference between the anode and the cathode in the corrosion cell causes a current flow. This DC current flows from the anode through the surrounding environment (electrolyte) to the cathode and back to the anode through the metallic path (wire) completing the circuit. The corrosion current is carried through the metallic path by electrons and while in the electrolyte, it is carried by ions. The electrode that discharges current to the electrolyte (anode) corrodes, while the metal receiving the current from the environment (cathode) remains uncorroded. In general

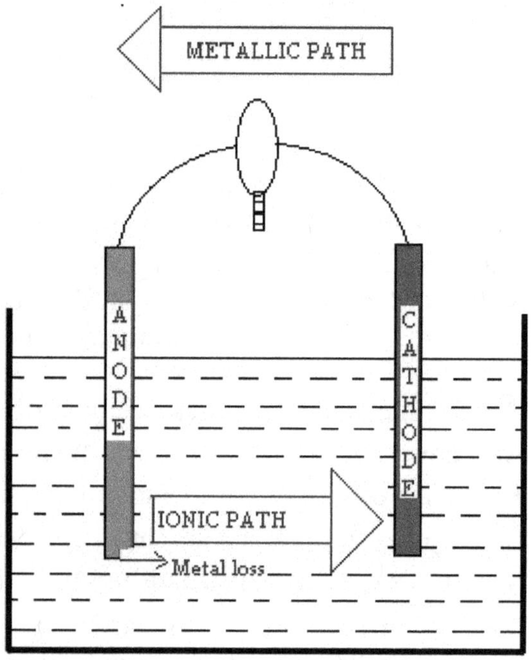

Fig. 119. Corrosion cell

Corrosion of metals occurs either by direct chemical attack or by electro chemical attack on the metal by the corrosive environment.

Dry corrosion involves direct attack of atmospheric gases on metal (In the absence of moisture). Wet corrosion involves electro chemical attack on metals in aqueous environments. Wet corrosion is more prevalent than dry corrosion.

Effects of corrosion

The effects of corrosion are many and varied and the effects of these on the safe, reliable and efficient operation of equipment or structures are often more serious than the simple loss of a mass of metal. Failures of various kinds and the need for expensive replacements may occur even though the amount of metal destroyed is quite small. Some of the major harmful effects of corrosion can be summarised as follows:

1. Reduction of metal thickness leading to loss of mechanical strength and structural failure or breakdown. When the metal is lost in localised zones so as to give a crack-like structure, very considerable weakening may result from quite a small amount of metal loss.
2. Hazards or injuries to people arising from structural failure or breakdown (e.g. bridges, cars, aircraft).

3. Loss of time in availability of profile-making industrial equipment.
4. Reduced value of goods due to deterioration of appearance.
5. Contamination of fluids in vessels and pipes (e.g. beer goes cloudy when small quantities of heavy metals are released by corrosion).
6. Loss of technically important surface properties of a metallic component. These could include frictional and bearing properties, ease of fluid flow over a pipe surface, electrical conductivity of contacts, surface reflectivity or heat transfer across a surface.
7. Mechanical damage to valves, pumps, etc, or blockage of pipes by solid corrosion products.

Theories of corrosion

Dry Corrosion (Direct Chemical Attack)

The Dry corrosion occurs in the absence of a liquid phase i.e, conducting (aqueous) medium. The corrosive media include vapours, gases, etc. The example of dry corrosion is the reaction between metal and oxygen (atmosphere) at elevated temperatures in perfectly dry conditions. Oxygen is mainly responsible for the corrosion of most metallic substances when compared to other gases and chemicals. According to this theory, corrosion on the surface of a metal is due to direct reaction of atmospheric gases like oxygen, halogens, oxides of sulphur, oxides of nitrogen, hydrogen sulphide and fumes of chemicals with metal.

There are three main types of dry corrosion.

(i) Oxidation corrosion (*Reaction with oxygen*): Some of the metals directly react with oxygen in the absence of moisture. Alkali and alkaline earth metals react with oxygen at room temperature and form corresponding oxides, while some metals react with oxygen at higher temperature. Noble metals like Ag, Au and Pt are not easily oxidisable. The metal oxidelayer formed at the surface of the metal can be protective to the metal. If diffusion of either oxygen or metal is across this layer, further corrosion is possible. Thus, the layer of metal oxide plays an important role in the process of corrosion. Oxides of Pb, Al and Sn are stable and hence inhibit further corrosion. They form a stoichiometric, nonporous, stable, tightly adhering oxide film. In case of porous oxide film, atmospheric gases pass through the pores and react with the metal and the process of corrosion continues to occur till the entire metal is converted into oxide. Porous oxide layer is formed by al- kali and alkaline earth metals. Molybdenum forms a volatile oxide film of MoO_3 which accelerates corrosion.

Au, Ag, Pt form unstable oxide layer which decomposes soon after the for- mation, thereby preventing further corrosion.

(ii) Corrosion by other gases such as Cl_2, SO_2, H_2S, NOx: In dry atmosphere, these gases react with metal and form corrosion products which may be protective or non-protective. Dry Cl_2 reacts with Ag and forms AgCl which is a protective layer, while $SnCl_4$ is volatile. In petroleum industries at high temperatures, H_2S attacks steel forming FeS scale which is porous and interferes with normal operations.

(iii) Liquid metal corrosion: In several industries, molten metal passes through metallic pipes and causes corrosion due to dissolution or due to internal penetration. For example, liquid metal mercury dissolves most metals by forming amalgams, thereby corroding them.

Wet Corrosion (Electrochemical theory of corrosion)

Wet corrosion usually occurs in conducting liquid (aqueous) phase. This type of corrosion occurs when the metal comes in contact with a conducting liquid or when two dissimilar metals are immersed or dipped partly in a solution.

According to the electrochemical theory of corrosion, the corrosion of a metal, such as Iron, exposed to the environment may be thought of as three steps process.

1) Formation of large number of galvanic cells i.e., the Development of anodic and cathodes regions on the same metal. The electrons released at the anode are absorbed at cathode.
2) Corrosion (oxidation) takes place at the anodic region.
3) Reduction of O_2 (air) in presence of water to hydroxyl [OH^-] ions takes place at cathodic region.

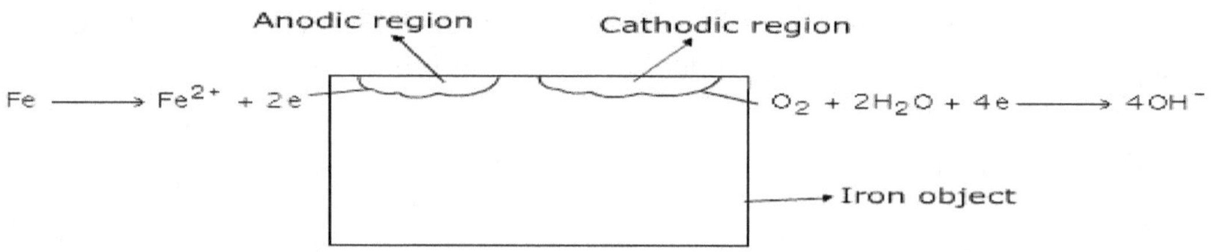

Corrosion Reactions

Anodic reaction: Metal undergoes oxidation (corrosion) with +ve release of electrons

$$Fe \longrightarrow Fe^{2+} + 2e$$

At the cathodic region, electrons are absorbed and cause reduction of the constituents.

Cathodic reactions:

Cathodic reactions are dependent on the constituents of the corrosion medium. There are three possible ways in which reduction takes place.

i) If the solution is aerated and almost neutral, oxygen is reduced in presence of H_2O to OH^- ions.

$$O_2 + 2H_2O + 4e \longrightarrow 4OH^-$$

ii) If the solution is deaerated and almost neutral. H_2 is liberated along with OH^- ions.

$$2H_2O + 2e \longrightarrow H_2 + 2OH^-$$

iii) If the solution is deaerated and acidic, H^+ ions are reduced to hydrogen gas

$$2H^+ + 2e \longrightarrow H_2$$

The metal ions formed at the anode combine with hydroxyl ions and form the corresponding metallic hydroxide $Fe(OH)_2$ which further gets oxidized to hydrated ferric oxide [rust].

$$2\,Fe^{2+} + 4\,OH^{-} \longrightarrow 2\,Fe(OH)_2$$

$$2\,Fe(OH)_2 + O_2 + (x-2)H_2O \longrightarrow \underset{Rust}{Fe_2O_3 \cdot x.H_2O}$$

In the presence of limited oxygen, Black rust is formed as follows:

$$3\,Fe(OH)_2 + \frac{1}{2}O_2 \longrightarrow \underset{\text{Black rust}}{Fe_3O_4 \cdot 3H_2O}$$

Differences between Dry corrosion and wet corrosion

	Dry corrosion	Wet corrosion
1	Corrosion occurs in the absence of moisture	Corrosion occurs in the presence of conducting medium
2	It involves direct attack of chemicals on the materials surface	Electrochemical (galvanic) cells are formed on the materials surface
3	It is a slow process	It is a rapid process
4	Corrosion products are produced at the site of corrosion	Corrosion occurs at anode but rust is deposited at cathode
5	The corrosion is uniform.	It depends on the size of the anodic part of metal

Types of Electrochemical Corrosion

1) Differential metal corrosion

2) Differential aeration corrosion - water line and pitting corroison

3) Crevice corrosion

1) Differential metal corrosion (Galvanic Corrosion)

Differential metal corrosion occurs when two dissimilar metals are in contact with each other in a corrosive environment. The metal with low reduction potential (Anode) undergo oxidation (corrosion) while metal with high reduction potential (cathode) involves reduction. The Greater the electrode potential difference, the higher the corrosion rate. Thus it is anodic metal, which always undergoes corrosion, and cathodic metal is unattacked.

Consider a bimetallic couple of Iron and Copper.

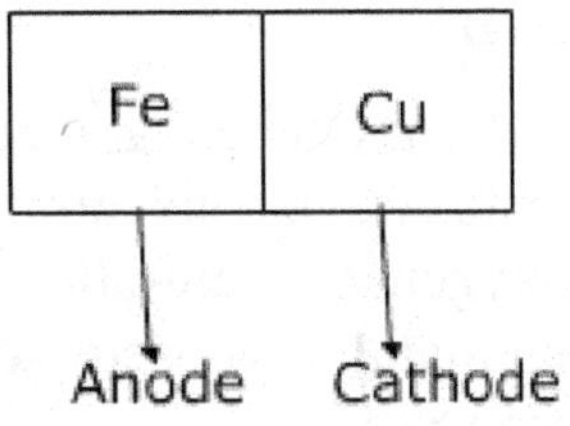

Mg → - 2.52 V

Zn → -0.76V

Cu → 0.34 V

Ag → 0.80 V

Iron is placed above copper in electrochemical series therefore Iron is anodic to copper i.e, Iron acts as anode and undergoes corrosion where as copper acts as cathode and remain unattacked.

Reactions: $Fe \longrightarrow Fe^{2+} + 2e$

$$\frac{1}{2}O_2 + H_2O + 2e \longrightarrow 2OH^-$$

$$2Fe^{2+} + 4OH^- \longrightarrow 2Fe(OH)_2$$

In a bimetallic sample of Iron and Zinc

Zinc is placed above Iron in electrochemical series. Zinc is anodic to Fe, therefore Zinc undergo corrosion. Therefore Iron corrodes in contact with copper or silver where as Zinc and Magnesium corrode in contact with Iron. The rate of Differential metal corrosion depends upon the potential difference between the two metals, higher the difference, faster is the rate of corrosion.

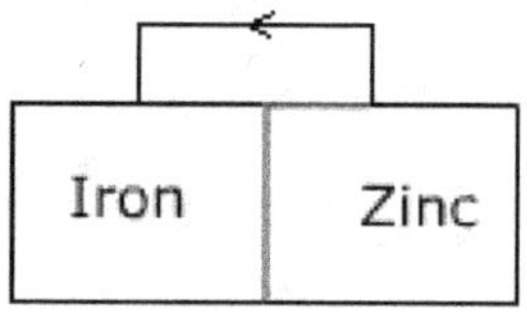

Consider another bimetallic couple of Steel and Copper.

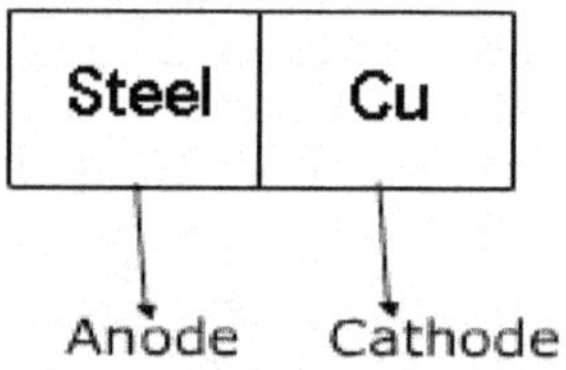

Steel is anodic in nature thus undergoes corrosion. The rate of corrosion depends on the potential difference between the two metals (materials).

Examples of Differential metal corrosion (galvanic corrosion)

1. Buried iron pipeline connected to zinc bar
2. Steel pipe connected to copper plumbing
3. Steel propeller shaft in bronze bearing
4. Zinc coating on mild steel

Preventive measure

1. Place a thin layer of an insulator between two metals or materials
2. Materials selected should have very less potential difference.

2) Differential aeration corrosion

Differential aeration corrosion occurs when a metal is exposed to differential air concentrations or oxygen concentrations. The part of the metal exposed to higher oxygen concentration acts as cathodic region and the part of the metal exposed to lower oxygen concentration acts as anodic region.

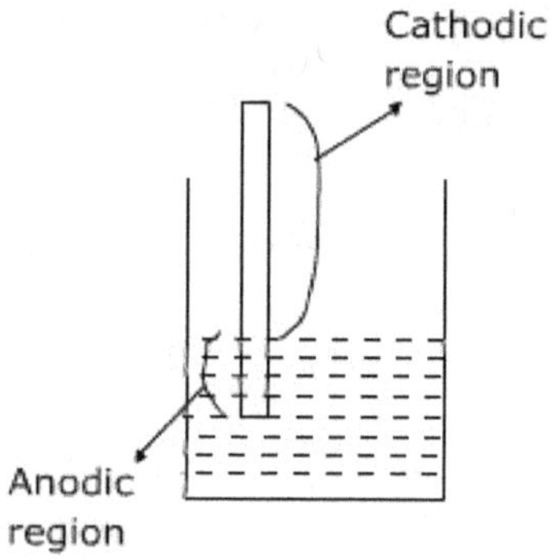

For example consider a plate of pure iron which is partially immersed in an aerated solution of sodium chloride.

Since cathodic reaction requires O_2. The part of the metal exposed to higher oxygen concentration always acts as cathodic region. The portion of the metal immersed in solution is anodic region.

Anodic reaction $Fe \rightarrow Fe^{2+} + 2e$

[bottom portion]

Cathodic reaction $\frac{1}{2}O_2 + H_2O + 2e \longrightarrow 2OH^-$

[Near the water line]

a) Waterline corrosion

Waterline corrosion is a case of differential aeration corrosion. Ocean going ships and water storage steel tanks involve waterline corrosion.

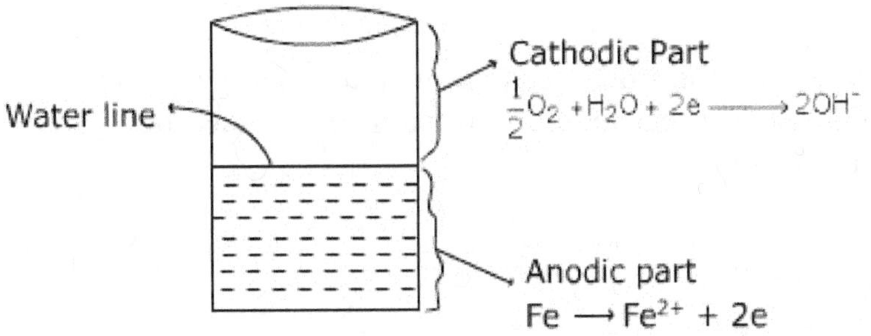

The metal portion just below the water line is more anodic to portion above the water line.

∴ The metal just below the waterline undergoes corrosion.

Water line corrosion is common in water lifting pipes under water. Ocean going ships also suffer from differential aeration corrosion. Ships sunk under water for several years do not undergo corrosion. For the height of the ship sunk under water, the difference in O_2 concentration between the top and bottom of ship is negligibly small. Thus ship sunk under sea, is exposed to almost uniform concentration of air and doesn't undergo differential aeration corrosion.

b) Pitting corrosion

Pitting corrosion involves localized attack usually in chloride medium. Pitting corrosion arises when small portion of the metallic surface is either defective(with cracks) or occupied by dust or scale or sand or water drop (Fig. 120).

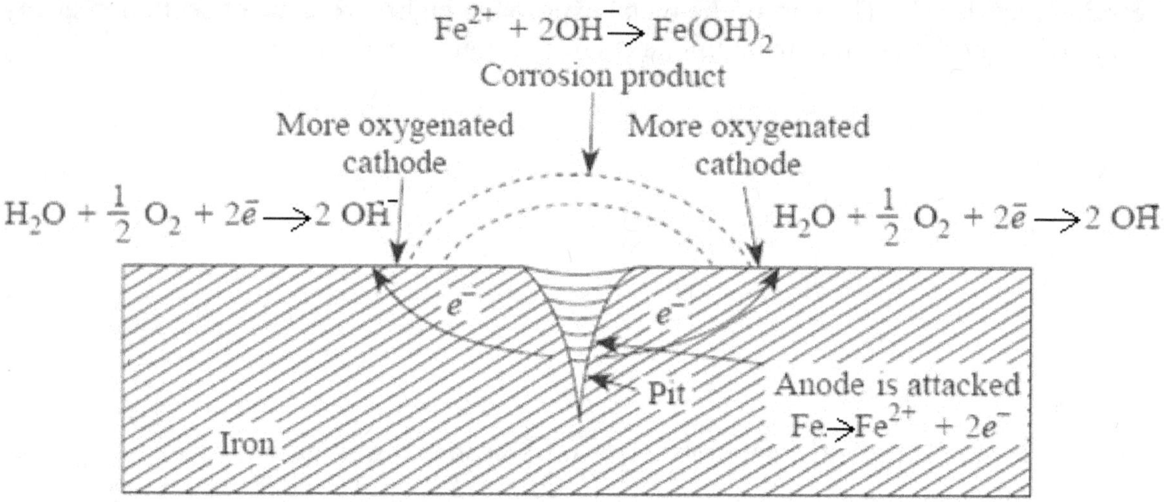

Fig. 120. Pitting corrosion at the surface of iron

The metal below the dust (deposit particle) is exposed to lower O_2 concentration and undergoes corrosion. Once a pit is formed, the rate of corrosion increases. The accelerated corrosion takes place below the deposit because of small anodic area and large cathodic area.

Examples of Differential aeration corrosion

1. Half immersed ion plate in aqueous solution
2. Steel pipe carrying any liquid exposed to atmosphere
3. Ocean going ships
4. Steel storage tanks

Preventive measures

1. Use of metallic coating
2. Use of electroless plating method
3. Use of chemical conversion method
4. Maintaining the material's own protective film.
5. Control of the chemistry of fluids and use of inhibitors

3) Crevice corrosion

Crevice corrosion is often associated with small volume of stagnant solution or electrolyte trapped in crevices of joints, rivets, bolts, gasket surfaces, lap joints, nuts, washers or in surface deposits. An example of this is crevice corrosion which occurs when oxygen cannot penetrate a crevice and a differential aeration cell is set up. Corrosion occurs rapidly in the area with less oxygen. If surface of painted metal is scratched, it will undergo corrosion. The scratched portion acts as small anode and the rest part will act as cathode forming a local cell.

Mechanism of crevice corrosion

In general crevice corrosion occurs rapidly in the area with less oxygen. Increase in Acidity & Chloride Content of Crevice Solution initiates passive film breakdown process and there by enhances rate of corrosion reaction. Finally corrosion propagation continues for further deterioration of material.

The Fontana and Greene model describes crevice corrosion mechanism. This model consists of four stages. Stage 1: Corrosion occurs as normal both inside and outside the crevice: Anodic reaction: $M \rightarrow M^{n+} + ne$, cathodic reaction: $O_2 + 2H_2O + 4e = 4OH^-$. The positively charged metallic ions are electrostatically counterbalanced by OH-. Stage 2: at this stage, the cathodic reaction inside the crevice consumed most of the oxygen available. Stage 3: Cl- and OH- diffuse into the crevice to maintain a minimum potential energy. Metal chloride is formed. Hydrolysis of metal chloride lowers pH and MCln + nH_2O = M(OH)n + nHCl. Stage 4: More Mn+ ions attack more Cl- leads to lower pH inside crevice, metal dissolution accelerates and more Mn+ ions will be produced that will lower pH.

Preventive measures

1. Use alloys sufficiently high in chromium, molybdenum and/or nitrogen to ensure resistance.
2. Prepare surfaces to best possible finish. Mirror-finish resists pitting best.
3. Remove all contaminants, especially free-iron, by passivation
4. Design and fabricate to avoid crevices.
5. Design and fabricate to avoid trapped and pooled liquids
6. Weld with correct consumables and practices and inspect to check for inadvertent crevices.
7. Close crevices in existing lap joints by continuous welding, caulking, or soldering.

Other Types of Corrosion

1. Stress Corrosion

Mechanical forces within the metal bring about stress corrosion. The internal stress of metal arises when metal is subjected to mechanical operations such as pressing, hammering, bending, welding or riveting. The stress can also be external stress acting on the metal during service conditions.

The metal atoms under stress are always at higher energy levels as compared to atoms free from stress. The stressed part of the metal therefore becomes more reactive than the stress free part. As a result, Galvanic cell is formed with the stressed part, acting as anode and stress free part acting as cathode under specific corrosion environments, the stressed part undergoes corrosion, initiating the crack. In the presence of continuously acting stress and corrosion medium the crack propagates through the metal. Cracking of brass and caustic embrittlement of steel boilers are two examples of stress corrosion.

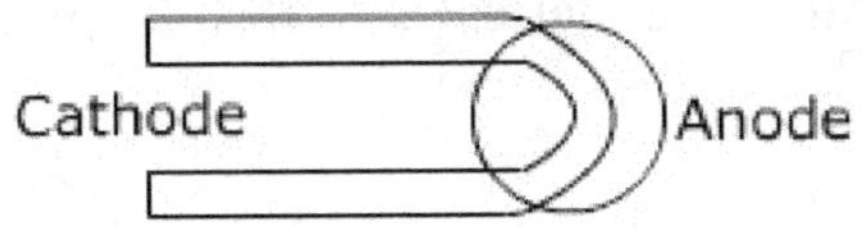

Caustic embrittlement of steel boilers

Mild steel boilers undergo corrosion at the stressed portion when the operating pressures are between 10 to 20 atmospheres. Fine cracks may be present at the stressed portion of the boiler. Boiler water containing alkaline impurities pass into the cracks by capillary action. Water evaporates leaving behind caustic soda in the cracks, concentration of which may increase to 10%. A galvanic cell is set up between the iron under stress and the iron in the main body.

Iron under stress	Concentrat ed NaOH (in the cracks)	Dilute NaOH in boiler water	Iron (main body)

Iron under stress acts as the anode and gets corroded resulting the boiler failure. Boiler water has Na_2CO_3; it will be hydrolysed at high temperature to give NaOH. It flows into hair cracks and crevices. There it reacts with iron and forms Na2FeO2 (sodium ferroate) which decomposes to give Fe_3O_4 and NaOH.
Reaction

$$3Na_2FeO_2 + 3H_2O\ Fe_3O_4 + H_2 + 6NaOH$$

NaOH thus formed further reacts with iron to cause corrosion. It is called caustic embrittlement. Addition of Na2SO4 to boiler water in addition to tannin and lignin to boiler water prevents caustic cracking.

Preventive measures

1. By applying protective coatings

2. Using corrosion inhibitors

3. By stress relief heat treatments

4. Reducing the overall stress level and designing out stress concentrations

2. Inter-granular corrosion

This corrosion is observed in case of alloys. Intergranular corrosion is an attack on the grain boundaries of a metal or alloy. The corrosion product is observed at the boundaries of grains. Externally, it is not seen. There is a sudden failure of material due to this Corrosion. For example, during the welding of stainless steel (an alloy of Fe, C, Cr), chromium carbide is precipitated at the grain boundaries and the region adjacent to grain boundaries becomes depleted of chromium composition and is made anodic with respect to solid solution within the grains richer in chromium. Rapid quenching after heat treatment of a metal is the remedy of inter-granular corrosion.

When stainless steel (18% Cr, 8% Ni) containing more than 0.1% carbon is heated to high temperature and cooled slowly or held at 650°C for a short interval of time a rapid reaction occurs between carbon and chromium. Chromium carbide, thus formed, precipitates at the grain boundaries and results in the formation of galvanic cells.

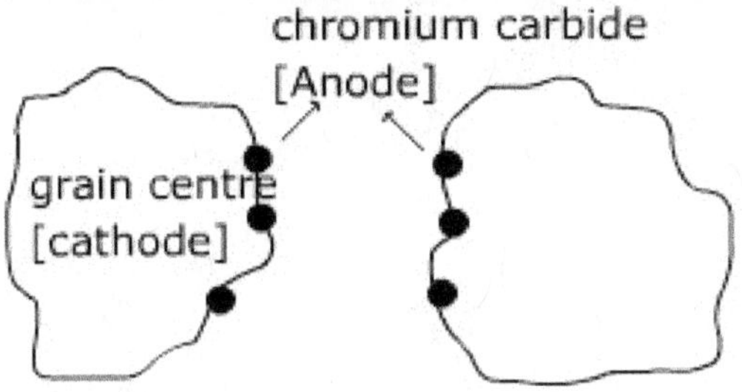

Chromium rich grain centers acts as cathode. Grain boundaries acts as anode and undergoes corrosion. This type of corrosion is said to be grain boundary corrosion.

Preventive measures

1. Select an alloy type that is resistant to intergranular corrosion.

2. Avoid heat treatments or service exposure that makes a material susceptible. Normally this occurs with austenitic stainless steels when they are held for some time in the sensitizing temperature range of 470 to 915 °C

3. Application of protective coating.

3. Microbiological corrosion

Microbial corrosion involves degradation of materials by bacteria, moulds and fungi or their by-products. Some types of bacteria consume oxygen and cause differential aeration type of system which results in corrosion. The corrosion occurs at the portion poor in oxygen concentration. Ex: The bacillus, algae diatoms.

It can occur by a range of actions such as:

a. Attack of the metal or protective coating by acid by-products, sulphur, hydrogen sulphide or ammonia

b. Direct interaction between the microbes and metal which sustains attack.

For example, corrosion of steel pipelines in waterlogged neutral ground containing little oxygen is often much more severe than might be expected. Cast iron pipes of 6 mm wall may perforate in less than five years. *Sulphate-reducing bacteria* such as *Desulphovibrio* which flourish in oxygen-free conditions assist the corrosion of iron apparently by consuming hydrogen produced in the cathodic reaction. H^+ + e H thus enhancing the rate of the cathodic reaction. The bacteria make use of the H^+ to reduce sulphate $SO_4^{2-} \longrightarrow$

The overall corrosion cell reaction is

$8H^+ + 8e \longrightarrow 8H$

$SO_4^{2-} + 8H \longrightarrow S^{2-} + 4\,H_2O$

producing iron sulphide FeS as the corrosion product.

Preventive measure

□Selection of resistant materials

□Frequent cleaning

□Control of chemistry of surrounding media and removal of nutrients

□Use of biocides

□Cathodic protection.

3. **Soil corrosion**
   Underground pipes, cables, etc. corrode due to soil corrosion.

It is caused due to moisture, pH of soil and micro-organisms. The differential aeration is also the cause of corrosion.

4. **Erosion corrosion**

Due to mechanical wear and tear, corrosion occurs on the surface of a metal and is called erosion corrosion

FACTORS INFLUENCING THE RATE OF CORROSION

Primary factors (Related to metal)	**Secondary factors** (Related to Environment)
1) Nature of the Metal	1) pH
2) Electrodes potential difference	2) Temperature
3) Surface state of the metal	3) Anodic and Cathodic area effect
4) Hydrogen over voltage	4) Polarisation of electrodes.
5) Formation of protective films by metals	5) Conductance
	6) Humidity

Primary factors (Related to metal)

1) Nature of the Metal The tendency of metal to undergo corrosion is dependent on the nature of the metal. Metals with low reduction potential exhibits high reactivity and hence they are more susceptible for undergoing corrosion. Thus the tendency of a metal to undergo corrosion decreases with increase in electrode potential (Galvanic potential). Metals such as Li, Na, Mg, Zn, etc have low reduction potentials and therefore undergo severe corrosion in corrosive environment. The noble metals with high reduction potential such as Ag, Au, Pt, etc are less susceptible for undergoing corrosion. However few metals deviate from the normal tendency, due to their ability to develop passive layer.

2) Electrode potential difference

Potential difference between the anodic region and cathodic region is known as open circuit potential difference, larger the open circuit potential, higher is the rate of corrosion.

For example, the potential difference between Iron and Silver is 1.14V, which is more than that between Iron and Tin [0.3 V]. Therefore Iron corrodes faster when in contact with Ag than with tin. Thus farther apart the metals stand from each other in the electrochemical series, greater is the intensity of corrosive attack on the anodic metal. On this account the use of dissimilar metals should be avoided wherever possible.

3) Surface state of the metal.

State or condition of the surface of the metal affects the corrosion process rate. For example if part of the surface is covered by oil or dust particles then severe pitting corrosion occurs at the metal surface. Corrosion at the surface is also accelerated by presence of water droplet on the surface. Heterogeneity at the surface of the metal results in high corrosion rate.

4) Hydrogen overvoltage.

In most of the electrochemical reactions, the competing cathodic reaction is the hydrogen ion reduction to liberate hydrogen gas. Hydrogen overvoltage is the measure of the tendency of the electrode to liberate hydrogen gas. A metal with low hydrogen overvoltage on its surface is more susceptible for corrosion. Low hydrogen over voltage results in faster cathodic reaction. This increases even anodic reaction rate and hence the overall rate of corrosion also increases. If hydrogen overvoltage is high for a given metal , then cathodic reaction rate is slower and hence the rate of corrosion is less.

5) Formation of protective films by metals

A few metals exhibit the tendency to develop passive, but protective films or layers on exposure to corrosive medium. The nature of such a passive layer largely decides the rate of corrosion. The passive layer [usually metal oxide] formed on the surface of the metal may act as a protective layer and prevents further corrosion. Metals like Al, Ti & Cr develop such layer on their surface and become passive to corrosion. In these cases the passive layer formed is stable, highly insoluble with low conductivities. Thus it acts as a protective film & prevents further corrosion of metal.

On the other hand, if the oxide layer is unstable, soluble, non uniform and Porous then corrosion continues to takes place. Example Fe & Zn metal oxides layers cannot control corrosion at their metal surfaces as the access to the corrosive environment cannot be completely prevented by the porous layer.

Secondary factors (Related to Environment)

1) pH

In general lower the pH of the corrosion medium, higher is the corrosion rate. At pH greater than 10, corrosion of iron practically ceases due to the formation of a protective coating of hydrous oxides. However metals like Al, Zn undergo fast corrosion in highly alkaline solution. If the pH is lower than 3, severe corrosion occurs even in the absence of air due to the continuous evolution of hydrogen at the cathodic region.

2) Temperature

The rate of any chemical reaction increases with rise in temperature. Therefore rate of corrosion increases as the temperature increases. Increase in temperature increases the conductance of the medium and reduces passivity of the metal and thereby increases the rate of corrosion.

3) Anodic and Cathodic area effect

The rate of corrosion is greatly influenced by the relative sizes of Anodic and Cathodic areas.

1) If the ratio of Anodic to Cathodic region is small i.e., smaller the anodic region and larger the Cathodic region then the corrosion is more intensive and faster is the rate of corrosion at anodic region. Example: Tin coated Iron with some pinholes or uncovered areas on the surface. Tin is cathodic to Iron.

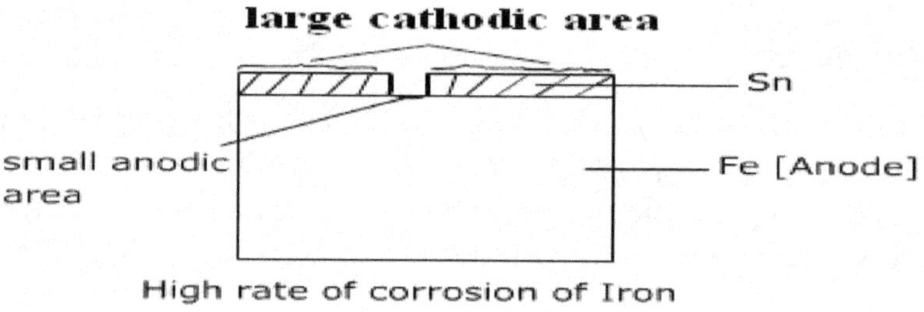

ii) If the ratio of Anodic to Cathodic region is large i.e., larger the anodic region and smaller the cathodic region, then the corrosion at anodic region occurs very slowly.

Example: Zinc coated Iron articles with some pinholes on the surface. Zinc is anodic to Iron.

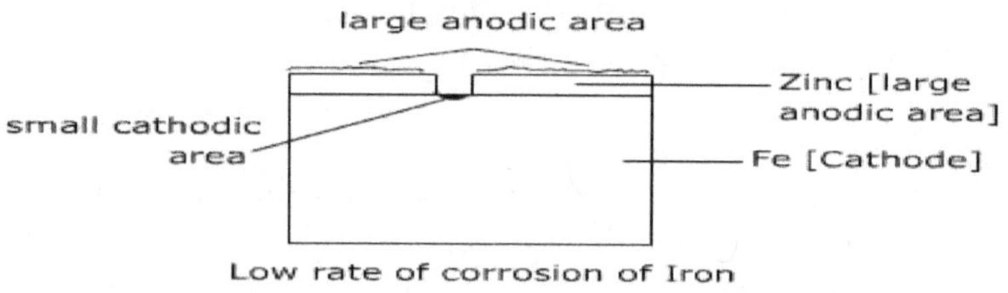

4) Polarisation of electrodes

Polarisation is defined as the departure of electrode potentials from their equilibrium values that is from the open circuit potentials causing a decrease in the current density.

Polarisation at the electrodes is due to concentration changes in the electrode region. Polarization causes the potential of the cathode to become more anodic and that of anode to become more cathodic.

The anodic and cathodic reactions must go in step with one another, if corrosion is to be sustained. The overall rate of electrochemical corrosion may be principally governed either by the Anodic or by Cathodic reaction rate. For metals such as Fe and Zn in aqueous solution, the rate of the cathodic reaction is the rate - determining step and the reaction is said to be under cathodic control [cathode polarisation] in which corrosion current is under cathodic control.

If the anode alone undergoes, the rate of corrosion is controlled by anodic polarization.

The following diagrams represent the polarization curves for corrosion processes proceeding by (a) Cathodic control (b) Anodic control and (c) Mixed control

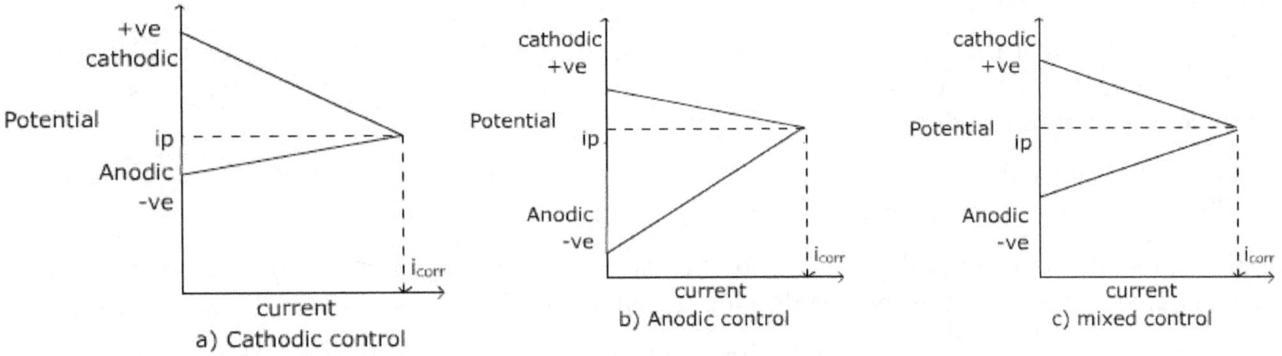

a) Cathodic control
b) Anodic control
c) mixed control

In these diagrams the polarisation curve has a steeper slope for the electrode that is polarized to a greater extent.

5) Conductance of the medium

Corrosion is an electrochemical reaction, which involves electrons transfer [(Conduction)] conducting species in the atmosphere increase the corrosion rate. The rate of corrosion in a wet atmosphere (more conducting) is more than the dry atmosphere.

5) Conductance of the medium

Corrosion is an electrochemical reaction, which involves electrons transfer [(Conduction)] conducting species in the atmosphere increase the corrosion rate. The rate of corrosion in a wet atmosphere (more conducting) is more than the dry atmosphere.

6) Humidity

Rate of corrosion is low at low relative humidity. Corrosion rate increases with rise in humidity upto a certain value called critical humidity but corrosion abruptly increases above critical humidity.

1.14 CORROSION CONTROL METHODS

Corrosion control is more realistic than corrosion prevention since corrosion of a metal is natural process by which metal is converted into a more stable compound state. There are varieties of corrosion control techniques as the corrosion depends on the nature of environment. The important corrosion control methods are discussed below.

1. Design and material selection
2. Protective coatings
   a) Inorganic coatings
      i) Metal coatings → anodic and cathodic metal coatings
      ii) Surface conversion coatings → Anodizing, phosphating and chromating.
   b) Organic coatings → Paints and enamels
3. Cathodic protection
4. Anodic protection
5. Corrosion inhibitors

1. Design and Material Selection

i) In any design, the greater the number of angles, corners, edges and internal surface, the more difficult it becomes for efficient surface treatment. Hence, as far as possible, L, T & U shaped profiles in construction are to be minimized.

ii) No corrosion occurs without the presence of moisture. The design of a structure should be such that retention of moisture should be as low as possible.

iii) Differential metal corrosion can be reduced by the use of single metal rather than dissimilar metals. If their use is unavoidable, position of the metals chosen shall be as close as possible in the galvanic series.

An insulator should be placed between two dissimilar metals.

iv) The rate of corrosion depends on the relative sizes of anode and cathode. Small anode and large cathode should be avoided as far as possible.

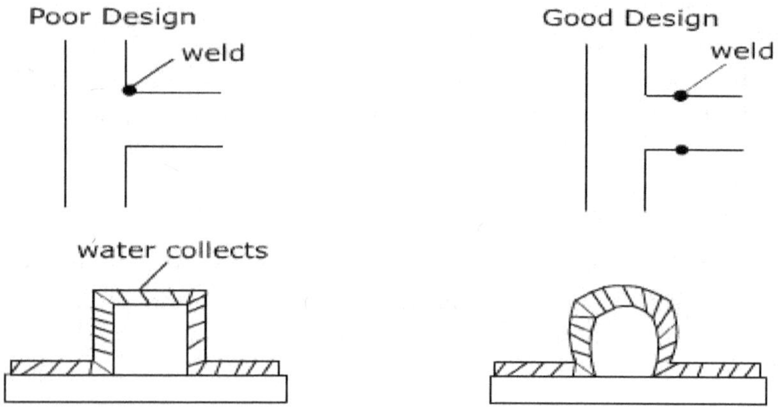

2. Protective Coatings.

Protective coatings are used to prevent corrosion at the surfaces of materials. Protective coatings includes metallic coatings, chemical conversion coatings, ceramic coatings and organic coatings

3. Cathodic Protection

The corrosion of metal takes place at the anodic region where as at the cathodic region, metal is unaffected. The principle of cathodic protection involves the elimination of anodic sites and conversion of entire metal into cathode area. This can be achieved by providing electrons from an external source so that the specimen always remains cathode.

The technique of offering protection to a specimen against corrosion by providing electrons from an external source is called cathodic protection.

Cathodic protection can be achieved by the following two methods.

a) Sacrificial anode method

b) Impressed current method

a) Sacrificial anode method

In this method, metal structure is converted into a cathode by connecting into a more active metal. This active metal acts as an auxiliary anode. Zn, Mg and Al are the common auxiliary anodes. These metals being more active, acts as anode and undergo preferential corrosion, protecting the metal structure. Since the anodic metals are sacrificed to protect the metal structure. The method is known as sacrificial anode method. New ones are replacing exhausted sacrificial anodes as and when required.

Examples of sacrificial anode methods are

- A magnesium block connected to a buried pipelines
- Magnesium bars are fixed to the sides of the ships

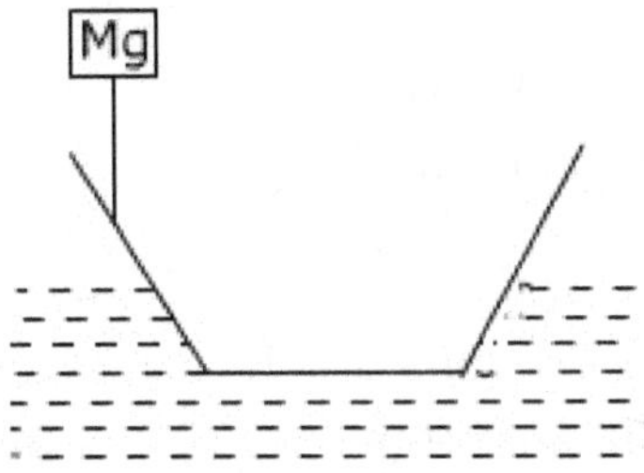

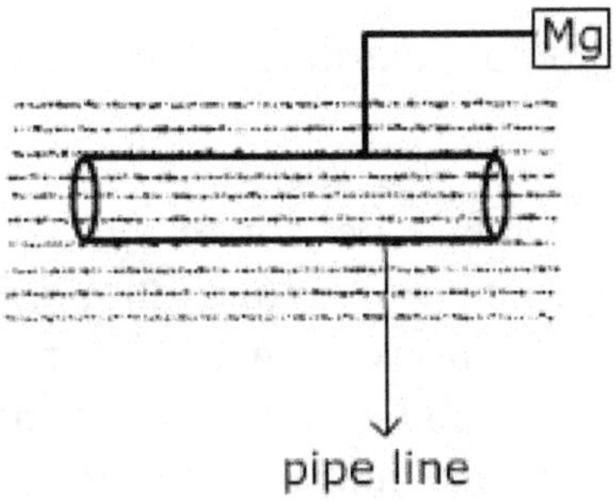

b) Impressed current method

Another method of providing cathodic protection is by applying a direct current larger than the corrosion current. The protected metal is made cathodic by connecting it to the negative terminal of a d.c. source. Positive terminal is connected to inert anode such as Graphite. The metal structure being cathode, doesnot undergo corrosion.

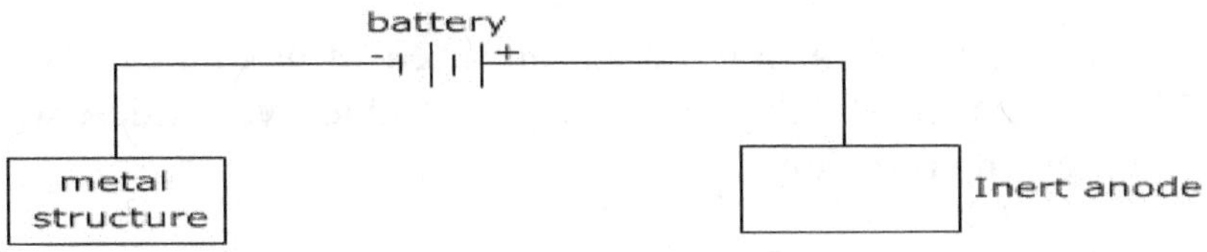

4. Anodic protection

The prevention of corrosion by impressed anodic current method is called Anodic protection. Few metals like Ti and alloys such as steel, when made as anode exhibits passivity by developing their oxide layers. In case of passive metals such as Ti, Ni, Cr and their alloys, application of suitable anodic current makes them passivated and decreases the rate of dissolution of metals. The potential required to protect the metal can be obtained from potential - current curve as shown below.

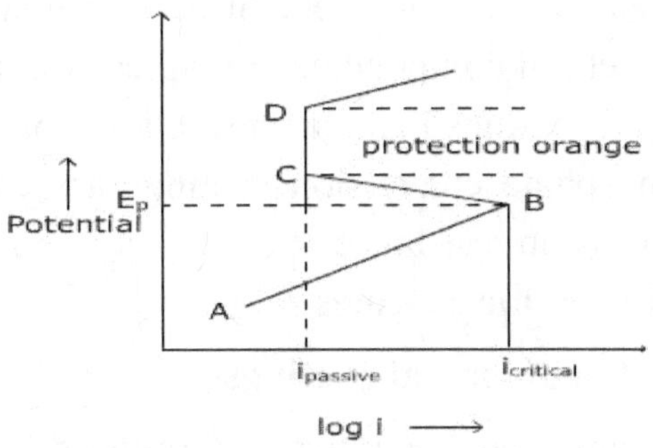

As the potential is increased, initially the current also increases [AB], indicating the dissolution of metal. This trend continues until the current reaches a critical value (i_{crit}) and passivation due to the development of oxide layers sets in this potential is called passivating potential (E_P). Above E_P, the current flow decreases and reaches a minimum value called the passivating current $i_{passive}$. It is the minimum protective current density to maintain passivation. If the potential is further increased, metal remains unattacked upto a particular potential is reached [CD]. This potential range in which anodic protection can be achieved is called protection range.

The advantage of the method is that it requires a small current. However the corrosion rate cannot be reduced to zero as in cathodic protection. A drawback of the method is that it cannot be applied to metals, which don't passivate.

This method of anodic protection is utilized in the transportation of concentrated acids.

5. Corrosion Inhibitors

Corrosion inhibitors reduce corrosion by retarding either anodic or cathodic reactions.

i) Anodic inhibitors

Anodic reaction is the oxidation of metal i.e. for ex., in rusting of Fe, Fe^{2+} ions are formed at anode. Anodic inhibitors prevent the formation of Fe^{2+} ions [anodic reaction] & thereby retard corrosion process. This is achieved by the addition of anions such as chromate, tungstate, molybdate etc. These anions combine with metal ions formed at anodic region, forming sparingly soluble respective salts. These compounds formed are deposited on the anodic sites forming protective films which acts as barriers between metal surface and corrosion medium and prevents further anodic reaction and hence corrosion.

ii) Cathodic inhibitors

Cathodic inhibitors act by inhibiting the cathodic reactions which involve the liberation of hydrogen in acidic solutions and OH^- ions in alkaline solution.

Organic cathodic inhibitors such as amines, mercaptans, thioureas, sulphoxides, form a protective layer on cathodic regions and prevent the evolution of hydrogen thus decreasing the rate of corrosion.

Inorganic cathodic inhibitors such as sulphates of Mg, Mn, Ni and Zn are used in neutral or alkaline medium. These inhibitors react with OH^- ions liberated at cathode forming insoluble hydroxides, which form protective film over cathodes areas and prevent corrosion.

Protective coatings

Protective coatings are used to prevent corrosion at the surfaces of materials. Protective coatings include metallic coatings, chemical conversion coatings, ceramic coatings and organic coatings.Underbody structural components are typically coated to provide a first line of defense against corrosion. Commonly used coatings include conversion, hot melt wax, electrocoat, metallic,organic, autodeposition and powder. Phosphate conversion coatings are employed to enhance paint adhesion, thereby indirectly enhancing corrosion resistance.

Inorganic coatings

1. Anodic metal coatings

Anodic metal coatings are produced by coating a base metal with more active metals

[low E^o value], which are anodic to the base metal.

Example: Al, Mg, Zn and Cd coatings on Iron.

A characteristic feature of anodic coating is that, even if the coating falls off at some places, the base metal doesn't undergo corrosion because base metal is cathodic in nature.Anodic metal coatings are carried out by four methods such as cladding, hotdipping, spraying and concentration.

a) Cladding

Cladding is used mainly with metal sheets. The iron sheet to be protected is cladded (sandwiched) between two sheets of the anodic metal and squeezed by rolling under pressure and heat.

Ex: Cladding of Aluminium by Duralumin. [Alloy of Al, Mg & Cu]. These corrosion resistant materials are used in aircraft industry.

b) **Hot dipping [Galvanisation]**

Hot dipping involves dipping the base metal in molten anodic metal.

Ex: Galvanising is coating Zinc on Iron by hot dipping.

Galvanisation involves pretreatment of Iron article with organic solvents followed by dilute H_2SO_4. Iron is then treated with $ZnCl_2$ and NH_4Cl and dipped in molten zinc at 430 - 470°C

c) Metal spraying

Metal spraying is done through flame spraying technique where metal particles are sprayed by an air blast on the surface to be coated.

Ex: Al, Zn, Cd and Zn are coated by this technique.

d) Cementation

Cementation is a process of obtaining a coating by heating the metal object in a powder of the coating metal. Coating of Iron by Zinc in this way is called sheardizing.

Coating Al on Fe → Colorising

Cr and Al coating on alloys → Chromaluminising.

2. Cathodic metal coatings

Cathodic metal coatings are produced by coating a base metal with more noble metals [High E^o values] which are cathodic to the base metal.
Cathodic coatings are effective only when the coating is continuous, nonporous and free from gap otherwise rapid corrosion of the base metal takes place as a result of the formation of large cathodic and small anodic areas.

Tinning is most commonly used cathodic coating on steel surfaces.

Tinning

Tinning is a process of coating base metal with tin.

Tinning involves pretreatment of Iron with organic solvents followed by dilute H_2SO_4. Iron is then passed through Zinc chloride and ammonium chloride flux and then through Tin. Finally it is dipped in palm oil to prevent the oxidation of Tin.

3. Chemical conversion coatings

a)Anodizing

Anodized coating is generally produced on metals like Al, Zn, Mg and their alloys by anodic oxidation process, in which base metal is made as anode, in an electrolytic bath and by passing direct current. Al exhibits high anodic passivity by developing it's oxide layer and there by prevents corrosion when coated on other metals.

The metal article, after pretreatment is made as Anode and steel or copper is made the cathode. The electrolyte consists of 5 - 10% chromic acid. The temperature of bath is maintained at 35°C. Anodic current density is maintained at 10 - 20 mA/cm^2. An oxide layer of 2 - 8 □ m thickness is obtained.

The anodic oxide film formed on Al in the bath of phosphoric acid tends to be porous and provides good adherence for paints and dyes. Anodized articles are used as soap boxes, tiffin carriers, window frames etc.

b)Phosphating

Phosphate coatings are produced by chemical reaction of aqueous solution of phosphates of Fe, Mn and Zn in dilute phosphoric acid solution, with the base metal such as Fe, steel or Zn.

Phosphating involves the following steps

i) Dissolution of metal as metal ions

ii) Metal ions reacting with the phosphate ions to form a metal phosphate

iii) Deposition of the metal phosphate on the surface of the metal.

Phosphating offers improved corrosion resistance and also imparts the surface a good adhesion quality phosphating is a standard pretreatment before painting of automotive bodies and other steel sheet fabrications such as refrigerators and washing machines.

4. Organic Coating

Organic coatings are materials that are applied to protect or beautify a surface. Coatings include paints, varnishes, stains, industrial maintenance coatings and they can be applied to stationary and/or mobile sources. When applied, the solvents in organic coatings can evaporate into the atmosphere and contribute to volatile organic compound and airborne toxic emissions.

Coating of metal surface with organic paints is the most widely used anticorrosion coating. Organic paint is a dispersion of one or more finely divided solids in a medium. The medium consists of non-volatile, film-forming materials like drying oils and resins in thinners. When a properly treated metal surface is painted, the thinner evaporates leaving behind the film-former and the pigment on the surface.

5. Ceramic coatings

Hard materials suitable for thin film ceramic coatings are usually carbides, nitrides, borides and silicides of the IVth, Vth and VIth groups of the periodic table. The ceramic coatings are formed by introducing nitrogen, hydrocarbon, or silicide during the sputtering process. The ceramic compound is thought to form at the surface of the substrate as this is the most energetically favourable.

The most common ceramic coatings available are TiN, CrN, TiCN, and TiAlN.

Alloyed elements improve oxidation resistance of ceramic coatings
The properties of the ceramic coatings, such as oxidation resistance, have been improved by the addition of elements such as aluminium, vanadium, yttrium, chromium etc. and the development of multilayers and nanocomposite coatings.

Multilayers and Superlattices improve ceramic coatings
Further improvement to the properties of hard PVD ceramic coatings are achieved by depositing multilayers and superlattices. These are thin films formed by alternately depositing two different components to form layered structures. Multilayers become superlatices when the period of the different layers is less than 100Å.

1.15 EXERCISE QUESTIONS

TYPE I QUESTIONS

1. State Faraday's 2nd law of electrolysis.
2. Chlorine gas will oxidize Br^- but will not oxidize F^-. Justify. The reduction half-reactions and standard electrode potentials are,

   $Br_2(l) + 2e^- \rightarrow 2Br^-(aq)$ $\quad E° = 1.07$ V, $\quad Cl_2(g) + 2e^- \rightarrow 2Cl^-(aq)$ $\quad E° = 1.36$ V

   $F_2(g) + 2e^- \rightarrow 2F^-(aq)$ $\quad E° = 2.87$ V
3. Predict whether $Cr_2O_7^{2-}$ will oxidize Sn^{2+} ions in acidic solution under standard conditions. The reduction half-reactions and standard potentials are,

   $$Sn^{4+}(aq) + 2e^- \rightarrow Sn^{2+}(aq) \quad E° = 0.15 \text{ V}$$

   $$Cr_2O_7^{2-}(aq) + 14H^+(aq) + 6e^- \rightarrow 2Cr^{2+}(aq) + 7H_2O(l) \quad E° = 1.33 \text{ V}$$

4. Write the salient features of Debye-Huckel theory?
5. Define the term equivalent conductivity. Give its S I unit.
6. Distinguish between galvanic cell and electrolytic cell
7. Calculate the Gibbs free energy change, $\Delta G°$, in J/mol at 25°C for the following reaction:

   $3\,Sn^{4+} + 2Cr \rightarrow 3Sn^{2+} + 2Cr^{3+}$

   ($Sn^{4+} + 2e^- \rightarrow Sn^{2+}$ $\quad E° = +0.15$V

   $Cr^{3+} + 3e^- \rightarrow Cr$ $\quad E° = -0.74$V)
8. What are ion selective electrodes? Give an example.
9. Explain the following characteristics of a battery:

   i) Energy efficiency ii) Shelf life
10. Pin holes on tin coated iron articles are more disastrous to iron than pin holes on zinc coated iron articles. Justify.

TYPE II QUESTIONS

1. Define standard electrode potential? List any four differences between galvanic and electrolytic cell.
2. Derive Nernst equation for single electrode potential?
3. Explain the construction and working of a calomel electrode.
4. Explain briefly the experimental method of determination of pH of a solution using glass electrode.
5. The emf of the cell Cd /$CdSO_4$ (0.0093M)//$CdSO_4$ (xM)/Cd is 0.086 V at 25°C. Find the value of x
6. What are ion selective electrodes? Explain the construction and working of Glass electrode.
7. Explain the following characteristics of a battery:
   i) Cycle life ii) Shelf life
8. A galvanic cell is constructed by coupling Ag and Cd electrodes dipped in 0.5 M $AgNO_3$ and 0.25 M $CdSO_4$ respectively at 25°C. Write the cell reaction, and calculate emf of the cell. Given that SRPs of Ag and Cd are +0.80 V and – 0.40 V respectively.
9. Explain the construction and working of lead - acid battery with charging and discharging reactions. Mention its applications.
10. Explain the construction and working of Ni-MH battery with charging and discharging reactions
11. What are fuel cells? Distinguish between fuel cells from Batteries.
12. Explain the construction and working of the methanol – oxygen fuel cell.
13. Define the term Corrosion. Discuss the corrosion of iron based on electrochemical theory.
14. Explain how the following factors influence the arte of corrosion,
    i) Nature of the electrode
    ii) Ratio of anodic to cathodic areas
15. Explain sacrificial anodic method of corrosion control.
16. Balance the following equation in basic medium,
    $Cr_2O_7{}^{2-}$ (aq) + 2NO (g) $\rightarrow$ $2Cr^{3+}$ (aq) + $2NO_3^-$ (aq)
17. Balance the following redox reaction in consideration of the redox potential of each reaction and predict the feasibility of reactions;

$$\mathbf{K_2Cr_2O_7 + H_2SO_4 + KBr \rightarrow K_2SO_4 + Cr_2(SO_4)_3 + Br_2 + H_2O}$$

Potentials of the reactions are as given below,

$\mathbf{Cr_2O_7{}^{2-}(aq) + 14H^+ + 6\bar{e} \rightarrow 2Cr^{3+}(aq) + 7H_2O}$ $\quad$ $\mathbf{E^0 = +1.33\ V}$

$\mathbf{Br_2{}^0 + 2\bar{e} \rightarrow 2Br^{1-}}$ $\quad$ $\mathbf{E^0 = +1.09\ V}$

18. In an electrolysis experiment, electric current was passed for 4 hours through two cells connected in series. The first cell contains a copper salt and the second contains Ag_2SO_4 solution. 22 g of silver were deposited in the first cell. Find the amount of copper deposited in the second cell. Also calculate the magnitude of current in amperes. (At. wt of Cu = 63.5, At. wt of Ag = 108)
19. Define electrochemical equivalence. What volume of oxygen gas (measured at STP) is produced by the oxidation of water at the anode in the electrolysis of copper (II) sulfate passage of 2.50 amps of current through a solution of copper (II) sulfate for 50.0 minutes?
20. Explain the conductivity in metals, semiconductors and insulators based on band theory of solids with the band structure.

21. An electro chemical cell is formed from Ni and Pb electrodes having 0.01M NiSO4 and 0.5M PbSO4 electrolytes. The standard electrode potentials of Ni and Pb electrodes are -0.24V and -0.13V respectively. Write the cell scheme, cell reaction and calculate emf of the cell at 298K.
22. Explain the construction and working of calomel electrode. Mention its advantages.

23. Use the standard cell potential to calculate the value of equilibrium constant, K at 25°C for the following reaction,

    $2Cu + PtCl_6^{2-} \rightarrow 2Cu^+ + PtCl^{4-} + Cl^-$

    $(Cu^+ + e \rightarrow Cu^+$ $\quad E^o = 0.521$ V

    $PtCl_6^{2-} \rightarrow PtCl^{4-} + 2Cl^-$ $\quad E^o = 0.680$ V)
24. Explain the construction and working of Zn-MnO_2 battery
25. What is meant by corrosion? Explain differential aeration corrosion with suitable examples.
26. Explain the cathodic protection method of corrosion control.

TYPE III QUESTIONS

1. Which one of the following reactions is not feasible?
   a) Oxidation of magnesium by ferrous chloride b) Oxidation of copper by silver nitrate
   c) Oxidation of gold by copper nitrate d) Reduction of silver ions by copper

2. Which one of the following reactions is feasible?
   a) Oxidation of ferrous hydroxide to ferric hydroxide by oxygen
   b) Reduction of ferric hydroxide to ferrous hydroxide by oxygen
   c) Oxidation of iodide ions by chromium ions
   d) Reduction of manganese ions by ammonia

3. Molten $AlCl_3$ is electrolyzed for 5.0 hours with a current of 0.40 amperes. Metallic aluminum is produced at one electrode and chlorine gas, Cl_2, is produced at the other. How many liters of Cl_2 measured at STP are produced when the electrode efficiency is only 65%?

   a) 0.55 L b) 0.63 L c) 0.84 L d) 1.0 L

4. How many faradays are required to reduce 1.00 g of aluminum (III) to the aluminum metal?
   a) 1.00 b) 1.50 c) 3.0 d) 4.0
5. A current of 0.2 ampere is passed in aqueous solution of copper sulphate for one hour. The mass of copper deposited at the cathode is, [Given: equivalent mass of Cu = 31.75]
   a) 1.2369 g b) 0.2369 g c) 2.369 g d) 2.2369
6. The number of electrons gained during electrolysis of 10 g of Zn^{2+} ions(Eq. wt. 32.6)
   a) 1.85×10^{23} b) 1.85×10^{21} c) 1.85×10^{24} d) 1.85×10^{25}
7. What current strength in ampere will be required to liberate 10 g of chlorine from sodium chloride solution in one hour? a) 7.55 b) 6.81 c) 10 d) 38
8. Which of the following statements is true about Debye length?

a) Debye length decreases as the concentration of ions decreases
b) Debye length decreases as the valency of ions decreases
c) Debye length decreases as the temperature decreases
d) Debye length decreases as the concentration of ions increases

9. The unit of equivalent conductance is,
a) $S\ cm^2\ eq^{-1}$ b) $S\ cm^2\ eq^{-2}$ c) $S\ cm\ eq^{-1}$ d) $S\ cm^{-1}\ eq^{-1}$

10. 15 ml of 0.13M of $NiSO_4$ is electrolyzed using a current of 360 milliamperes for 55.3 minutes. How much of the metal would have been plated out if current efficiency was only 50%.
a) 1.55 g b) 0.135 g c) 1.35 d) 0.155 g

11. An increase in equivalent conductance of a strong electrolyte with dilution is mainly due to
a) Increase in both i.e. number of ions and ionic mobility of ions b) increase in number of ions
c) increase in ionic mobility of ions d) 100% ionization of electrolyte at normal dilution

12. A concentration cell is constructed by placing identical Cu electrodes in two Cu^{2+} solutions. If the concentrations of the two Cu^{2+} solutions are 1.0 M and 0.0020 M, then the potential of the cell is,
a) 0.020 V b) 0.030 V c) 0.080 V d) 1.0 V

13. For the reduction of silver ions with copper metal the standard cell potential was found to be +0.46V at 25°C. The value of standard Gibbs energy, ΔG° will be ($F = 96500\ C\ mol^{-1}$),
a) -44.5 kJ b) -98.0 kJ c) -89.0 kJ d) -89.0 J

14. The Photovoltaic cell is,
a) Storage cell b) Rechargeable cell c) Energy conversion device d) Fuel cell

15. The conventional solar cell efficiency is, a) 90% b) 60% c) 50% d) 25%

16. E1, E2 and E3 are the emf values of the three galvanic cells respectively.

(a) $Zn\,|\,Zn^{+2}_{1M}\,||\,Cu^{+2}_{0.1M}\,|\,Cu$ (b) $Zn\,|\,Zn^{+2}_{1M}\,||\,Cu^{+2}_{1M}\,|\,Cu$ (c) $Zn\,|\,Zn^{+2}_{0.1M}\,||\,Cu^{+2}_{1M}\,|\,Cu$

Which one of the following is true?

a) E2 > E3 >E1 b) E3 > E2 > E1 c) E1 > E2 > E3 d) E1 > E3> E2

17. The potential of a hydrogen electrode at pH = 10 is
a) 0.59 V b) 0.00 V c) -0.59 V d) -0.059 V

18. The standard reaction Gibbs energy for the electrochemical reaction
Cr2O72- + 2 Fe + 14 H+ → 2 Cr3+ + 2 Fe3+ 7 H2O is -793 kJ mol-1. The standard cell potential is,
a) +1.37 V b) +4.11 V c) +2.74 V d) +2.05 V

19. Which electrode develop potential with respect to specific ions, ignoring the other ions,
a) Calomel electrode b) Redox electrode c) Glass electrode d) Gas electrode

20. The equilibrium constant for the following system at 25oC, 3 Mg^{2+} + 2Al $\rightleftharpoons$ 3Mg + 2Al^{3+} is
    a) 1069 b) 1024 (or 10-24) c) 10-36 d) 1072
21. During charging of lead storage battery,
    a) PbSO4 anode is reduced to Pb b) PbSO4 cathode is reduced to Pb
    c) PbSO4 anode is oxidised to Pb d) PbSO4 anode is oxidised to PbO_2
22. What causes the cathode to be "eaten away" in the primary cell while the cell is discharging?
    a) The material of the cathode combines with the negative ions to form a new substance.
    b) The material of the cathode dissolves in the electrolyte.
    c) The material of the cathode leaves the negative termina and goes through the anode.
    d) Bacteria in the electrolyte erodes the material in the cathode.
23. Which of the following factors adversely affect the capacity of secondary battery
    a) Temperature b) Specific gravity of electrolyte c) rate of discharge d) All of these
24. One advantage of using hydrogen fuel cells
    a) the potential for the production of greenhouse gas b) the ease of storage
    c) Low impact on Ozone d) Nonpolluting nature

Chapter 2

ENVIRONMENTAL CHEMISTRY

2.1 INTRODUCTION

Environmental science is an interdisciplinary academic field that integrates physical and biological sciences, (including physics, chemistry, biology, soil science, geology, and geography) to the study of the environment, and the solution of environmental problems. Environmental science provides an integrated, quantitative, and interdisciplinary approach to the study of environmental systems.

Related areas of study include environmental studies and environmental engineering.
Environmental studies incorporate more of the social sciences for understanding human relationships, perceptions and policies towards the environment.

Environmental engineering focuses on design and technology for improving environmental quality.
Environmental scientists work on subjects like:

- the understanding of earth processes,
- evaluating alternative energy systems,
- pollution control and mitigation,
- natural resource management,
- the effects of global climate change.

Environmental issues almost always include an interaction of physical, chemical, and biological processes. Environmental scientists bring a systems approach to the analysis of environmental problems. Key elements of an effective environmental scientist include the ability to relate space, and time relationships as well as quantitative analysis.

Environmental science came alive as a substantive, active field of scientific investigation in the 1960s and 1970s driven by **(a)** the need for a multi-disciplinary approach to analyze complex environmental problems, **(b)** the arrival of substantive environmental laws requiring specific environmental protocols of investigation and **(c)** the growing public awareness of a need for action in addressing environmental problems.

Environmental scientists investigate and seek out natural solutions to environmental problems. To solve water quality issues, they look at the sources of pollution and fine new ways to stop the pollution at its source or how to use natural systems, like wetlands, to clean and filter the water. They investigate the health and sustainability of our soils to grow food, and look at ways to restore ecosystems and conserve our national natural treasures.

Components of Environmental Science

1. Atmospheric sciences focuses on the Earth's atmosphere, with an emphasis upon its interrelation to other systems. Atmospheric sciences can include studies of meteorology, greenhouse gas phenomena, atmospheric dispersion modeling of airborne contaminants, sound propagation phenomena related to noise pollution, and even light pollution.

Taking the example of the global warming phenomena, physicists create computer models of atmospheric circulation and infra-red radiation transmission, chemists examine the inventory of atmospheric chemicals and their reactions, biologists analyze the plant and animal contributions to carbon dioxide fluxes, and specialists such as meteorologists and oceanographers add additional breadth in understanding the atmospheric dynamics.

2. Ecology. An interdisciplinary analysis of an ecological system which is being impacted by one or more stressors might include several related environmental science fields. For example, one might examine an estuarine setting where a proposed industrial development could impact certain species by water and air pollution.

For this study, biologists would describe the flora and fauna, chemists would analyze the transport of water pollutants to the marsh, physicists would calculate air pollution emissions and geologists would assist in understanding the marsh soils and bay muds.

3. Environmental chemistry is the study of chemical alterations in the environment. Principal areas of study include soil contamination and water pollution. The topics of analysis include chemical degradation in the environment, multi-phase transport of chemicals (for example, evaporation of a solvent containing lake to yield solvent as an air pollutant), and chemical effects upon biota.

As an example study, consider the case of a leaking solvent tank which has entered the habitat soil of an endangered species of amphibian. As a method to resolve or understand the extent of soil contamination and subsurface transport of solvent, a computer model would be implemented.

Chemists would then characterize the molecular bonding of the solvent to the specific soil type, and biologists would study the impacts upon soil arthropods, plants, and ultimately pond-dwelling organisms that are the food of the endangered amphibian.

4. Geosciences include environmental geology, environmental soil science, volcanic phenomena and evolution of the Earth's crust. In some classification systems this can also include hydrology, including oceanography.

As an example study of soils erosion, calculations would be made of surface runoff by soil scientists. Fluvial geomorphologists would assist in examining sediment transport in overland flow. Physicists would contribute by assessing the changes in light transmission in the receiving waters. Biologists would analyze subsequent impacts to aquatic flora and fauna from increases in water turbidity.

5.Atmospheric dispersion modeling

6. Biogeochemistry

7. Ecological sanitation

8. Environmental chemistry

9. Environmental toxicology

10. Environmental Impact Statement

11. Environmental monitoring

12. Environmental planning

13. Environmental statistics

14. Environmental studies

15. Environmental informatics

Usual mistake - In common usage, "environmental science" and "ecology" are often used interchangeably, but technically, ecology refers only to the study of organisms and their interactions with each other and their environment.

Ecology could be considered a subset of environmental science, which also could involve purely chemical or public health issues (for example) ecologists would be unlikely to study. In practice, there is considerable overlap between the work of ecologists and other environmental scientists.

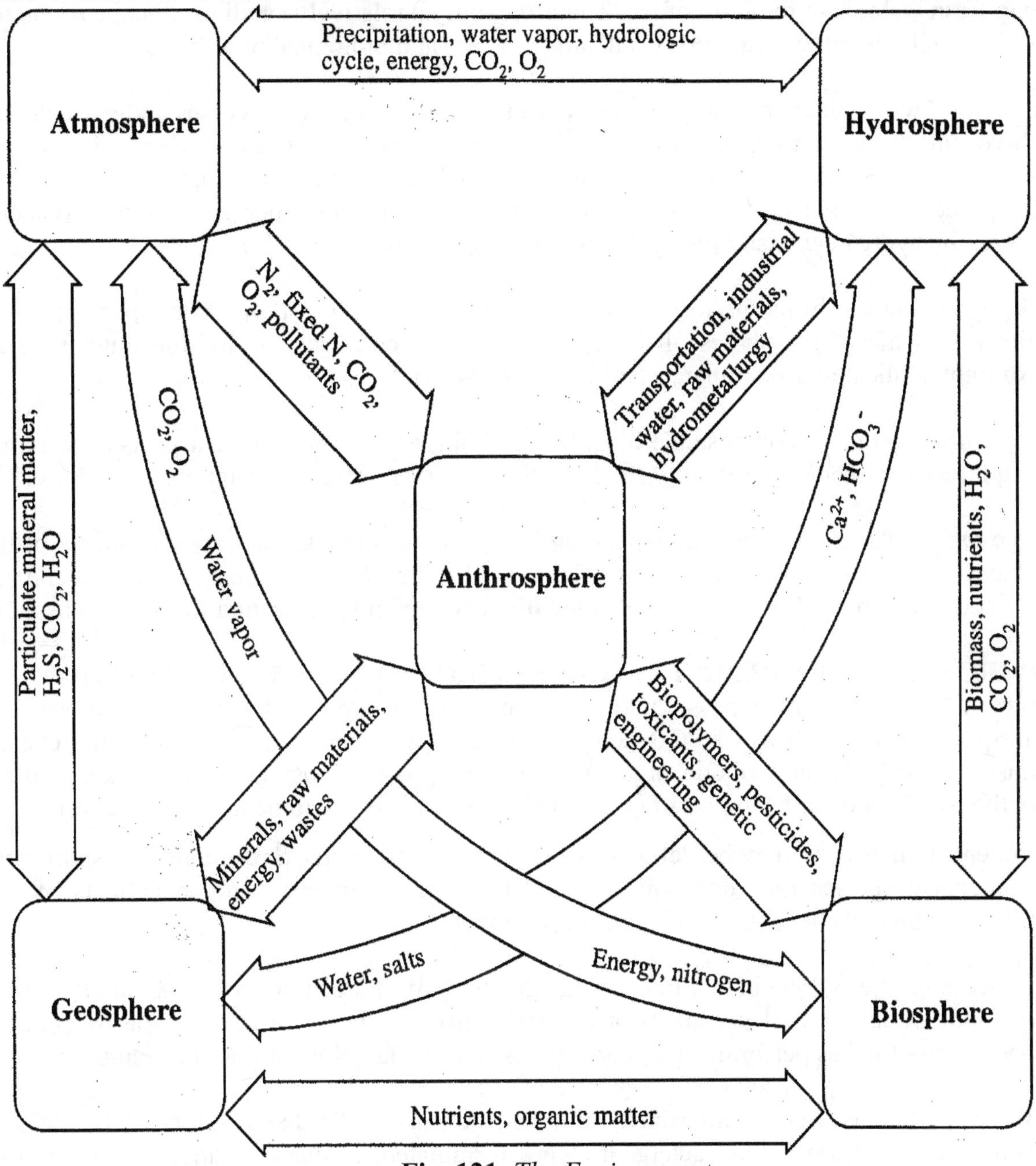

Fig. 121. *The Environment*

Illustration of the close relationships among the air, water, and earth environments with each other and with living systems, as well as the tie-in with technology (the anthrosphere).

Traditionally, environmental science has been divided among the study of the atmosphere, the hydrosphere, the geosphere, and the biosphere. The **atmosphere** is the thin layer of gases that cover Earth's surface. In addition to its role as a reservoir of gases, the atmosphere moderates Earth's temperature, absorbs energy and damaging ultraviolet radiation from the sun, transports energy away from equatorial regions, and serves as a pathway for vapor-phase movement of water in the hydrologic cycle. The **hydrosphere** contains Earth's water. Over 97% of Earth's water is in oceans, and most of the remaining fresh water is in the form of ice. Therefore, only a relatively small percentage of the total water on Earth is actually involved with terrestrial, atmospheric, and biological processes.

Exclusive of seawater, the water that circulates through environmental processes and cycles occurs in the atmosphere, underground as groundwater, and as surface water in streams, rivers, lakes, ponds, and reservoirs. The **geosphere** consists of the solid earth, including soil, which supports most plant life. The part of the geosphere that is directly involved with environmental processes through contact with the atmosphere, the hydrosphere, and living things is the solid **lithosphere**.

The lithosphere varies from 50 to 100 km in thickness. The most important part of it insofar as interactions with the other spheres of the environment are concerned is its thin outer skin composed largely of lighter silicate-based minerals and called the **crust**.

All living entities on Earth compose the **biosphere**. Living organisms and the aspects of the environment pertaining directly to them are called **biotic**, and other portions of the environment are **abiotic**.

To a large extent, the strong interactions among living organisms and the various spheres of the abiotic environment are best described by cycles of matter that involve biological, chemical, and geological processes and phenomena. Such cycles are called **biogeochemical cycles**, and we will discuss it further.

2.2 ENVIRONMENTAL CHEMISTRY AND ENVIRONMENTAL BIOCHEMISTRY

Environmental chemistry encompasses many diverse topics. It may involve a study of Freon reactions in the stratosphere or an analysis of PCB deposits in ocean sediments. It also covers the chemistry and biochemistry of volatile and soluble organometallic compounds biosynthesized by anaerobic bacteria. Literally thousands of other examples of environmental chemical phenomena could be given.

Environmental chemistry may be defined as the study of the sources, reactions, transport, effects, and fates of chemical species in water, soil, air, and living environments, and the effects of technology thereon. Environmental chemistry is not a new discipline.

Excellent work has been done in this field for the greater part of a century. Until about 1970, most of this work was done in academic departments or industrial groups other than those primarily concerned with chemistry. Much of it was performed by people whose basic education was not in chemistry.

Thus, when pesticides were synthesized, biologists observed firsthand some of the less desirable consequences of their use. When detergents were formulated, sanitary engineers were startled to see sewage treatment plant aeration tanks vanish under meter-thick blankets of foam, while limnologists wondered why previously normal lakes suddenly became choked with stinking cyanobacteria.

Despite these long standing environmental effects, and even more recent and serious problems, such as those from hazardous wastes, relatively few chemists have been exposed to material dealing with environmental chemistry as part of their education.

Environmental Chemistry and the Environmental Chemist

An encouraging trend is that in recent years many chemists have become deeply involved with the investigation of environmental problems. Academic chemistry departments have found that environmental chemistry courses appeal to students, and many graduate students are attracted to environmental chemistry research.

Help-wanted ads have included significant numbers of openings for environmental chemists among those of the more traditional chemical subdisciplines. Industries have found that well-trained environmental chemists at least help avoid difficulties with regulatory agencies, and at best are instrumental in developing profitable pollutioncontrol products and processes.

Some background in environmental chemistry should be part of the training of every chemistry student. The ecologically illiterate chemist can be a very dangerous species. Chemists must be aware of the possible effects their products and processes might have upon the environment. Furthermore, any serious attempt to solve environmental problems must involve the extensive use of chemicals and chemical processes.

There are some things that environmental chemistry is not. It is not just the same old chemistry with a different cover and title. Because it deals with natural systems, it is more complicated and difficult than "pure" chemistry. Students sometimes find this hard to grasp, and some traditionalist faculty find it impossible.

Accustomed to the clear-cut concepts of relatively simple, well-defined, though often unrealistic systems, they may find environmental chemistry to be poorly delineated, vague, and confusing. More often than not, it is impossible to come up with a simple answer to an environmental chemistry problem. But, building on an ever-increasing body of knowledge, the environmental chemist can make educated guesses as to how environmental systems will behave.

Chemical Analysis in Environmental Chemistry

One of environmental chemistry's major challenges is the determination of the nature and quantity of specific pollutants in the environment. Thus, chemical analysis is a vital first step in environmental chemistry research. The difficulty of analyzing for many environmental pollutants can be awesome. Significant levels of air pollutants may consist of less than a microgram per cubic meter of air.

For many water pollutants, one part per million by weight (essentially 1 milligram per liter) is a very high value. Environmentally significant levels of some pollutants may be only a few parts per trillion. Thus, it is obvious that the chemical analyses used to study some environmental systems require a very low limit of detection.

However, environmental chemistry is not the same as analytical chemistry, which is only one of the many subdisciplines that are involved in the study of the chemistry of the environment. Although a "brute-force" approach to environmental control, involving attempts to monitor each environmental niche for every possible pollutant, increases employment for chemists and raises sales of analytical instruments, it is a wasteful way to detect and solve environmental problems, degenerating into a mindless exercise in the collection of marginally useful numbers.

Those responsible for environmental protection must be smarter than that. In order for chemistry to make a maximum contribution to the solution of environmental problems, the chemist must work toward an understanding of the nature, reactions, and transport of chemical species in the environment. Analytical chemistry is a fundamental and crucial part of that endeavor.

Environmental Biochemistry

The ultimate environmental concern is that of life itself. The discipline that deals specifically with the effects of environmental chemical species on life is **environmental biochemistry**. A related area, **toxicological chemistry**, is *thechemistry of toxic substances with emphasis upon their interactions with biologic tissue and living organisms*. Toxicological chemistry deals with the chemical nature and reactions of toxic substances and involves their origins, uses, and chemical aspects of exposure, fates, and disposal.

2.3 WATER, AIR, EARTH, LIFE, AND TECHNOLOGY

Water and the Hydrosphere

Water, with a deceptively simple chemical formula of H_2O, is a vitally important substance in all parts of the environment. Water covers about 70% of Earth's surface. It occurs in all spheres of the environment—in the oceans as a vast reservoir of saltwater, on land as surface water in lakes and rivers, underground as groundwater, in the atmosphere as water vapor, in the polar icecaps as solid ice, and in many segments of the anthrosphere such as in boilers or municipal water distribution systems.

Water is an essential part of all living systems and is the medium from which life evolved and in which life exists. Energy and matter are carried through various spheres of the environment by water. Water leaches soluble constituents from mineral matter and carries them to the ocean or leaves them as mineral deposits some distance from their sources. Water carries plant nutrients from soil into the bodies of plants by way of plant roots. Solar energy absorbed in the evaporation of ocean water is carried as latent heat and released inland.

The accompanying release of latent heat provides a large fraction of the energy that is transported from equatorial regions toward Earth's poles and powers massive storms.
Water is obviously an important topic in environmental sciences. Its environmental chemistry will be discussed further.

Air and the Atmosphere

The atmosphere is a protective blanket which nurtures life on the Earth and protects it from the hostile environment of outer space. It is the source of carbon dioxide for plant photosynthesis and of oxygen for respiration. It provides the nitrogen that nitrogen-fixing bacteria and ammonia-manufacturing industrial plants use to produce chemically-bound nitrogen, an essential component of life molecules.

As a basic part of the hydrologic cycle, the atmosphere transports water from the oceans to land, thus acting as the condenser in a vast solarpowered still. The atmosphere serves a vital protective function, absorbing harmful ultraviolet radiation from the sun and stabilizing Earth's temperature. **Atmospheric science** deals with the movement of air masses in the atmosphere, atmospheric heat balance, and atmospheric chemical composition and reactions.

Earth

The **geosphere**, or solid Earth is that part of the Earth upon which humans live and from which they extract most of their food, minerals, and fuels. The earth is divided into layers, including the solid, iron-rich inner core, molten outer core, mantle, and crust. Environmental science is most concerned with the **lithosphere**, which consists of the outer mantle and the **crust**.

The latter is the earth's outer skin that is accessible to humans. It is extremely thin compared to the diameter of the earth, ranging from 5 to 40 km thick. **Geology** is the science of the geosphere. As such, it pertains mostly to the solid mineral portions of Earth's crust. But it must also consider water, which is involved in weathering rocks and in producing mineral formations; the atmosphere and climate, which have profound effects on the geosphere and interchange matter and energy with it; and living systems, which largely exist on the geosphere and in turn have significant effects on it.

Geological science uses chemistry in the form of geochemistry to explain the nature and behavior of geological materials, physics to explain their mechanical behavior, and biology to explain the mutual interactions between the geosphere and the biosphere.3 Modern technology, for example the ability to move massive quantities of dirt and rock around, has a profound influence on the geosphere.

The most important part of the geosphere for life on earth is **soil** formed by the disintegrative weathering action of physical, geochemical, and biological processes on rock. It is the medium upon which plants grow, and virtually all terrestrial organisms depend upon it for their existence. The productivity of soil is strongly affected by environmental conditions and pollutants.

Life

Biology is the science of life. It is based on biologically synthesized chemical species, many of which exist as large molecules called *macromolecules*. As living beings, the ultimate concern of humans with their environment is the interaction of the environment with life. Therefore, biological science is a key component of environmental science and environmental chemistry. The role of life in environmental science is discussed in numerous scientific discussions.

The Anthrosphere and Technology

Technology refers to the ways in which humans do and make things with materials and energy. In the modern era, technology is to a large extent the product of engineering based on scientific principles. Science deals with the discovery, explanation, and development of theories pertaining to interrelated natural phenomena of energy, matter, time, and space.

Based on the fundamental knowledge of science, engineering provides the plans and means to achieve specific practical objectives. Technology uses these plans to carry out the desired objectives. It is essential to consider technology, engineering, and industrial activities in studying environmental science because of the enormous influence that they have on the environment.

Humans will use technology to provide the food, shelter, and goods that they need for their well-being and survival. The challenge is to interweave technology with considerations of the environment and ecology such that the two are mutually advantageous rather than in opposition to each other.

Technology, properly applied, is an enormously positive influence for environmental protection. The most obvious such application is in air and water pollution control. As necessary as "end-of-pipe"

measures are for the control of air and water pollution, it is much better to use technology in manufacturing processes to prevent the formation of pollutants.

Technology is being used increasingly to develop highly efficient processes of energy conversion, renewable energy resource utilization, and conversion of raw materials to finished goods with minimum generation of hazardous waste by-products. In the transportation area, properly applied technology in areas such as high speed train transport can enormously increase the speed, energy efficiency, and safety of means for moving people and goods.

Until very recently, technological advances were made largely without heed to environmental impacts. Now, however, the greatest technological challenge is to reconcile technology with environmental consequences. The survival of humankind and of the planet that supports it now requires that the established two-way interaction between science and technology become a three-way relationship including environmental protection.

Ecology and the Biosphere

The **biosphere** is the name given to that part of the environment consisting of organisms and living biological material. Virtually the entire biosphere is contained by the geosphere and hydrosphere in the very thin layer where these environmental spheres interface with the atmosphere. There are some specialized life forms at extreme depths in the ocean, but these are still relatively close to the atmospheric interface.

The biosphere strongly influences, and in turn is strongly influenced by, the other parts of the environment. It is believed that organisms were responsible for converting Earth's original reducing atmosphere to an oxygen-rich one, a process that also resulted in the formation of massive deposits of oxidized minerals, such as iron in deposits of Fe_2O_3.

Photosynthetic organisms remove CO_2 from the atmosphere, thus preventing runaway greenhouse warming of Earth's surface. Organisms strongly influence bodies of water, producing biomass required for life in the water and mediating oxidation-reduction reactions in the water.

Organisms are strongly involved with weathering processes that break down rocks in the geosphere and convert rock matter to soil. Lichens, consisting of symbiotic (mutually advantageous) combinations of algae and fungi, attach strongly to rocks; they secrete chemical species that slowly dissolve the rock surface and retain surface moisture that promotes rock weathering.

The biosphere is based upon plant photosynthesis, which fixes solar energy (*hv*) and carbon from atmospheric CO_2 in the form of high-energy biomass.

$$CO_2 + H_2O \rightarrow \{CH_2O\} + O_2(g)$$

In so doing, plants and algae function as autotrophic organisms, those that utilize solar or chemical energy to fix elements from simple, nonliving inorganic material into complex life molecules that compose living organisms. The opposite process, biodegradation, breaks down biomass either in the presence of oxygen (aerobic respiration):

$$\{CH_2O\} + O_2(g) \rightarrow CO_2 + H_2O$$

or absence of oxygen (anaerobic respiration):

$$2\{CH_2O\} + CO_2(g) \rightarrow CH_4(g)$$

Both aerobic and anaerobic biodegradation get rid of biomass and return carbon dioxide to the atmosphere. The latter reaction is the major source of atmospheric methane. Nondegraded remains of these processes constitute organic matter in aquatic sediments and in soils, which has an important influence on the characteristics of these solids. Carbon that was originally fixed photosynthetically forms the basis of all fossil fuels in the geosphere.

There is a strong interconnection between the biosphere and the anthrosphere. Humans depend upon the biosphere for food, fuel, and raw materials. Human influence on the biosphere continues to change it drastically. Fertilizers, pesticides, and cultivation practices have vastly increased yields of biomass, grains, and food. Destruction of habitat is resulting in the extinction of vast numbers of species, in some cases even before they are discovered.

Bioengineering of organisms with recombinant DNA technology and older techniques of selection and hybridization are causing great changes in the characteristics of organisms and promise to result in even more striking alterations in the future. It is the responsibility of humankind to make such changes intelligently and to protect and nurture the biosphere.

Ecology

Ecology is the science that deals with the relationships between living organisms with their physical environment and with each other. Ecology can be approached from the viewpoints of (1) the environment and the demands it places on the organisms in it or (2) organisms and how they adapt to their environmental conditions.

An **ecosystem** consists of an assembly of mutually interacting organisms and their environment in which materials are interchanged in a largely cyclical manner. An ecosystem has physical, chemical, and biological components along with energy sources and pathways of energy and materials interchange. The environment in which a particular organism lives is called its **habitat**. The role of an organism in a habitat is called its **niche**.

For the study of ecology it is often convenient to divide the environment into four broad categories. The **terrestrial environment** is based on land and consists of **biomes**, such as grasslands, savannas, deserts, or one of several kinds of forests. The **freshwater environment** can be further subdivided between *standing-water habitats* (lakes, reservoirs) and *running-water habitats* (streams, rivers). The oceanic **marine environment** is characterized by saltwater and may be divided broadly into the shallow waters of the continental shelf composing the **neritic zone** and the deeper waters of the ocean that constitute the **oceanic region**. An environment in which two or more kinds of organisms exist together to their mutual benefit is termed a **symbiotic environment**.

A particularly important factor in describing ecosystems is that of **populations** consisting of numbers of a specific species occupying a specific habitat. Populations may be stable, or they may grow exponentially as a **population explosion**. A population explosion that is unchecked results in resource depletion, waste accumulation, and predation culminating in an abrupt decline called a **population crash**.

Behavior in areas such as hierarchies, territoriality, social stress, and feeding patterns plays a strong role in determining the fates of populations.

Two major subdivisions of modern ecology are **ecosystem ecology**, which views ecosystems as large units, and **population ecology**, which attempts to explain ecosystem behavior from the properties of individual units. In practice, the two approaches are usually merged.

Descriptive ecology describes the types and nature of organisms and their environment, emphasizing structures of ecosystems and communities, and dispersions and structures of populations.

Functional ecology explains how things work in an ecosystem, including how populations respond to environmental alteration and how matter and energy move through ecosystems. An understanding of ecology is essential in the management of modern industrialized societies in ways that are compatible with environmental preservation and enhancement.

Applied ecology deals with predicting the impacts of technology and development and making recommendations such that these activities will have minimum adverse impact, or even positive impact, on ecosystems.

2.4 ENERGY AND CYCLE OF ENERGY

Biogeochemical cycles and virtually all other processes on Earth are driven by energy from the sun. The sun acts as a so-called **blackbody** radiator with an effective surface temperature of 5780 K. It transmits energy to Earth as electromagnetic radiation with a maximum energy flux at about 500 nanometers, which is in the visible region of the spectrum. A 1-square-meter area perpendicular to the line of solar flux at the top of the atmosphere receives energy at a rate of 1,340 watts, sufficient, for example, to power an electric iron.

This is called the **solar flux**.

Energy Flow and Photosynthesis in Living Systems

Whereas materials are recycled through ecosystems, the flow of useful energy is essentially a one-way process. Incoming solar energy can be regarded as high-grade energy because it can cause useful reactions to occur, such as production of electricity in photovoltaic cells or photosynthesis in plants.

As shown in Figure 122, solar energy captured by green plants energizes chlorophyll, which in turn powers metabolic processes that produce carbohydrates from water and carbon dioxide.

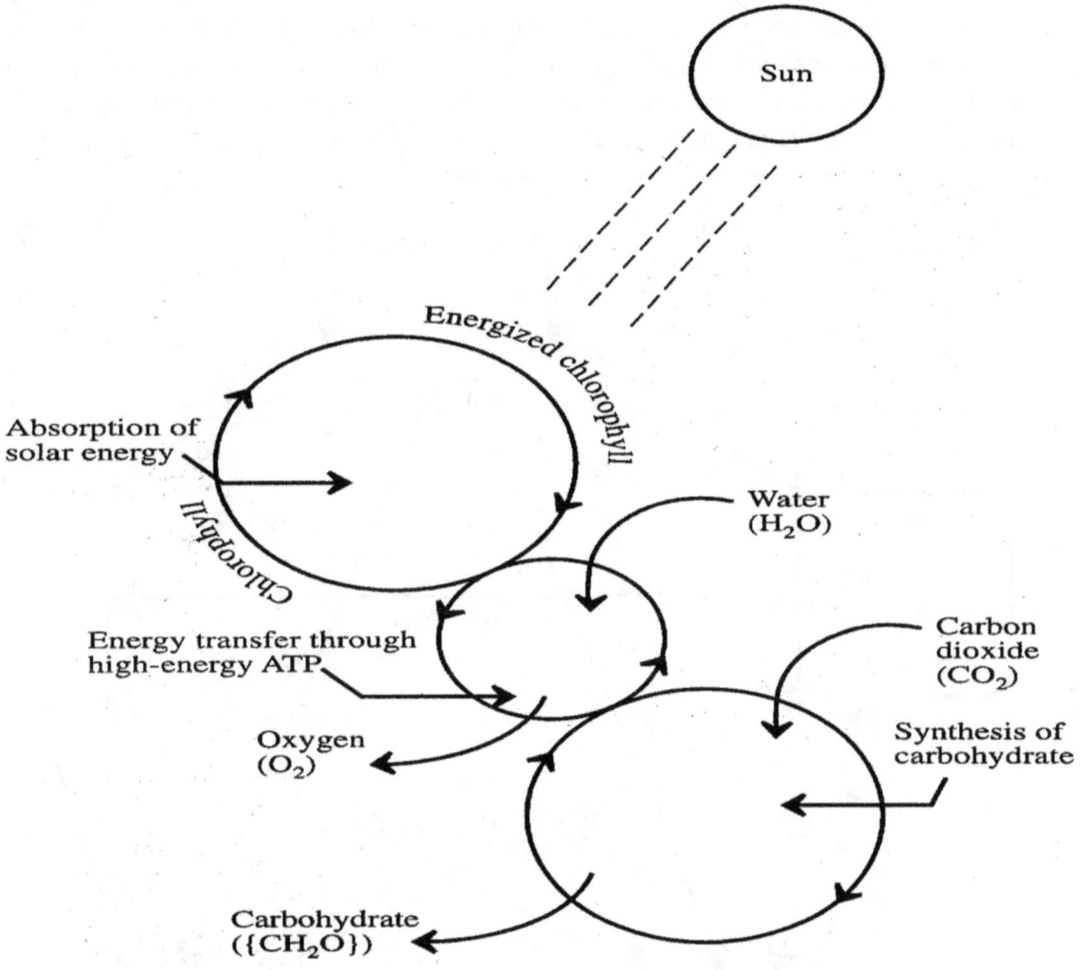

Fig. 122. *Energy conversion and transfer by photosynthesis.*

These carbohydrates are repositories of stored chemical energy that can be converted to heat and work by metabolic reactions with oxygen in organisms. Ultimately, most of the energy is converted to low-grade heat, which is eventually reradiated away from Earth by infrared radiation.

Energy Utilization

During the last two centuries, the growing, enormous human impact on energy utilization has resulted in many of the environmental problems now facing humankind. This time period has seen a transition from the almost exclusive use of energy captured by photosynthesis and utilized as biomass (food to provide muscle power, wood for heat) to the use of fossil fuel petroleum, natural gas, and coal for about 90 percent, and nuclear energy for about 5 percent, of all energy employed commercially.

Although fossil sources of energy have greatly exceeded the pessimistic estimates made during the "energy crisis" of the 1970s, they are limited and their pollution potential is high. Of particular importance is the fact that all fossil fuels produce carbon dioxide, a greenhouse gas. Therefore, it will be necessary to move toward the utilization of alternate renewable energy sources, including solar energy and biomass. The study of energy utilization is crucial in the environmental sciences.

Matter and cycles of matter

Cycles of matter (Fig. 123), often based on elemental cycles, are of utmost importance in the environment. These cycles are summarized here and are discussed further in later chapters. Global geochemical cycles can be regarded from the viewpoint of various reservoirs, such as oceans, sediments, and the atmosphere, connected by conduits through which matter moves continuously. The movement of a specific kind of matter between two particular reservoirs may be reversible or irreversible.

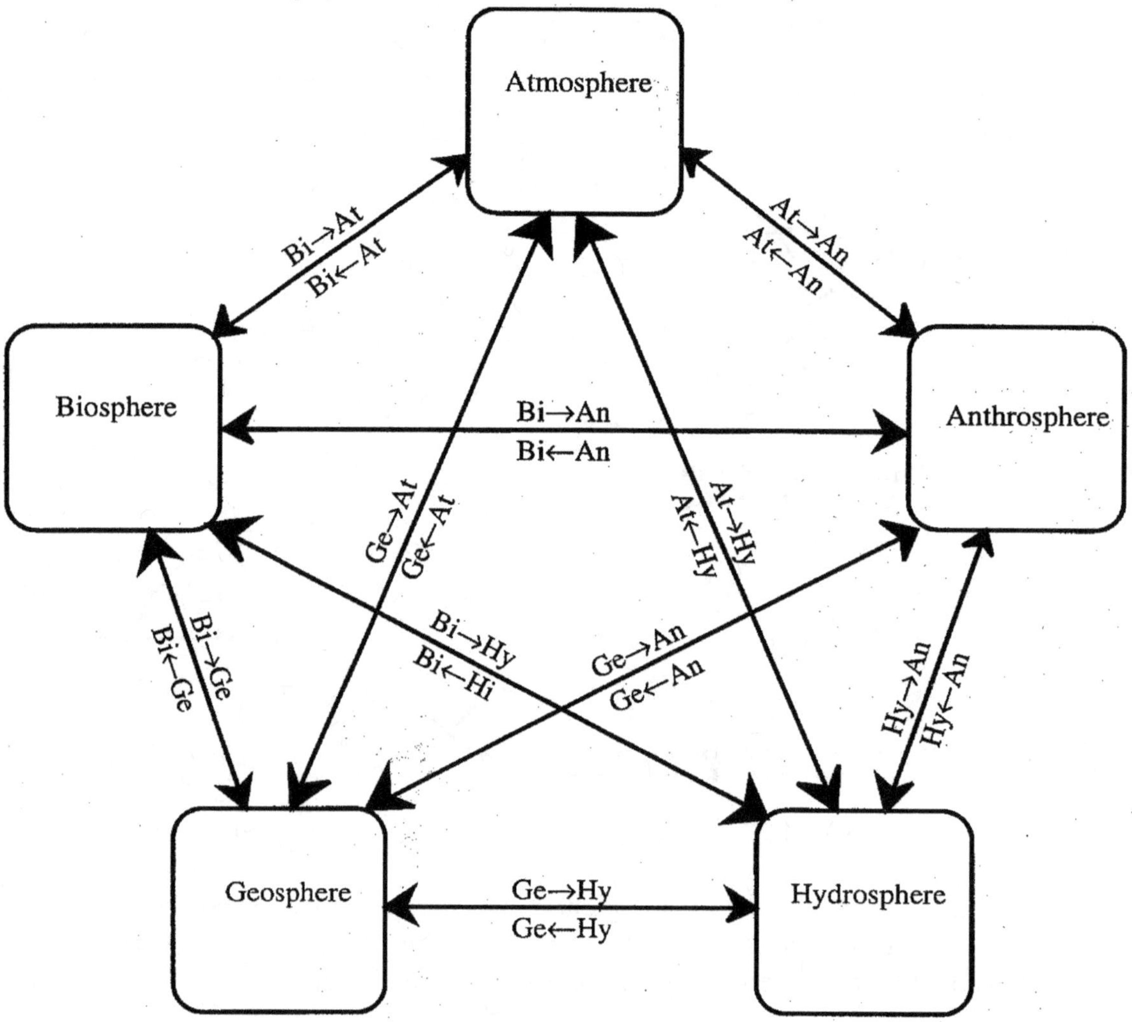

Fig. 123. *General cycle showing interchange of matter among the atmosphere, biosphere, anthrosphere, geosphere, and hydrosphere*

The fluxes of movement for specific kinds of matter vary greatly as do the contents of such matter in a specified reservoir. Cycles of matter would occur even in the absence of life on Earth but are strongly influenced by life forms, particularly plants and microorganisms. Organisms participate in **biogeochemical cycles**, which describe the circulation of matter, particularly plant and animal nutrients, through ecosystems. As part of the carbon cycle, atmospheric carbon in CO_2 is fixed as biomass; as part of the nitrogen cycle, atmospheric N_2 is fixed in organic matter. The reverse of these kinds of processes is **mineralization**, in which biologically bound elements are returned to inorganic states. Biogeochemical cycles are ultimately powered by solar energy, which is fine-tuned and directed by energy expended by organisms.

Most biogeochemical cycles can be described as **elemental cycles** involving **nutrient elements** such as carbon, nitrogen, oxygen, phosphorus, and sulfur. Many are exogenic cycles in which the element in question spends part of the cycle in the atmosphere—O_2 for oxygen, N_2 for nitrogen, CO_2 for carbon. Others, notably the phosphorus cycle, do not have a gaseous component and are endogenic cycles. All sedimentary cycles involve **salt solutions** or **soil solutions** that contain dissolved substances leached from weathered minerals; these substances may be deposited as mineral formations, or they may be taken up by organisms as nutrients.

Carbon Cycle

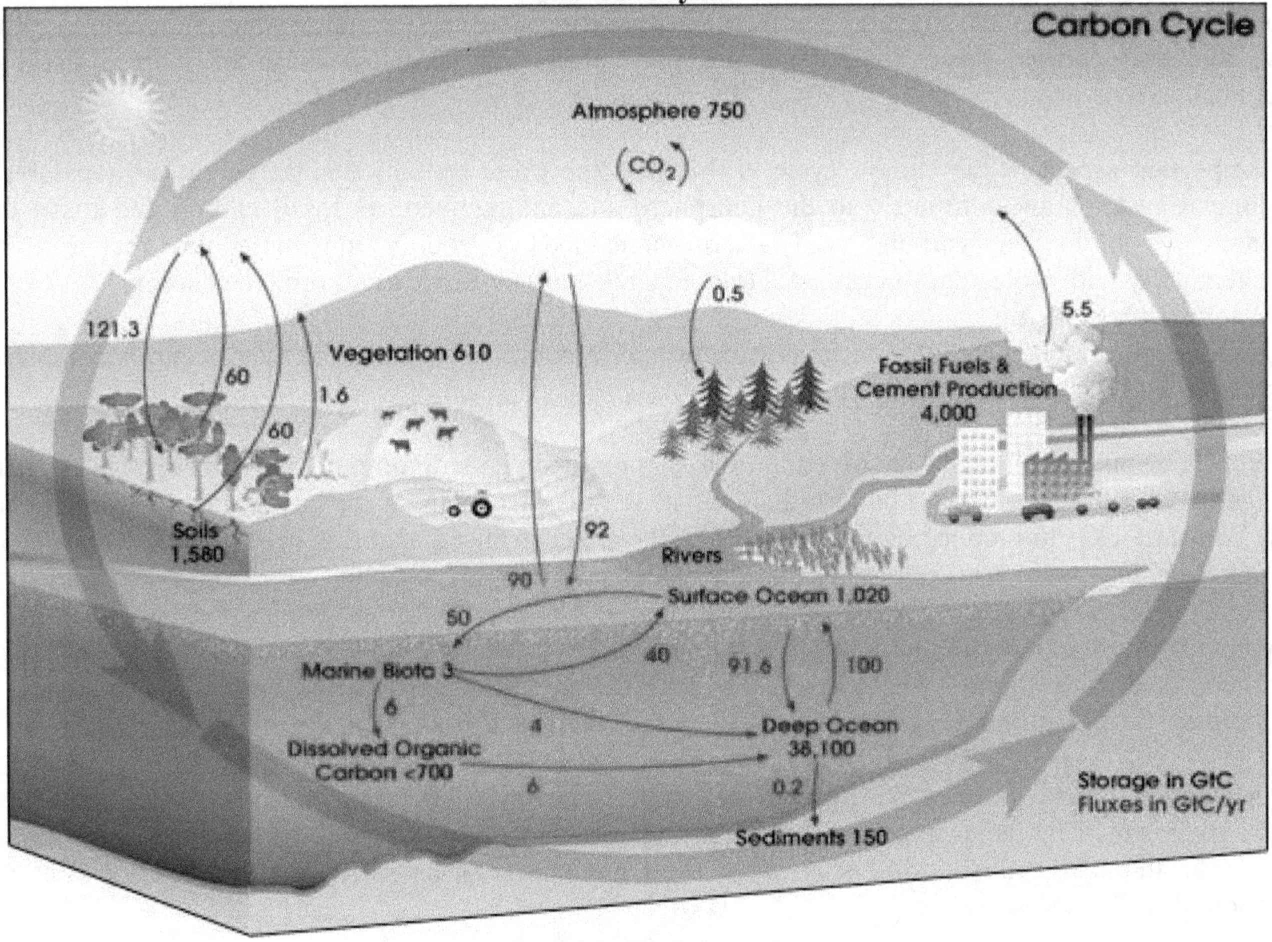

Fig. 124. *Carbon cycle*

Mineral carbon is held in a reservoir of limestone, $CaCO_3$, from which it may be leached into a mineral solution as dissolved hydrogen carbonate ion, HCO_3^-, formed when dissolved $CO_2(aq)$ reacts with $CaCO_3$. In the atmosphere carbon is present as carbon dioxide, CO_2. Atmospheric carbon dioxide is fixed as organic matter by photosynthesis, and organic carbon is released as CO_2 by microbial decay of organic matter.

Carbon is circulated through the **carbon cycle** (Fig. 124). This cycle shows that carbon may be present as gaseous atmospheric CO_2, constituting a relatively small but highly significant portion of global carbon. Some of the carbon is dissolved in surface water and groundwater as H_2CO_3 or molecular $CO_2(aq)$.

A very large amount of carbon is present in minerals, particularly calcium and magnesium carbonates such as $CaCO_3$. Photosynthesis fixes inorganic C as biological carbon, represented as $\{CH_2O\}$, which is a consituent of all life molecules. Another fraction of carbon is fixed as petroleum and natural gas, with a much larger amount as hydrocarbonaceous kerogen (the organic matter in oil shale), coal, and lignite, represented as C_xH_{2x}.

Manufacturing processes are used to convert hydrocarbons to xenobiotic compounds with functional groups containing halogens, oxygen, nitrogen, phosphorus, or sulfur. Though a very small amount of total environmental carbon, these compounds are particularly significant because of their toxicological chemical effects.

An important aspect of the carbon cycle is that it is the cycle by which solar energy is transferred to biological systems and ultimately to the geosphere and anthrosphere as fossil carbon and fossil fuels. Organic, or biological, carbon, $\{CH_2O\}$, is contained in energy-rich molecules that can react biochemically with molecular oxygen, O2, to regenerate carbon dioxide and produce energy.

This can occur biochemically in an organism through aerobic respiratio, or it may occur as combustion, such as when wood or fossil fuels are burned.

Microorganisms are strongly involved in the carbon cycle, mediating crucial biochemical reactions discussed later in this section. Photosynthetic algae are the predominant carbon-fixing agents in water; as they consume CO_2 to produce biomass the pH of the water is raised enabling precipitation of $CaCO_3$ and $CaCO_3 \bullet MgCO_3$.

Organic carbon fixed by microorganisms is transformed by biogeochemical processes to fossil petroleum, kerogen, coal, and lignite. Microorganisms degrade organic carbon from biomass, petroleum, and xenobiotic sources, ultimately returning it to the atmosphere as CO_2.

Hydrocarbons such as those in crude oil and some synthetic hydrocarbons are degraded by microorganisms. This is an important mechanism for eliminating pollutant hydrocarbons, such as those that are accidentally spilled on soil or in water. Biodegradation can also be used to treat carbon-containing compounds in hazardous wastes.

The Nitrogen Cycle

Nitrogen occurs prominently in all the spheres of the environment. The atmosphere is 78% elemental nitrogen, N_2, by volume and comprises an inexhaustible reservoir of this essential element. Nitrogen, though constituting much less of biomass than carbon or oxygen, is an essential constituent of proteins.

The N_2 molecule is very stable so that breaking it down into atoms that can be incorporated with inorganic and organic chemical forms of nitrogen is the limiting step in the nitrogen cycle. This does occur by highly energetic processes in lightning discharges that produce nitrogen oxides.

Elemental nitrogen is also incorporated into chemically bound forms, or **fixed** by biochemical processes mediated by microorganisms. The biological nitrogen is mineralized to the inorganic form during the decay of biomass.

Large quantities of nitrogen are fixed synthetically under high temperature and high pressure conditions according to the following overall reaction: $N_2 + 3H_2 \rightarrow 2NH_3$

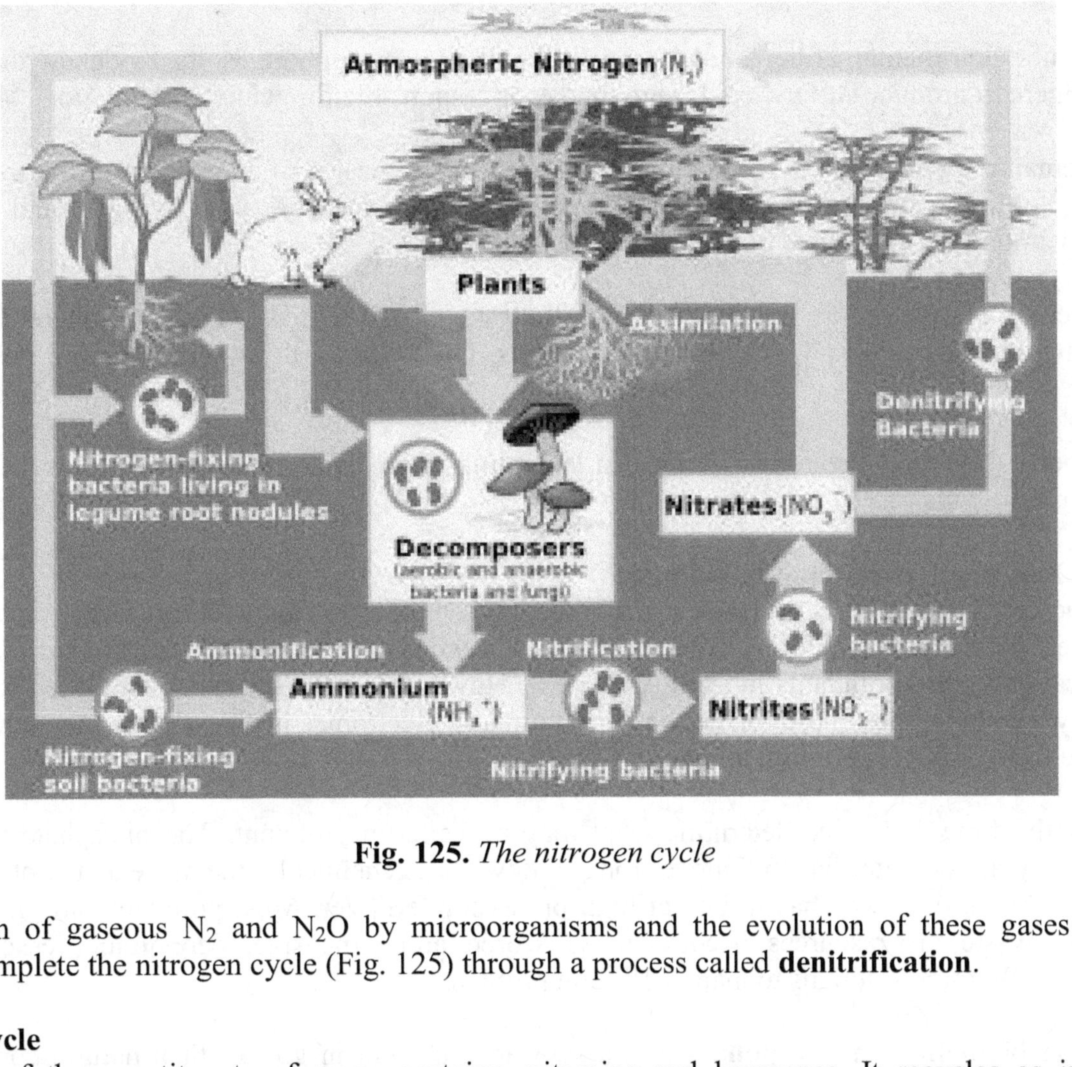

Fig. 125. *The nitrogen cycle*

The production of gaseous N_2 and N_2O by microorganisms and the evolution of these gases to the atmosphere complete the nitrogen cycle (Fig. 125) through a process called **denitrification**.

The Sulfur Cycle

Sulfur is one of the constituents of many proteins, vitamins and hormones. It recycles as in other biogeochemical cycles.

The essential steps of the sulfur cycle are:

- Mineralization of organic sulfur to the inorganic form, hydrogen sulfide: (H_2S).
- Oxidation of sulfide and elemental sulfur (S) and related compounds to sulfate (SO_4^{2-}).
- Reduction of sulfate to sulfide.
- Microbial immobilization of the sulfur compounds and subsequent incorporation into the organic form of sulfur.

The sulfur cycleis relatively complex in that it involves several gaseous species, poorly soluble minerals, and several species in solution. It is tied with the oxygen cycle in that sulfur combines with oxygen to form gaseous sulfur dioxide, SO_2, an atmospheric pollutant, and soluble sulfate ion, SO_4^{2-}.

Among the significant species involved in the sulfur cycle are gaseous hydrogen sulfide, H_2S; mineral sulfides, such as PbS, sulfuric acid, H_2SO_4, the main constituent of acid rain; and biologically bound sulfur in sulfur-containing proteins.

Insofar as pollution is concerned, the most significant part of the sulfur cycle is the presence of pollutant SO_2 gas and H_2SO_4 in the atmosphere. The former is a somewhat toxic gaseous air pollutant evolved in the combustion of sulfur-containing fossil fuels.

The major detrimental effect of sulfur dioxide in the atmosphere is its tendency to oxidize in the atmosphere to produce sulfuric acid. This species is responsible for acidic precipitation, "acid rain".

The demands of increasing population coupled with the desire of most people for a higher material standard of living are resulting in worldwide pollution on a massive scale. Environmental pollution can be divided among the categories of water, air, and land pollution.

All three of these areas are linked. For example, some gases emitted to the atmosphere can be converted to strong acids by atmospheric chemical processes, fall to the earth as acid rain, and pollute water with acidity.

Improperly discarded hazardous wastes can leach into groundwater that is ventually released as polluted water into streams.

Some Definitions Pertaining to Pollution

In some cases pollution is a clear-cut phenomenon, whereas in others it lies largely in the eyes of the beholder. Toxic organochlorine solvent residues leached into water upplies from a hazardous waste chemical dump are pollutants in anybody's view. However, loud rock music amplified to a high decibel level by the sometimes questionable miracle of modern electronics is pleasant to some people, and a very definite form of noise pollution to others.

Frequently, time and place determine what may be called a pollutant. The phosphate that the sewage treatment plant operator has to remove from wastewater is chemically the same as the phosphate that the farmer a few miles away has to buy at high prices for fertilizer. Most pollutants are, in fact, resources gone to waste; as resources become more scarce and expensive, economic pressure will almost automatically force solutions to many pollution problems.

A reasonable definition of a **pollutant** is a substance present in greater than natural concentration as a result of human activity that has a net detrimental effect upon its environment or upon something of value in that environment. **Contaminants**, which are not classified as pollutants unless they have some detrimental effect, cause deviations from the normal composition of an environment.

Every pollutant originates from a **source**. The source is particularly important because it is generally the logical place to eliminate pollution. After a pollutant is released from a source, it may act upon a receptor. The **receptor** is anything that is affected by the pollutant. Humans whose eyes smart from oxidants in the atmosphere are receptors.

Trout fingerlings that may die after exposure to dieldrin in water are also receptors. Eventually, if the pollutant is long-lived, it may be deposited in a **sink**, a long-time repository of the pollutant. Here it will remain for a long time, though not necessarily permanently. Thus, a limestone wall may be a sink for atmospheric sulfuric acid through the reaction:

$CaCO_3 + H_2SO_4 \rightarrow CaSO_4 + H_2O + CO_2$

which fixes the sulfate as part of the wall composition.

2.5 ATMOSPHERE

The Environment of Atmosphere

When the Earth first formed, it may have had a thin atmosphere of light gases, mainly hydrogen and helium, derived from the stellar nebula. If so, this atmosphere was lost when the Sun began to radiate, supplying the gas molecules with the energy they needed to escape the planet's gravitational attraction. Volcanism then released the gases which formed a new atmosphere.

No one knows the precise composition of that atmosphere, but probably it was rich in carbon dioxide and contained only traces of nitrogen and free oxygen. Our present atmosphere has evolved to its present state, partly (some would say mainly) as a result of biological processes.

These processes continue to maintain it. Nitrogen, for example, is chemically somewhat inert but, in the presence of oxygen and with a sufficient application of energy, it will oxidize to nitrate (NO_3). This reacts with water to form nitric acid (HNO_3) and is washed to the ground. Lightning supplies enough energy for the oxidation, and in the world as a whole it is estimated that there are about 1800 thunderstorms, with lightning, happening at any one time, delivering some 100 million tonnes of fixed nitrogen to the surface every year.

At this rate it would not take long to strip the atmosphere of its nitrogen, were it not for the activities of denitrifying bacteria, which utilize nitrogen compounds in the soil and release gaseous nitrogen as a metabolic by-product. Were the air to be seriously depleted of nitrogen the proportion (partial pressure) of oxygen would increase and were it to increase to more than about 25 per cent exposed carbon compounds would burn even more readily than they do. This would reduce the amount of oxygen, replacing it with carbon dioxide.

As it is, carbon dioxide is removed from the air by green plants as well as by dissolving in rain water. To a considerable extent, therefore, our present atmosphere has been constructed and is maintained by living organisms. It is an essential component of the environment. It is also the route by which many organisms are disseminated and most nutrients cycled. Air is a mixture of gases.

They are not combined, soair itself is not a chemical compound, although some of its minor constituent gases are. As the many practical uses for compressed air (in car and bicycle tyres, for example) testify, air is highly compressible. The total weight of the atmosphere compresses its lower layers, so the average density of the air decreases from about 1.2 kg m-3 at sea level to about 0.7 kg m^{-3} at a height of 5 km. Half of the total mass of the atmosphere lies below 5 km.

The pressure exerted by the atmosphere is usually expressed in millibars (mb) in weather forecasts or in pascals (Pa) by physicists (1 mb=100 N m-2=100 Pa). The average sealevel pressure is 1013.2 mb (or

101320 Pa). One consequence of its compressibility is that for all practical purposes the atmosphere is quite shallow.

Although it has no clearly defined upper boundary, merging imperceptibly with the solar atmosphere at an altitude of about 80000 km (ALLABY, 1992, p. 50), pressure and therefore density decrease logarithmically with height. At 30 km, air density is at only 0.02 per cent of its sea-level value.

The part of the atmosphere in which weather occurs is confined to the lowermost 8–16 km. None of the familiar meteorological phenomena occur above this height, but they are influenced by events in the lower reaches of the overlying layer, up to a height of about 30 km.
The air is warmed convectively from below, by contact with the surface. As it warms it expands and as it expands it cools. Temperature and density determine the amount of water vapour air can hold (its humidity), and the combination of the humidity, density, and temperature imposes a layered structure on the atmosphere. Figure 126 illustrates this structure, relating it to height, temperature, and pressure.

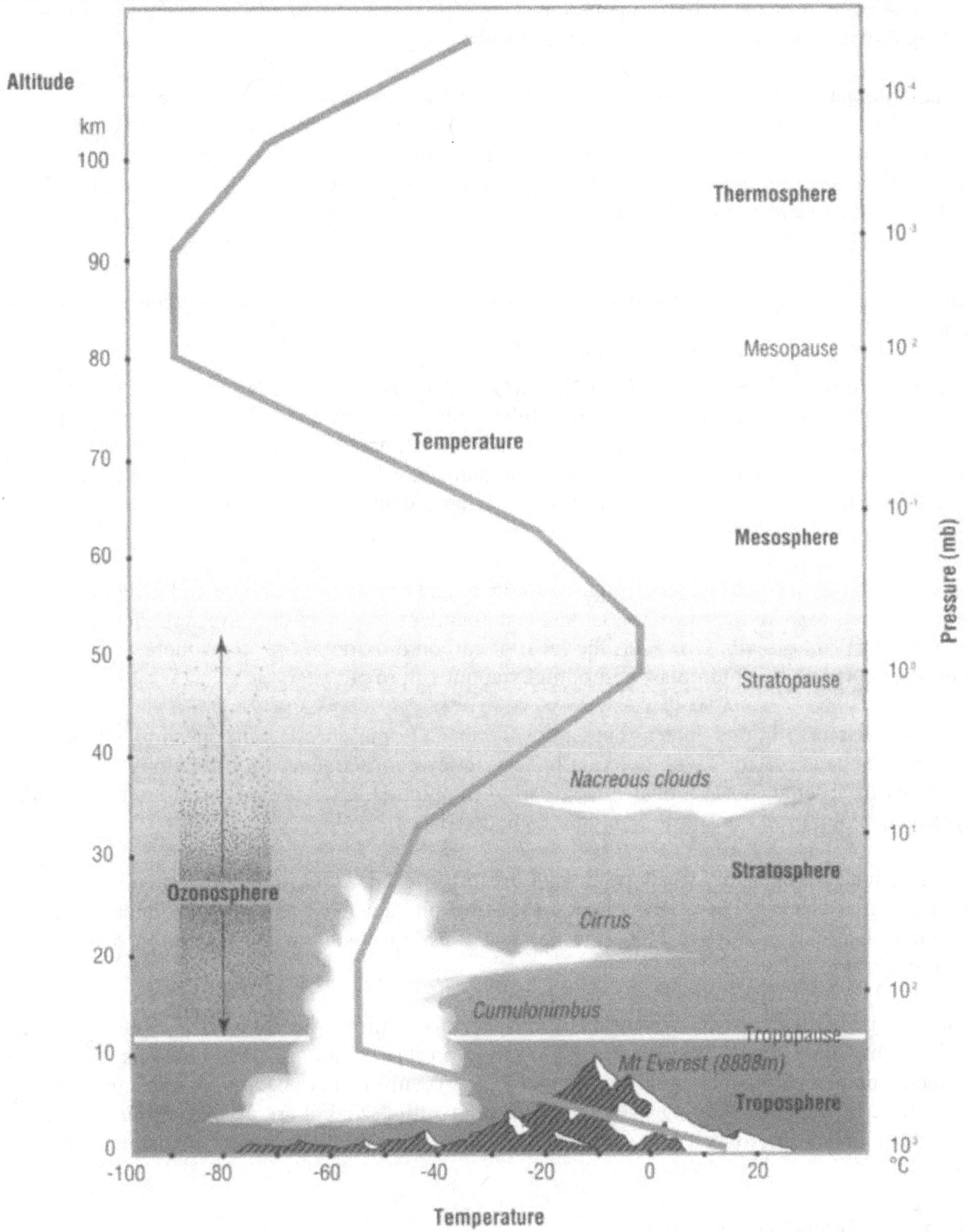

Fig. 126. *The athmosphere*

The most significant feature of atmospheric chemistry is the occurrence of **photochemical reactions** resulting from the absorption by molecules of light photons, designated *h*n. (The energy, E, of a photon of visible or ultraviolet light is given by the equation, E = *h*n, where *h* is Planck's constant and n is the frequency of light, which is inversely proportional to its wavelength.

Ultraviolet radiation has a higher frequency than visible light and is, therefore, more energetic and more likely to break chemical bonds in molecules that absorb it.) One of the most significant photochemical reactions is the one responsible for the presence of ozone in the stratosphere (see above), which is initiated when O2 absorbs highly energetic ultraviolet radiation in the wavelength ranges of 135-176 nanometers (nm) and 240-260 nm in the stratosphere:

$$O_2 + hn = O + O$$

The oxygen atoms produced by the photochemical dissociation of O_2 react with oxygen molecules to produce ozone, O_3,

$$O + O_2 + M = O_3 + M$$

where M is a third body, such as a molecule of N2 , which absorbs excess energy from the reaction. The ozone that is formed is very effective in absorbing ultraviolet radiation in the 220-330 nm wavelength range, which causes the temperature increase observed in the stratosphere. The ozone serves as a very valuable filter to remove ultraviolet radiation from the sun's rays. If this radiation reached the earth's surface, it would cause skin cancer and other damage to living organisms.

Gaseous Oxides in the Atmosphere

Oxides of carbon, sulfur, and nitrogen are important constituents of the atmosphere and are pollutants at higher levels. Of these, carbon dioxide, CO_2, is the most abundant. It is a natural atmospheric constituent, and it is required for plant growth. However, the level of carbon dioxide in the atmosphere, now at about 360 parts per million (ppm) by volume, is increasing by about 1 ppm per year.

As discussed in Chapter 14, this increase.in atmospheric CO_2 may well cause general atmospheric warming—the "greenhouse effect," with potentially very serious consequences for the global atmosphere and for life on earth. Though not a global threat, carbon monoxide, CO, can be a serious health threat because it prevents blood from transporting oxygen to body tissues.

The two most serious nitrogen oxide air pollutants are nitric oxide, NO, and nitrogen dioxide, NO_2, collectively denoted as "NOx." These tend to enter the atmosphere as NO, and photochemical processes in the atmosphere can convert NO to NO_2. Further reactions can result in the formation of corrosive nitrate salts or nitric acid, HNO3.

Nitrogen dioxide is particularly significant in atmospheric chemistry because of its photochemical dissociation by light with a wavelength less than 430 nm to produce highly reactive O atoms. This is the first step in the formation of photochemical smog (see below). Sulfur dioxide, SO_2, is a reaction product of the combustion of sulfur-containing fuels such as high-sulfur coal. Part of this sulfur dioxide is converted in the atmosphere to sulfuric acid, H_2SO_4, normally the predominant contributor to acid precipitation.

Hydrocarbons and Photochemical Smog

The most abundant hydrocarbon in the atmosphere is methane, CH_4, released from underground sources as natural gas and produced by the fermentation of organic matter. Methane is one of the least reactive atmospheric hydrocarbons and is produced by diffuse sources, so that its participation in the formation of pollutant photochemical reaction products is minimal.

The most significant atmospheric pollutant hydrocarbons are the reactive ones produced as automobile exhaust emissions. In the presence of NO, under conditions of temperature inversion, low humidity, and sunlight, these hydrocarbons produce undesirable **photochemical smog** manifested by the presence of visibility-obscuring particulate matter, oxidants such as ozone, and noxious organic species such as aldehydes.

Particulate Matter

Particles ranging from aggregates of a few molecules to pieces of dust readily visible to the naked eye are commonly found in the atmosphere and are discussed in detail in Chapter 10. Some atmospheric particles, such as sea salt formed by the evaporation of water from droplets of sea spray, are natural and even beneficial atmospheric constituents.

Very small particles called **condensation nuclei** serve as bodies for atmospheric water vapor to condense upon and are essential for the formation of rain drops. Colloidal-sized particles in the atmosphere are called **aerosols**. Those formed by grinding up bulk matter are known as **dispersion aerosols**, whereas particles formed from chemical reactions of gases are **condensation aerosols**; the latter tend to be smaller.

Smaller particles are in general the most harmful because they have a greater tendency to scatter light and are the most respirable (tendency to be inhaled into the lungs).

Much of the mineral particulate matter in a polluted atmosphere is in the form of oxides and other compounds produced during the combustion of high-ash fossil fuel. Smaller particles of **fly ash** enter furnace flues and are efficiently collected in a properly equipped stack system. However, some fly ash escapes through the stack and enters the atmosphere. Unfortunately, the fly ash thus released tends to consist of smaller particles that do the most damage to human health, plants, and visibility.

Importance of the atmosphere

The atmosphere is a protective blanket which nurtures life on the Earth and protects it from the hostile environment of outer space. The atmosphere is the source of carbon dioxide for plant photosynthesis and of oxygen for respiration.

It provides the nitrogen that nitrogen-fixing bacteria and ammonia-manufacturing plants use to produce chemically-bound nitrogen, an essential component of life molecules. As a basic part of the hydrologic cycle the atmosphere transports water from the oceans to land, thus acting as the condenser in a vast solar-powered still.

Unfortunately, the atmosphere also has been used as a dumping ground for many pollutant materials—ranging from sulfur dioxide to refrigerant Freon—a practice which causes damage to vegetation and materials, shortens human life, and alters the characteristics of the atmosphere itself.

In its essential role as a protective shield, the atmosphere absorbs most of the cosmic rays from outer space and protects organisms from their effects. It also absorbs most of the electromagnetic radiation

from the sun, allowing transmission of significant amounts of radiation only in the regions of 300-2500 nm (nearultraviolet, visible, and near-infrared radiation) and 0.01-40 m (radio waves).
By absorbing electromagnetic radiation below 300 nm, the atmosphere filters out damaging ultraviolet radiation that would otherwise be very harmful to living organisms. Furthermore, because it reabsorbs much of the infrared radiation by which absorbed solar energy is re-emitted to space, the atmosphere stabilizes the earth's temperature, preventing the tremendous temperature extremes that occur on planets and moons lacking substantial atmospheres.

Physical characteristics of the atmosphere

Atmospheric science deals with the movement of air masses in the atmosphere, atmospheric heat balance, and atmospheric chemical composition and reactions. In order to understand atmospheric chemistry and air pollution, it is important to have an overall appreciation of the atmosphere, its composition and physical characteristics.

Atmospheric Composition

Dry air within several kilometers of ground level consists of two **major components**
• Nitrogen, 78.08 % (by volume) • Oxygen, 20.95 %
two **minor components**
• Argon, 0.934 % • Carbon dioxide, 0.036 %
in addition to argon, four more **noble gases**,
• Neon, 1.818 x 10-3 % • Helium, 5.24 x 10-4 %
• Krypton, 1.14 x 10-4 % • Xenon, 8.7 x 10-6 %
and **trace gases**. Atmospheric air may contain 0.1–5% water by volume, with a normal range of 1–3%.

Energy transfer in the atmosphere

The physical and chemical characteristics of the atmosphere and the critical heat balance of the earth are determined by energy and mass transfer processes in the atmosphere. Incoming solar energy is largely in the visible region of the spectrum.

The shorter wavelength blue solar light is scattered relatively more strongly by molecules and particles in the upper atmosphere, which is why the sky is blue as it is viewed by scattered light. Similarly, light that has been transmitted through scattering atmospheres (Fig. 127) appears red, particularly around sunset and sunrise, and under circumstances in which the atmosphere contains a high level of particles.

The solar energy flux reaching the atmosphere is huge, amounting to 1.34 x 1000 watts per square meter (19.2 kcal per minute per square meter) perpendicular to the line of solar flux at the top of the atmosphere. This value is the **solar constant**, and may be termed **insolation**, which stands for "incoming solar radiation."

If all this energy reached the earth's surface and was retained, the planet would have vaporized long ago. As it is, the complex factors involved in maintaining the Earth's heat balance within very narrow limits are crucial to retaining conditions of climate that will support present levels of life on earth.
The great changes of climate that resulted in ice ages during some periods, or tropical conditions during others, were caused by variations of only a few degrees in average temperature. Marked climate changes within recorded history have been caused by much smaller average temperature changes.

The mechanisms by which the earth's average temperature is retained within its present narrow range are complex and not completely understood, but the main features are explained here.

About half of the solar radiation entering the atmosphere reaches the earth's surface either directly or after scattering by clouds, atmospheric gases, or particles. The remaining half of the radiation is either reflected directly back or absorbed in the atmosphere, and its energy radiated back into space at a later time as infrared radiation.

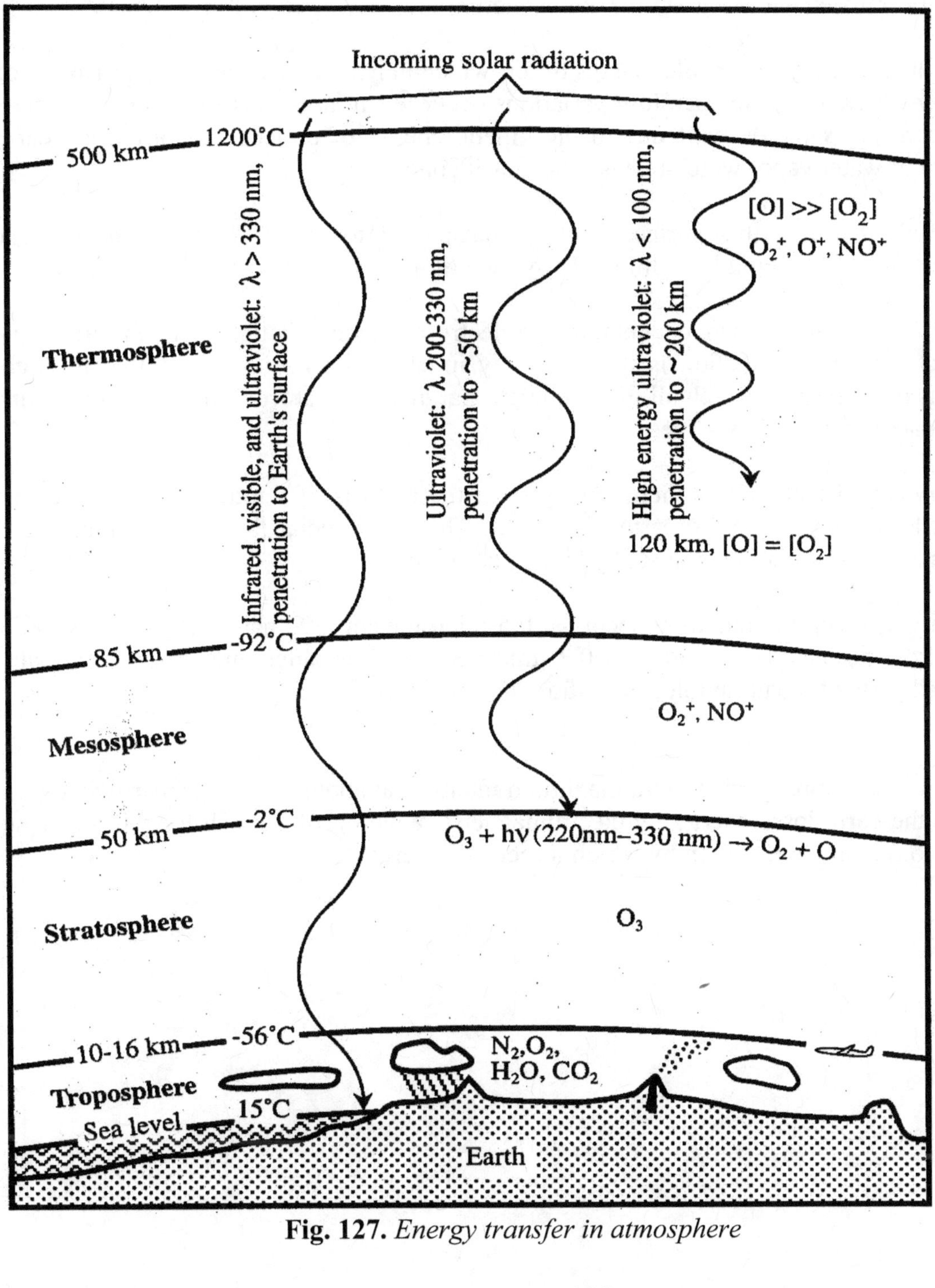

Fig. 127. *Energy transfer in atmosphere*

Most of the solar energy reaching the surface is absorbed and it must be returned to space in order to maintain heat balance. In addition, a very small amount of energy (less than 1% of that received from the sun) reaches the earth's surface by convection and conduction processes from the earth's hot mantle, and this, too, must be lost.
Energy transport, which is crucial to eventual reradiation of energy from the earth, is accomplished by three major mechanisms. These are conduction, convection, and radiation. **Conduction** of energy occurs through the interaction of adjacent atoms or molecules without the bulk movement of matter and is a relatively slow means of transferring energy in the atmosphere.

Convection involves the movement of whole masses of air, which may be either relatively warm or cold. It is the mechanism by which abrupt temperature variations occur when large masses of air move across an area. As well as carrying **sensible heat** due to the kinetic energy of molecules, convection carries **latent heat** in the form of water vapor which releases heat as it condenses.

An appreciable fraction of the earth's surface heat is transported to clouds in the atmosphere by conduction and convection before being lost ultimately by radiation.

Radiation of energy in earth's atmosphere occurs through electromagnetic radiation in the infrared region of the spectrum. Electromagnetic radiation is the only way in which energy is transmitted through a vacuum; therefore, it is the means by which all of the energy that must be lost from the planet to maintain its heat balance is ultimately returned to space.

The electromagnetic radiation that carries energy away from the earth is of a much longer wavelength than the sunlight that brings energy to the earth (Fig. 128). This is a crucial factor in maintaining the earth's heat balance, and one susceptible to upset by human activities.

The maximum intensity of incoming radiation occurs at 0.5 micrometers (500 nanometers) in the visible region, with essentially none outside the range of 0.2 μm to 3 μm. This range encompasses the whole visible region and small parts of the ultraviolet and infrared adjacent to it.

Outgoing radiation is in the infrared region, with maximum intensity at about 10 μm, primarily between 2 μm and 40 μm. Thus the earth loses energy by electromagnetic radiation of a much longer wavelength (lower energy per photon) than the radiation by which it receives energy.

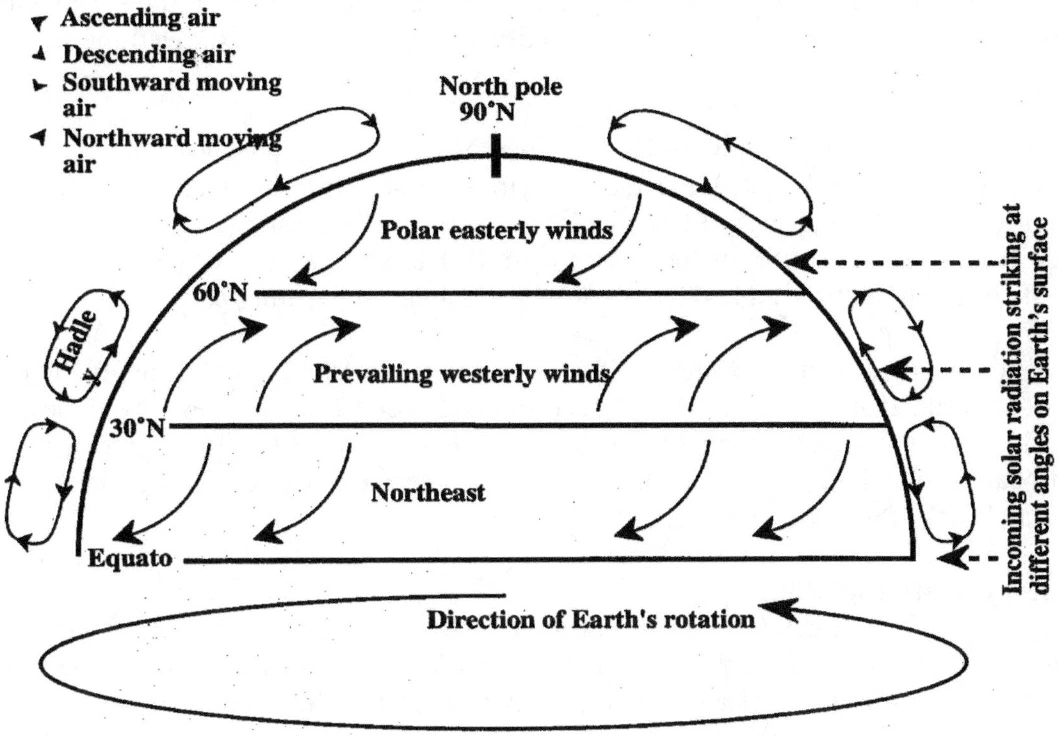

Fig. 128. *Global Weather*

The central feature of global weather is the redistribution of solar energy that falls unequally on earth at different latitudes (relative distances from the equator and poles). Consider 3. Sunlight, and the energy flux from it, is most intense at the equator because, averaged over the seasons, solar radiation comes in perpendicular to earth's surface at the Equator. With increasing distance from the equator (higher latitudes) the angle is increasingly oblique and more of the energyabsorbing atmosphere must be traversed, so that progressively less energy is received per unit area of earth's surface.

The net result is that equatorial regions receive a much greater share of solar radiation, progressively less is received farther from the Equator, and the poles receive a comparatively miniscule amount. The excess heat energy in the equatorial regions causes the air to rise.

The air ceases to rise when it reaches the stratosphere because in the stratosphere the air becomes warmer with higher elevation. As the hot equatorial air rises in the troposphere, it cools by expansion and loss of water, then sinks again. The air circulation patterns in which this occurs are called **Hadley cells**.

As shown in Figure 8, there are three major groupings of these cells, which result in very distinct climatic regions on earth's surface. The air in the Hadley cells does not move straight north and south, but is deflected by earth's rotation and by contact with the rotating earth; this is the **Coriolis effect**, which results in spiral-shaped air circulation patterns called cyclonic or anticyclonic, depending upon the direction of rotation.

These give rise to different directions of prevailing winds, depending on latitude. The boundaries between the massive bodies of circulating air shift markedly over time and season, resulting in significant weather instability.

The movement of air in Hadley cells combined with other atmospheric phenomena results in the development of massive **jet streams** that are, in a sense, shifting rivers of air that may be several kilometers deep and several tens of km wide. Jet streams move through discontinuities in the tropopause, generally from west to east at velocities around 200 km/hr (well over 100 mph); in so doing, they redistribute huge amounts of air and have a strong influence on weather patterns.

The air and wind circulation patterns described above shift massive amounts of energy over long distances on earth. If it weren't for this effect, the equatorial regions would be unbearably hot, and the regions closer to the poles intolerably cold. About half of the heat that is redistributed is carried as sensible heat by air circulation, almost 1/3 is carried by water vapor as latent heat, and the remaining approximately 20 % by ocean currents.

Solar energy in atmosphere

The Sun radiates in all directions and the Earth, being a very small target at a distance of 150 million km, intercepts 0.0005 per cent of the total. At the top of the Earth's atmosphere this amounts to about 1360 W m^{-2}, a value known as the 'solar constant'. Radiant heat and light are both forms of electromagnetic radiation, varying only in their wavelengths, and the Sun radiates across the whole electromagnetic spectrum.

According to Wien's law10, the wavelength at which a body radiates most intensely is inversely proportional to its temperature, so the hotter the body the shorter the wavelength at which it radiates most intensely. This is not surprising, because electromagnetic radiation travels only at the speed of light (beyond the Earth's atmosphere, in space, about 300000 km s-1) and the only way its energy can increase is by reducing the wavelength.

Very short-wave (high-energy) gamma (10-4–10-8 μm) and X (10^{-3}–10^{-5} μm) solar radiation is absorbed in the upper atmosphere and none reaches the surface. Radiation with a wavelength between 0.2 and 0.4 μm is called 'ultraviolet' (UV); at wavelengths below 0.29 μm, most UV is absorbed by stratospheric oxygen (O_2) and ozone (O3).

The wavelengths between 0.4 and 0.7 μm are what we see as visible light, with violet at the short-wave end of the spectrum and red at the long-wave end. These are the wavelengths at which the Sun radiates most intensely, with an intensity peak at around 0.5 μm in the green part of the spectrum. It is the part of the spectrum to which our eyes are sensitive, for the obvious reason that the most intense radiation is also the most useful, although some animals have eyes receptive to slightly shorter or longer wavelengths.

Beyond the red end of the visible spectrum lie the infra-red wavelengths (0.7 μm to 1 mm) and, with increasing wavelengths, microwaves and radio waves, the longest of which have wavelengths up to about 100 km.

The atmosphere is transparent to wavelengths longer than 0.29 μm, although water vapour absorbs energy in several narrow bands between 0.9 and 2.1 μm. When radiant energy, as light or heat, strikes the surface of land or water its energy is absorbed and the surface is warmed. The equator faces the Sun, which is always directly overhead at noon. Consequently, its radiation is most intense at the equator. With increasing distance from the equator, the Sun is lower in the sky at all seasons and its radiation covers a larger area less intensely.

Although latitude is obviously important, and places in high latitudes tend to receive less solar energy than those in low latitudes, cloudiness modifies the general distribution quite strongly. The equatorial region does not receive the most intense insolation, because for much of the time clouds shade the surface, reflecting incoming sunlight.

Tropical and subtropical deserts, where skies are mainly clear, receive 50 to 100 per cent more insolation than the equator and the dry interiors of North America and Eurasia are much sunnier than maritime regions.

Rather less than half of the solar radiation reaching the top of the atmosphere penetrates all the way to the surface. Most of the 'lost' incoming radiation is reflected directly back into space, and about 10 per cent is absorbed or scattered by ozone, water vapour, and particulate matter in the troposphere. It is scattering that gives the sky its colour. Radiation bounces off particles (mainly molecules) of a particular size in relation to its wavelength. All that changes is the direction of the radiation.

There is no loss of energy, but shorter wavelengths scatter more than longer ones. This is called Rayleigh scattering, after Lord Rayleigh (1842–1919) who discovered it, and it reflects radiation in all directions. When the Sun is high in a clear sky, violet light is scattered and absorbed very high in the atmosphere and blue below it. Scattering diffuses the blue light evenly and so the sky appears blue.

If the sky is hazy, dust particles scatter light of all wavelengths and the sky appears white. When the Sun is low, dust particles scatter light in the orange and red wavelengths, but shorter wavelengths are absorbed during the much longer passage of the light through the air, and the sky appears orange or red. Spherical particles larger than those responsible for Rayleigh scattering (more than about 0.1 μm) scatter light of all wavelengths, mainly without changing its direction.

This is Mie scattering, discovered by Gustav Mie in 1908, and it tends to darken the sky colour by counteracting the effect of Rayleigh scattering; it makes the sky a darker blue after rain has washed out solid particles.

Once warmed, the Earth also behaves as a black body, radiating energy in the long, infra-red waveband. All the received energy is reradiated. All the portion which is captured by green plants and subsequently passed to animals that eat the plants is converted back into heat by the process of respiration and escapes from the Earth.

This must be so, because if captured energy were retained permanently the Earth would grow continually hotter, and it does not. Overall, the amount of radiation received from the Sun is equal to the amount radiated into space from the surface of the Earth, but a proportion of the outgoing energy is retained for a time in the atmosphere. This produces the 'greenhouse effect'.

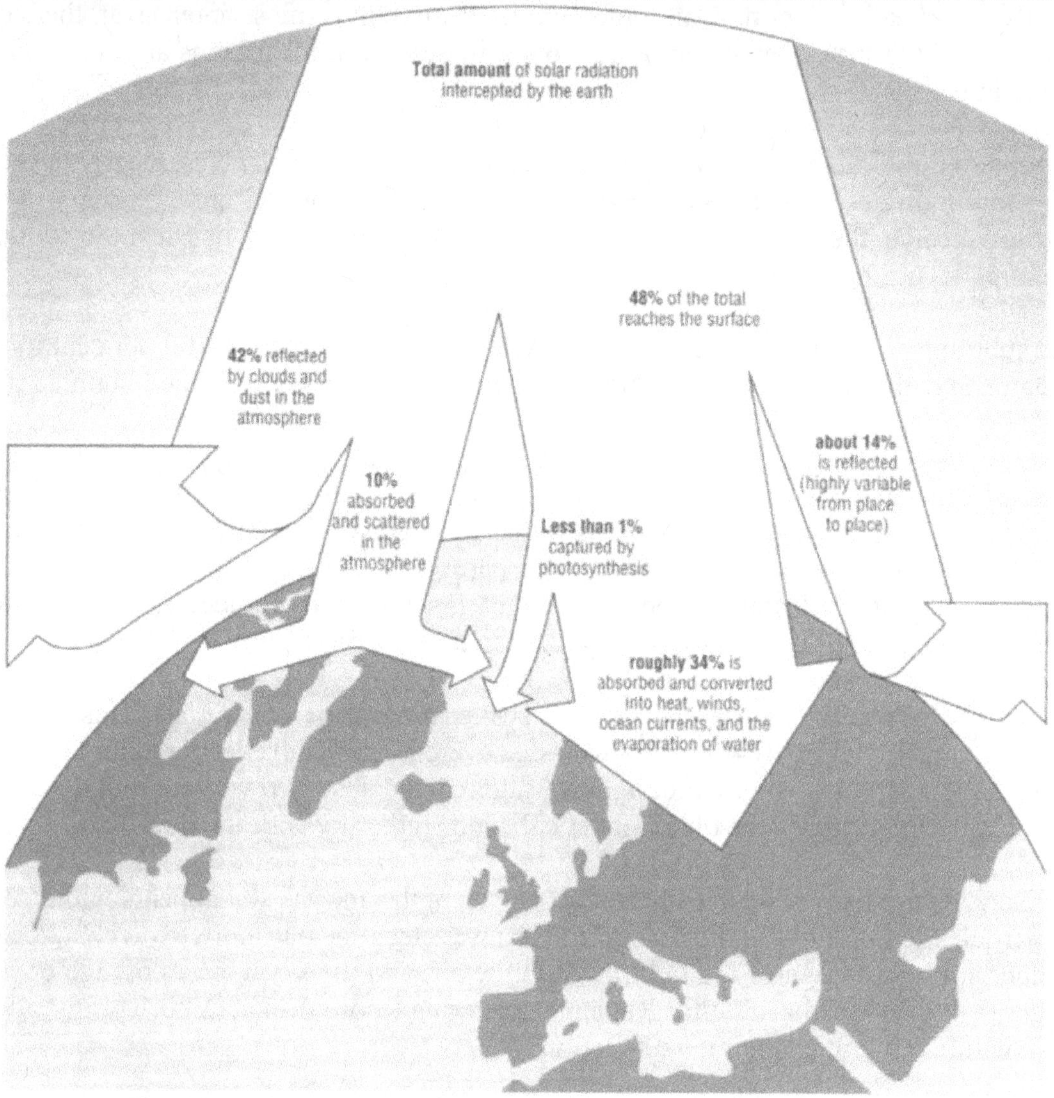

Fig. 129. *Absorption, reflection, and utilization of solar energy*

Albedo and heat capacity

The proportion of light reflected by a surface is called the 'reflection coefficient', or more usually 'albedo' of that surface (Fig. 130). It can be measured and is usually expressed as a fraction or a percentage. Albedo varies widely from one surface to another. About 70 per cent of the Earth's surface is covered by water.

Its albedo varies according to the angle at which sunlight reflects from it. When the Sun is low in the sky much more light is reflected and the water looks brighter than when the Sun is high; in the latter case most of its light penetrates the water and is absorbed, and the water looks dark. Early and late in the day, when the water is calm, the occupants of open boats can develop sunburn quite quickly, even in cool weather.

Reflected radiation does not warm the surface and so albedo has an important climatic effect, and one that can be modified by human intervention, although the relationship is more complex than it may seem. The clearance of tropical rain forest to grow field crops, for example, involves little change, but if that change is from forest to pasture for feeding livestock, the albedo could double.

In this case, the ground would absorb less heat, so there would be less evaporation of water and less cloud would form. This would reduce the average cloud albedo, however, so increasing the amount of radiation reaching the surface and warming it again, evaporating more water and increasing cloudiness once more, but not necessarily to its original value.

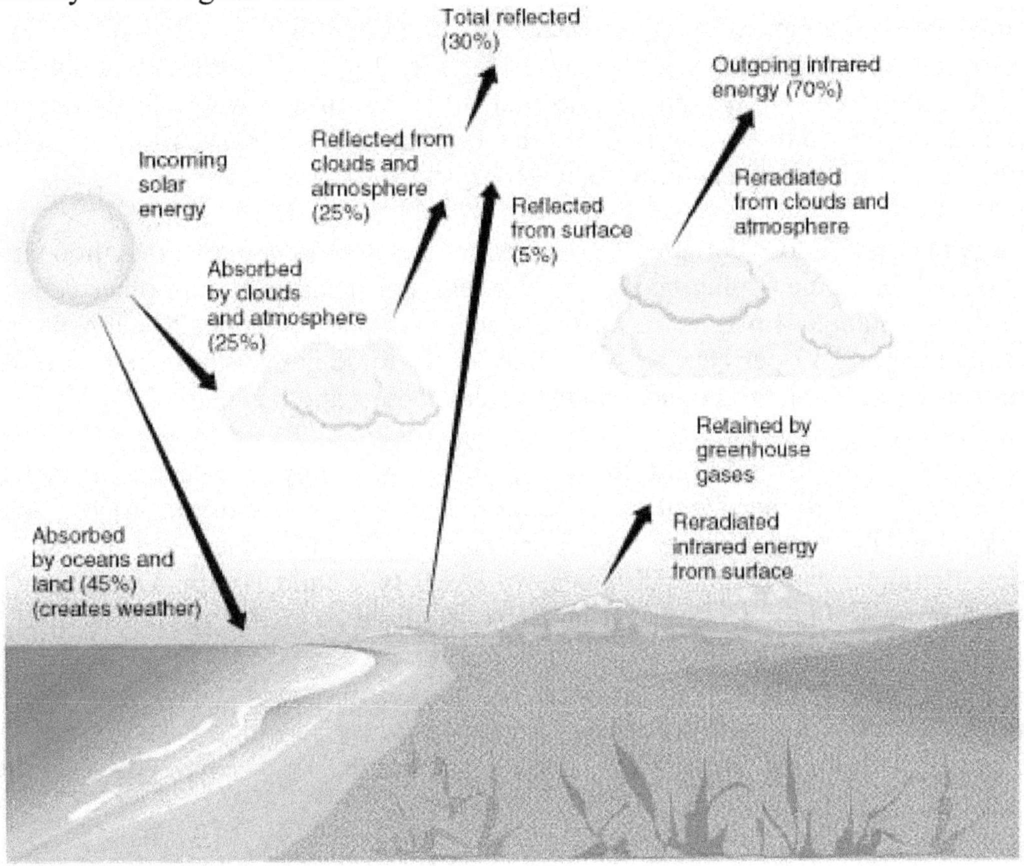

Fig. 130. *Albedo*

Clean air reflects no light, but air is seldom clean. It contains very small particles, called 'aerosols'. In the upper troposphere and at lower levels over the open sea, concentrations range from about 100 to 600 per cm^3, but at low level over continents they are much higher and in industrial regions can reach millions.

They vary in size from 10^{-3} to 10^2 μm and gravity has little effect on them, because they are so small. They tend to reflect short-wave radiation, thus increasing planetary albedo, and in saturated air water vapour will condense on aerosols smaller than about 0.5 μm, encouraging cloud formation and also increasing albedo (and removing the aerosols in precipitation).

Increasing the albedo has a cooling effect, but aerosols absorb radiation at infra-red wavelengths, so those in the lower atmosphere also have a warming influence; for those reaching the stratosphere, on the other hand, the cooling influence of increased albedo is dominant. To complicate matters further, aerosols settling on to clean snow 'dirty' it, reducing its albedo.

2.6 THE GREEN HOUSE EFFECT

Since the solar constant is known it is possible to calculate what the 'black-body' temperature of the Earth should be: it is 250 K (-23°C) . This is what the average temperature at the surface would be were it not for the absorption of long-wave radiation by the atmosphere, which delays the loss of heat and thus warms the planet. The absorption of radiation modifies the climate, or 'forces' it into a (warmer) state than it would be otherwise.

The actual average surface temperature is about 288 K (+15°C), a difference of 38°C. The forcing that achieves this difference is called the greenhouse effect (Fig. 131) and clearly, without it life on Earth would be very uncomfortable, if it were possible at all with surface water, including the oceans to considerable depth, frozen solid and the liquid water beneath the ice markedly more saline because it would contain the salt removed from solution as water crystallized to form ice.

Nitrogen and oxygen are almost completely transparent to electromagnetic radiation at wavelengths greater than 0.29 μm, but some of the incoming solar radiation is absorbed by other constituents of the atmosphere. Of the total, about 4 per cent is absorbed by stratospheric ozone, 20 per cent in the infra-red band by carbon dioxide and 13 per cent by water vapour (in three narrow infra-red wavebands at about 1.5 μm, 2.0 μm, and 2.5–4.5 μm), and 6 per cent by water droplets and dust.

At wavelengths greater than about 4.0 μm, however, several atmospheric gases absorb radiation, each in particular wavebands related to the size of its molecules. The significance of this arises from the fact that the Earth, warmed by the Sun, behaves like a black body at a temperature of 288 K and emits electromagnetic radiation at 4–100 μm, with a peak of intensity around 10 μm. More than 90 per cent of this outgoing long-wave radiation is absorbed in the atmosphere (BARRY AND CHORLEY, 1982, pp. 33–35).

The remainder, about 6 per cent, with wavelengths between those at which it can be absorbed, escapes into space. These 'gaps' in the absorption bands, at about 8.5 μm and 13.0 μm, are called the 'atmospheric window'.

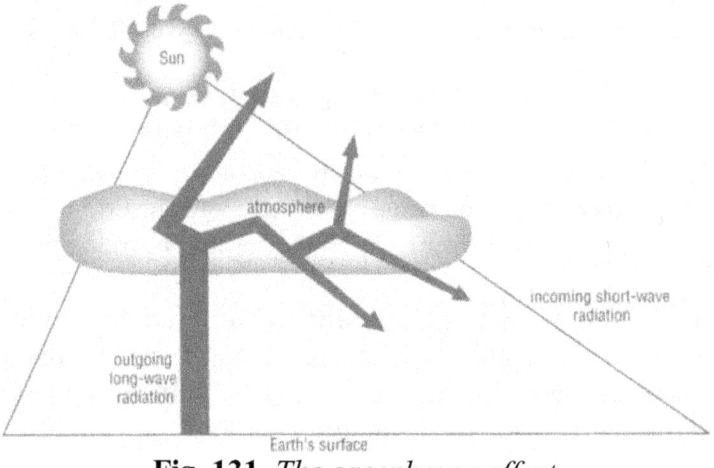

Fig. 131. *The greenhouse effect*

Molecules which absorb radiation reradiate it in all directions. Some returns to the surface, some is absorbed by other atmospheric molecules and some is radiated upwards, out into space. Of course, the surface is warmed by the Sun only during daytime, but its heat is radiated away by night as well as by day.

Eventually, as much energy leaves the Earth as reaches it from the Sun. It must do, because otherwise the atmospheric and surface temperatures would either rise or fall steadily over time; the overall energy budget must balance, and it does, although the transfer of energy is complicated.

The 'greenhouse' metaphor is colourful but a little misleading. It is true that the glass of a greenhouse is transparent to short-wave radiation and partly opaque to infra-red radiation, so its action is similar to that of the absorbing gases in the atmosphere, but the temperature difference inside and outside a greenhouse is due mainly to the fact that air inside is prevented from being cooled by mixing with air outside. With this minor qualification, however, the atmospheric greenhouse effect is real and important, and the gases which cause it are justly known as 'greenhouse gases'.

Carbon dioxide is the best-known greenhouse gas, because it is the most abundant of those over which we can exert some control, but it is not the only one. Methane, produced naturally, for example by termites, but also by farmed livestock and from wet-rice farming (present concentration about 1.7 ppm), nitrous oxide (0.31 ppm) and tropospheric ozone (0.06 ppm), products from the burning of fuels in furnaces and car engines, and the industrially manufactured compounds CFC-11 (0.00026 ppm) and CFC-12 (0.00044 ppm) are also important.

The most important of all, however, is water vapour. This enters into the calculations only indirectly, because its concentration varies greatly from place to place and from day to day and because it is strongly affected by temperature. Its influence, therefore, tends to add to those of the other gases and generally varies as they do. The anticipated changes in concentration for carbon dioxide, methane, and CFC-12, which is one of the family of CFC compounds.

These increases are based on the (uncertain) assumption that industrial and vehicle emissions are the only source of carbon dioxide. All greenhouse-gas effects are usually expressed as 'global warming potentials' (GWPs) which relate them to carbon dioxide. GWPs take account of the wavelengths at which particular molecules absorb, some of which overlap, and the length of time they remain in the atmosphere before decomposing or being deposited at the surface.

On this basis, over a 100-year period, with carbon dioxide given a value of 1, methane has a value of 11 (i.e. it is 11 times more effective than carbon dioxide, molecule for molecule), nitrous oxide 270, CFC-11 3400, and CFC-12 7100.

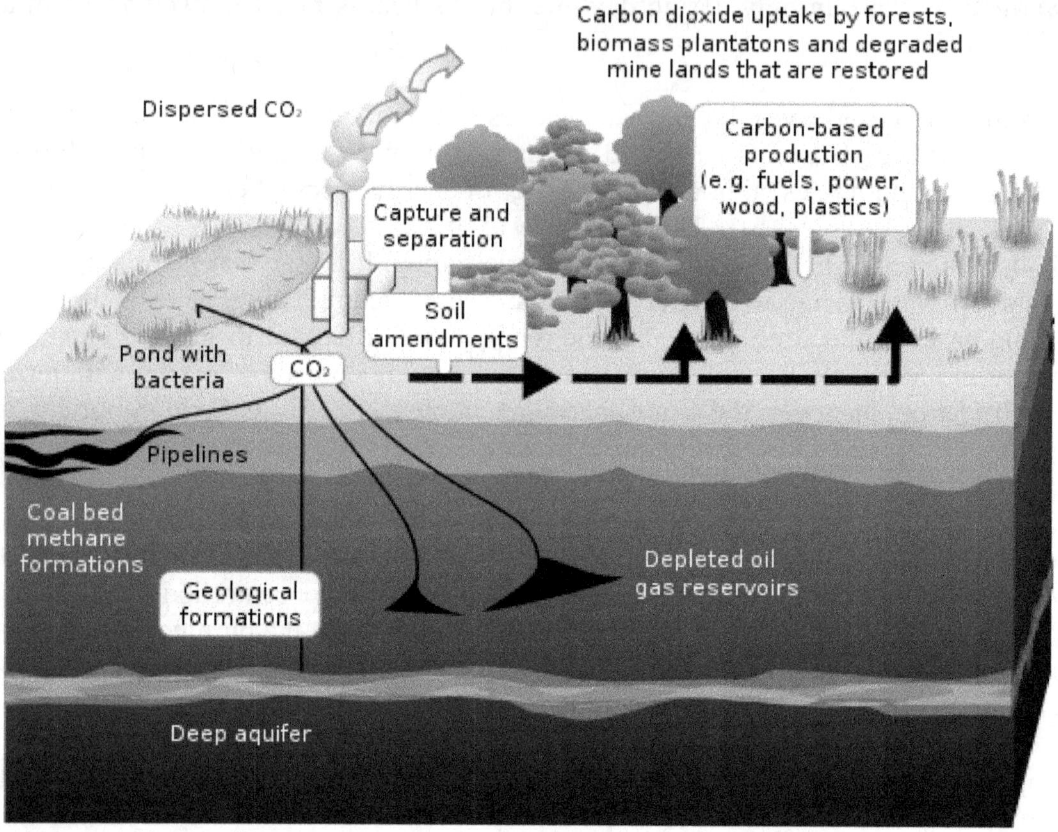

Fig. 132. *Sensitivity of atmosphere to green house effect*

The estimates of future climatic warming are based on the consequences calculated for a doubling of the carbon dioxide concentration, which includes the GWPs for all the relevant gases. Figure 132 shows that, depending on the sensitivity of the atmosphere to greenhouse forcing, a doubling of carbon dioxide would raise the average global temperature by 1.5–4.5°C, with a 'best estimate' of 2.5°C, and at the current rate of increase in greenhouse-gas concentrations these temperatures would be reached by around 2100. During the same period, warming of the oceans is calculated to cause them to expand, producing a rise in sea level of 2–4 cm per decade.

Problems of global warming

Global warming is the increase in the average temperature of Earth's near-surface air and oceans since the mid-20th century and its projected continuation. According to the 2007 Fourth Assessment Report by the Intergovernmental Panel on Climate Change (IPCC), global surface temperature increased 0.74 ± 0.18 °C (1.33 ± 0.32 °F) during the 20th century. Most of the observed temperature increase since the middle of the 20th century has been caused by increasing concentrations of greenhouse gases, which result from human activity such as the burning of fossil fuel and deforestation.

Global dimming, a result of increasing concentrations of atmospheric aerosols that block sunlight from reaching the surface, has partially countered the effects of warming induced by greenhouse gases.

Climate model projections summarized in the latest IPCC report indicate that the global surface temperature is likely to rise a further 1.1 to 6.4 °C (2.0 to 11.5 °F) during the 21st century. The uncertainty in this estimate arises from the use of models with differing sensitivity to greenhouse gas concentrations and the use of differing estimates of future greenhouse gas emissions.

2.7 POLLUTION

What is pollution and why is it important? Why does pollution occur, and is it harmful at all levels? What happens to pollutants in the environment? What are the root causes of pollution? Let us talk about these questions.

Human domination of Earth's ecosystems, ''Between one-third and one-half of the land surface has been transformed by human action; the carbon dioxide concentration in the atmosphere has increased by nearly 30% since the beginning of the Industrial Revolution; more atmospheric nitrogen is now fixed2 by humanity than by all natural terrestrial sources combined; more than half of all accessible surface fresh water is put to use by humanity; and about one-quarter of the bird species on Earth have been driven to extinction . . . All . . . trace to a single cause, the growing scale of the human enterprise.

The rates, scales, kinds, and combinations of changes occurring now are fundamentally different from those at any other time in history; we are changing Earth more rapidly than we are understanding it. In a very real sense, the world is in our hands and how we handle it will determine its composition and dynamics, and our fate."

In the past, we often did not even consider that we were changing our environment, let alone how that could affect us. In the twentieth century, many people willingly ignored gross pollution if its source was a factory on which the community depended for employment. ''That's the smell of money" they might say. This still occurs in some places in the world. If it took so long to recognize that pollution could directly affect human health, think how difficult it is to recognize our total dependence on the environment.

Ecosystems provide many services; a few of these services are outlined in the following. Vegetation and trees absorb the greenhouse gas carbon dioxide, while releasing the oxygen necessary to our lives. The atmosphere's stratospheric-ozone layer protects us from the sun's strongest ultraviolet radiation. Worms and other organisms, and vegetation enhance the fertility of soils that we need for agriculture. Healthy ecosystems provide insects, birds, and other animals that pollinate plants -- including crop plants.

Almost any substance, synthetic or natural, can pollute, but it is synthetic and other industrial chemicals that most concern people. If we learn that industrial chemicals in a water body are obviously impairing the ability of birds to reproduce, or are associated with fish tumors we all agree that the water is polluted.

But what if only tiny amounts of industrial chemicals are present and living creatures apparently unaffected? Is the water polluted? Some would say ''yes," arguing that *chronic* effects could result; that is, adverse effects resulting from long-term exposure to even very low concentrations, or that even largely unnoticed effects could be negative over time.

The word ''waste" differs from pollutant, although waste can pollute. Waste often refers to garbage or trash. Examples include the garbage discarded by households or restaurants, or the construction debris discarded by builders, or material that has reached the end of its useful life.

Category	Examples
Organic chemicals	Polychlorinated biphenyls (PCBs), oil, many pesticides
Inorganic chemicals	Salts, nitrate, metals and their salts
Organometallic chemicals	Methylmercury, tributyltin, tetraethyl lead
Acid	Sulfuric, nitric, hydrochloric, acetic
Physical	Eroded soil, trash
Radioactive	Radon, radium, uranium
Biological	Microorganisms, pollens

Pollution may be less obvious if you live in a wealthy country where the twentieth century brought cleaner air and drinking water, sewage treatment, safe food laws, and food refrigeration. But it took many years and hundreds of billions of dollars to reach those positive results. And wealth does not guarantee a healthy environment.

Read a 1994 description of Hong Kong, one of the Earth's wealthiest spots, ''Beaches that once were crowded by fun seekers are laden with industrial debris or too polluted for swimming. Rivers not foaming from pollutants often are purple from industrial dyes; others are clogged with hazardous metals from scores of electroplating plants. Some territorial waterways have been contaminated by untreated livestock wastes, and close to 25% of its 5.5 million inhabitants suffer from respiratory problems, many due to high levels of sulfur dioxide, nitrogen oxides, and particulate emissions from vehicles . . ." Moreover, over-fishing and pollution left its waters almost devoid of fish.

To make Hong Kong harbor less of an eyesore to tourists, boats collect many tons of trash a day. But its stench could not be disguised – until recently, 70% of the 1.7 million tons (1.5 million tonnes) of human sewage produced each day in Hong Kong was not treated before discharge. Only now are modern sewage-treatment plants being built as Hong Kong begins to confront seriously its many environmental problems.

Why does pollution happen?

Unless you assume that people and industry deliberately pollute, the question arises - why does pollution occur? Pollution happens because no process is 100% efficient. Consider your body - it cannot use 100% of the food you eat. For example, the fiber in food is not broken down in the gastrointestinal (GI) tract, and is excreted with the feces as solid waste. Enzymes in the gut break down other foods to molecules that can cross the GI wall into the bloodstream.

The blood carries these nutrient molecules to your organs. But organs cannot use 100% of the nutrient value, and a portion is excreted in urine as chemical waste. Likewise your body cannot convert all the potential energy in food into useful energy. Part becomes waste energy. No natural or human process, such as manufacturing or fuel burning, is 100% efficient.

Each process produces pollution or waste and waste energy. Carelessness or poor technology aggravates the amount of pollution produced, as do poorly designed processes.

A gallon of gasoline
Gasoline contains hydrocarbons (composed of hydrogen and carbon) along with smaller amounts of contaminants. During combustion, the chemicals in gasoline are converted into the products shown below, and are released through the vehicle's exhaust pipe. Notice the involvement of oxygen (O2) in each reaction. Waste energy is released as heat.

Hydrocarbon combustion
- Carbon reacts with atmospheric $O_2 \rightarrow$ carbon dioxide (a greenhouse gas).
- Hydrogen reacts with atmospheric $O_2 \rightarrow$ water (hydrogen oxide).

Combustion is not 100% efficient
- Hydrocarbons react with atmospheric $O_2 \rightarrow$ carbon dioxide + water.

However, unless excess O_2 is present some hydrocarbons end up as *incomplete products of combustion*. These include polycyclic aromatic hydrocarbons (PAHs), organic vapors, and soot. (Soot is mostly composed of fine black particles of carbon that has not reacted with O2 at all.) Although this does not ordinarily happen, excess O_2 can allow combustion to be almost 100% efficient; i.e., little or no incomplete products of combustion form.

Contaminants in gasoline react with O_2 too:
- Metals react with atmospheric $O_2 \rightarrow$ metal oxides (particulate pollutants).
- Sulfur reacts with atmospheric $O_2 \rightarrow$ sulfur dioxide (a gaseous pollutant).

Gasoline contains very little nitrogen, but at high combustion temperatures
- Atmospheric nitrogen reacts with atmospheric $O_2 \rightarrow$ nitrogen oxides.

Consider two natural laws. One tells us that *matter* is neither created nor destroyed. The matter in gasoline does not disappear. It becomes the pollutants shown above.

- The O2 that reacted with all these substances is conserved too. Another natural law tells us that *energy* is neither created nor destroyed. As gasoline burns to produce energy, only a portion of its energy powers the vehicle's engine – much is "lost" as heat to the environment. But, the energy is not "lost" although it is dissipated.

Natural pollutants
Further we will talk about human-generated pollutants, but natural chemicals pollute too. This happens most dramatically when a volcano erupts, spewing out huge quantities of ash, chlorine, sulfur dioxide, and other chemicals. Other natural chemicals may become pollutants too, but sometimes because man-made conditions allow them to build up to dangerous levels.

The radioactive chemical *radon* is produced from the radioactive uranium naturally found in rocks and soil around the world. Only small concentrations of radon are found in outside air. However, radon can seep up and into our constructed buildings from underlying soil and rocks.

Inside the building, concentrations build up to levels higher than those outside. Radon is associated with human lung cancer. The US Environmental Protection Agency (EPA) ranks radon second only to environmental tobacco smoke as an environmental health risk.

Arsenic is also a natural chemical. Until recently it was not a problem to people in Bangladesh and India. However, millions of wells were drilled to provide clean drinking water to Bangladeshis and Indians, freeing them from having to drink badly contaminated surface water. Unfortunately, arsenic in the rock and soil dissolves into the well water. The result is a massive ongoing poisoning event in which millions suffer from arsenic poisoning.

Governments regulate human exposure to the carcinogen *asbestos*, which is found in old insulation and tiles. However, asbestos is a natural substance and is found in unexpected places. In El Dorado County, California booming population growth has meant building homes in previously unoccupied regions, including those rich in asbestos deposits. Chronic asbestos exposure has been shown in certain regions of Turkey, where asbestos exposure is naturally high, to lead to respiratory diseases and cancer. It has also been a dangerous workplace pollutant.

Pollutant sources

"I am, therefore I pollute." That applies to any process:

- Motor vehicles including cars, buses, airplanes, ships, and off-road vehicles.
- Chemical and petroleum refineries.
- Manufacturing facilities.
- Commercial operations such as dry cleaners, bakeries, and garages.
- Plants that generate electric power by burning coal, oil, or natural gas.
- Agricultural operations growing crops or raising animals.
- Food processing operations.
- Mining operations.
- Construction operations.
- Military operations.
- Forestry operations.
- Construction and road building.
- Consumer product use.
- Municipal operations including drinking-water and wastewater treatment, and road maintenance.
- All activities occurring in commercial and municipal buildings, and in private dwellings.

As the population grows, consumption per individual grows. Technologies are becoming larger too. Thus, the scope of all the activities just mentioned grows too. Without concerted effort to prevent it, pollution and other forms of environmental degradation will also grow.

Pollutants move

Although pollutants seldom stay in one place, we often act as if they do. Countries, laws, and environmental agencies often have individual laws for air, water, and solid waste. But pollutants move through air, water, and soil, and may contaminate food as well. Pollutant effects are typically greater near their source, but pollutants often move and may have effects far from their sources too.

Pollution is greatest near the source

Many pollutants are detected far from the point of emission, and can exert adverse effects at a distance - acid deposition is an example. But the greatest effects typically arise near the emission's source. *Dioxins* emitted from an incinerator can also travel thousands of miles.

But, again the highest fallout occurs near the incinerator. Dioxins settle onto vegetation, crops, and soil. Cattle and other animals eat the contaminated forage or grain and store absorbed dioxins in their fat. Humans eating fatty meat, such as hamburgers, absorb dioxins into their own bodies and fat.

Some effects occur far from the source

Pollutants often move *transboundary*; that is across national boundaries via air currents, rivers, or, sometimes, with migrating animals such as whales. Damage may occur far from the point of emission. This complicates the ability of a government to reduce pollution within its borders.

Water movement

Chemicals spilled into a river in one country flow downstream into other countries: In the year 2000 a Romanian mining operation spilled cyanide and hazardous metals into a Romanian river which flowed into the Tisza River and later the Danube. The Associated Press reported that one Yugoslav mayor stated that 80% of the fish in the Tisza near his town died. Another mayor said ''The Tisza is a dead river. All life in it, from algae to trout, has been destroyed."

In a different accident at a Swiss facility, large quantities of chemicals washed into the Rhine River, which carried them into France and Germany killing fish and other aquatic life along the way.

Air movement

Sulfur dioxide and nitrogen oxides emitted to the atmosphere from sources burning fossil fuels can be blown many hundreds of miles. Converted to acidic substances as they travel, the result is acid deposition settling onto water and land over a whole region. Acid builds up over time in soil and water bodies as emissions continue. Forests and lakes in Sweden are harmed by acid originating in the European countries to its south. Japan's environment is damaged by coal burning in China.

The ''grasshopper effect" is a special case of pollutant movement. The insecticide *dichlorodiphenyltrichloroethane* (DDT) illustrates the grasshopper effect. When DDT is used in a Latin American country, it evaporates and the wind blows it north. When it reaches cooler air, allowing it to condense, it comes to Earth. On a warm day, it evaporates again. The process repeats itself, sometimes many times. Once it reaches the far North, it is too cold for DDT to evaporate again. The Arctic is a *sink* for DDT and other similar *persistent organic pollutants* (POPs).

Moreover, DDT and other POPs enter the Arctic food chain and build up in the fat of marine mammals. Inuit, the Arctic's indigenous people, eat the contaminated mammals and DDT builds up in their body fat to levels among the highest seen in the world.

Canada actively works for international treaties to cut pollutant flow from the south into the Canadian north. Sometimes, airborne pollutant movements are so prominent as to lead to *global change* such as stratospheric-ozone depletion, or global climate change. Even acid deposition, a regional phenomenon, is so widespread as to be global. At least one water pollutant, ''fixed" nitrogen, is also having such widespread regional effects as to be global.

2.8 AIR POLLUTION

Air pollution

The reality of outdoor air pollution is more than the words ''ambient air pollution". It is the eye-stinging pollution surrounding us in a city crowded with motor vehicles, the odor of ozone on a hot hazy day, the choking dust of a heavy dust storm, the smoke coming from wood or coal fires on a winter day, the fumes from an uncontrolled industrial facility, odor from uncontrolled sewage or an open dump. Many living in wealthy countries are spared the worst of these. Not so for the multitudes living in less-developed countries, who are exposed to these and more.

Criteria air pollutants

Remember that the higher the dose of a substance to which a living creature is exposed the greater is the possibility of an adverse effect. Now, apply this principle to criteria air pollutants. They are produced in large amounts, and we -- and other living creatures - are often exposed at levels high enough to exert adverse effects. Air pollution in Dar es Salaam, Mexico City, Istanbul, and many other cities, often causes painful breathing, eye irritation, and headaches. Chronic effects also occur, and trees and plants are also adversely affected. Combustion, especially fossil fuel combustion produces all six criteria pollutants. Each criteria pollutant is described below individually.

Polutant	**Characteristics or examples**
Criteria pollutants	**Six pollutants for which ambient air standards are set to protect human health and welfare**
Carbon monoxide (CO)	Produced by combustion of fossil fuel and biomass. All by itself, CO represents more than 50% of air pollution. Motor vehicles are the major source of CO, especially in cities.
Ozone (O_3)	Amajor component of photochemical smog formed from NOx, VOCs, and oxygen in the presence of sunlight and heat. Motor vehicles are major generators of NO_x and VOCs.
Sulfur dioxide (SO_2)	SO_2 is oxidized in air*b* to sulfuric acid under moist conditions, or to sulfate in dry conditions. Both are particulates and major components of haze. Fossil-fuel-burning power plants produce about two-thirds of the SO_2.
Nitrogen oxides (NOx)	NOx are oxidized in air*b* to acid under moist conditions or to nitrate in dry conditions. Both are particulates and components of haze. In cities, motor vehicles generate most NOx. Coal-burning facilities also produce significant quantities.
Lead (Pb)	Lead is emitted as a particulate during metal mining and processing, and during fossil fuel combustion.
Particulates (PM10 and PM2.5)	Tiny solid particles composed of one or several chemicals, and with many sources. Combustion is a major source of the tiniest particles. Notice that lead is emitted as a particulate and that SO2 and NOx can be converted to particulates; so can some organic VOCs.
Volatile organic pollutants (VOCs)	**Organic chemicals that evaporate easily. Some significantly contribute to smog. Motor vehicles are a major source.**
Hazardous air pollutants (HAPs)	**Each HAP has an emission control, no ambient air standard is set. HAPs are also called toxic air pollutants. About 70% are also VOCs.**
Organic chemicals	Examples are benzene, formaldehyde, and vinyl chloride.
Inorganic chemicals	Examples are asbestos and metals such as cadmium and mercury.

Carbon monoxide

Carbon monoxide (CO) is pervasive. All by itself, CO accounts for more than 50% of air pollution nationwide and worldwide. Worldwide, hundreds of millions of tons are emitted yearly. CO is a colorless, odorless, flammable gas; it is a product of incomplete combustion of carbon-containing material. Only under ideal conditions, with an excess of oxygen and optimal burning conditions, is carbon completely oxidized to carbon dioxide (CO_2).

Why carbon monoxide is of concern

Even levels of CO found in city traffic can aggravate heart problems. CO causes up to 11% of hospital admissions for congestive heart failure in elderly people. Examining how CO works will clarify how it adversely affects the heart, and causes other adverse effects. The blood protein *hemoglobin* contains an iron atom, which normally picks up oxygen in the lungs and transports it to the body's cells. There hemoglobin releases oxygen, exchanging it for the waste gas, carbon dioxide (CO_2).

It carries carbon dioxide back to the lungs, releases it, and once more picks up oxygen. CO is so toxic because it has 250 times greater affinity for the iron atom in hemoglobin than does oxygen - it displaces oxygen; so less oxygen reaches the heart. Even relatively small displacements affect heart function in sensitive people. The brain too demands a steady oxygen concentration for optimum functioning, so CO can cause headache, dizziness, fatigue, and drowsiness.

At higher doses, such as found in enclosed spaces with improperly operating combustion appliances CO may lead to coma and death. The US ambient air standard for CO is 9 ppm averaged over 8 hours.

Sources and sources of exposure

CO is formed anywhere that a carbon-containing material is burned, so CO exposure can happen anywhere that combustion occurs. In urban areas, up to 80 or 90% of CO is emitted by motor vehicles. Drivers stalled in traffic, or driving in highly congested areas, can have high exposure; so can traffic control personnel, mechanics working inside garages and parking garage attendants.

Cigarette smoke contains CO too. Individuals with CO exposure at work, and who also smoke, increase their risk of adverse effects. Facilities burning coal, natural gas, or biomass are CO sources. Biomass combustion (wood, dried manure, other dried vegetation) can lead to significant CO exposure in rural areas, and in impoverished locales where biomass is burned for cooking, heating, and even light. Atmospheric oxidation of methane gas and other hydrocarbons can produce CO too.

Reducing carbon monoxide emissions

In the 1950s, CO levels began to increase along with increasing combustion. The 1970s began a slow but steady downward trend as increasingly stringent emission controls were imposed in the United States and elsewhere. The following measures helped to achieve this reduction: The EPA established national standards for tailpipe emissions. About half of motor vehicle CO emissions in the United States come from only 10% of the vehicles. Inspection programs attempt to find such vehicles, and see to their repair or removal from the road. Owners need to maintain their vehicles to allow them to operate as cleanly as the designers intended. Facilities burning fossil fuels or wood are required to maintain high burning efficiencies to reduce emissions. Many places prohibit open burning of trash and garbage.

Ozone

Ozone (O3) has three oxygen atoms. It is related to the molecular oxygen (O_2) necessary to our lives, which has two. Many of us know the odor of ozone from lightning storms or from improperly maintained equipment such as photocopiers. O_3 is a summer pollutant. Ground-level O_3 is the same O_3 that is found in the stratosphere. However, ground-level (tropospheric) O_3 is a pollutant whereas in the stratosphere, O_3 depletion is the problem. Most ground-level O_3 comes from human activities. However, O_3 is formed naturally in areas remote from human activity, but at low levels, 0.02 to 0.05 ppm.

Smog and ozone

The word, smog, as used today refers to *photochemical smog* because sunlight plays a major role in its formation. O_3 is a major component of photochemical smog, but smog contains other photochemical oxidants too, including peroxyacetyl nitrate (PAN) and nitrogen dioxide. And smog contains particulate matter making it "air you can see."

The first use of the term "smog" came from London in 1905. It describes the *sm*oke and f*og* combination that then commonly obscured visibility. This smog results from the sulfur dioxide, soot, and tarry materials that are produced from uncontrolled burning of high-sulfur coal. This still happens today in places without the technology to capture sulfur dioxide. In the first half of the twentieth century, severe episodes in England and the United States caused thousands of deaths. ***Don't confuse*** ground-level (tropospheric) O_3 with stratospheric O_3.

Groundlevel O_3 is a serious pollutant, but in the stratosphere (the atmospheric layer above the troposphere), O_3 performs a vital function. It absorbs the sun's harmful ultraviolet rays, thereby protecting life on Earth.

Why care about ozone?

Oxygen (O_2) makes up about 20% of the air we breathe. Although essential to life, oxygen is reactive enough to sometimes harm us and other life. O_3 is much more reactive than O_2. The EPA considers O_3 the most serious and persistent air quality problem in the United States. It describes O_3 as the ''most . . . intractable air pollutant in urban air." Moreover, O3 is often present at levels known to have deleterious health and ecological effects. _ Effects on people.

The acute health effects of O_3 are to irritate eyes, nose, throat, and lungs and to decrease the ability of the lungs to function optimally. At 0.2 ppm young adults develop inflammation of the bronchial tubes and tissue deep within the lungs. Even at 0.08 ppm, O_3 adversely affects some people, including healthy individuals. Exercising people are especially susceptible, and so are advised to exercise early in the morning on days when O_3 levels are unhealthy. People with asthma or bronchitis, especially children, are also highly susceptible to O3; it can also increase susceptibility to infection. Chronic O_3 exposure can permanently damage lungs.

- Effects on plants and trees. In the Los Angeles of the 1940s, it was observed that O3 was greatly damaging vegetable crops. O_3 is primarily generated in the heavy traffic of urban areas, but its life span of weeks allows it time to spread over wide regions. The US Department of Agriculture reports that: ''ground-level ozone causes more damage to plants than all other air pollutants combined". Estimated crop losses in the United States are 5% to 10%. O_3 damages sensitive crops at 0.05 ppm whereas more resistant crops withstand 0.07 ppm or higher.

- Worldwide, perhaps 35% of crops grow in areas where O_3 levels exceed 0.05 to 0.07 ppm. If trends continue, by 2025 up to 75% of the world's crops will grow in areas with damaging O_3 levels. Trees are adversely affected too.

Many foresters consider O_3 the air pollutant most damaging to forests. In areas with acid precipitation and other air pollutants too, the combination may be more damaging than the effects of any one pollutant alone.

Sources of ozone
Remember that O_3 is not usually emitted as O_3; that is, it is not a primary pollutant. Rather O_3 is formed from *precursors*. Motor vehicles are a major source of the O_3 precursors, VOCs and NOx. In the summer's heat and the sun's strong ultraviolet rays, VOCs and NOx react with atmospheric oxygen (via several steps) to generate O_3. Knowing how ozone is formed, you may surmise, correctly, that in a city O3 builds up over the progress of a summer day: O_3 levels are typically low early in the morning.

Then vehicle exhausts from the morning traffic increase atmospheric levels of NOx and VOCs. As the day progresses, it becomes warmer, the sun's ultraviolet rays stronger, and O_3 is generated. Of course, off-road motor vehicles also contribute emissions; so do airplanes, construction equipment, lawnmowers and other garden equipment.

Yet other sources of O_3 precursors include facilities that burn fossil fuels and emit VOCs and NOx, especially coal-burning electric power plants and industrial facilities.
Reducing ground-level ozone.

We cannot reduce ground-level O_3 unless we reduce NOx and VOCs emissions; that is, the chemical precursors of ozone. Billions of dollars have been spent to do this, often with unsatisfactory results.
- **Motor vehicles.** *A major reason for our inabilityto lower O_3 is our inability to control motor vehicles*, which, in urban areas emit more than half of the NOx and 40% of the VOCs. Motor vehicles were designed to run much more cleanly after passage of the 1970 US Clean Air Act, and emissions fell 90% or more per gallon of fuel burned. Unfortunately, at the same time the number of motor vehicles continued to rise, as did the number of miles driven per vehicle.

So NOx and VOC emissions remain high. In the early twenty-first century the United States has a new O_3 standard, which primarily affects - not industry - but communities with heavy traffic. Many communities are out of compliance with even the older, higher standard of 0.12 ppm. Poorly maintained vehicles, even if quite new, have much greater NOx and VOC emissions than well-maintained ones.

- **Technical difficulties.** One difficulty in reducing ground-level O_3 is that regulations designed to reduce VOCs or NOx don't necessarily lead to the O_3 reductions anticipated. This happens because the reactions leading to O3 formation are not as straightforward as seems to indicate. Still, efforts have paid off. Within the United States, the number of cities exceeding the old O3 standard fell from 97 in 1990 to 56 in 1992. Nonetheless, in 1995 more than 70 million Americans still lived in areas not meeting the EPA's old 0.12 ppm O_3 standard.

- **Reducing other NOx and VOC sources.** Not only motor-vehicle emissions are being limited. There are controls too in industrialized countries on NOx emissions from power plants and industrial facilities.

There are also limitations on how factories can use solvents (with the purpose of limiting VOC emissions). Another measure is vapor-recovery nozzles on gasoline pumps that reduce VOC emissions during refueling; in addition, gasoline is being reformulated to burn more cleanly and thus to emit lower amounts of VOCs. Regular inspection of motor vehicles is also particularly important to make sure they are running cleanly.

- **Research.** Researchers found that not all VOCs are equally important to O_3 formation. The VOC formaldehyde contributes much more than some other VOCs. This knowledge may lead to efforts to reduce specifically emissions of VOCs that contribute most heavily to O_3 formation.
- **Non-technical approaches.** Technological changes to reduce emissions per vehicle or per factory are not enough. We must address social issues.

A changing ozone standard

This text has emphasized that a major difficulty in setting good health-based standards is that we lack good data. O_3 is an exception. There is abundant information on O3 effects in animals and people, including volunteers exposed to O_3 in enclosed chambers. Thus, when the EPA re-evaluated its previous standard (''safe" dose) of 0.12 ppm, it found evidence that the standard was not protective enough. It was harming humans, crops and trees. However, we can't really determine a safe dose for O_3 because it causes a biological response right down to background levels.

Further complicating the issue is the fact that the difference between adverse effects on children playing outside at the old 0.12 ppm standard, as compared to the new (0.08 ppm), may be small. Thus, the US EPA was told by its Clean Air Scientific Advisory Committee that setting an O_3 standard was more of a policy call than a scientific judgment. In 1997, the EPA did set the new standard at 0.08 ppm. But only in 2002, after legal challenges ended, could the EPA move forward to implement the standard. Meanwhile, in 2003, one-third of Americans continue to live in areas that do not comply with the old 0.12 ppm standard.

Sulfur dioxide

Sulfur dioxide (SO_2) is a colorless gas with a sharp irritating odor. It accounts for about 18% of all air pollution, making it second only to CO as the most common urban air pollutant.

Why care about sulfur dioxide?

- **Direct exposure to the gas, SO_2.** The SO_2 gas reacts with moisture in the eyes, lungs, and other mucous membranes to form strongly irritating acid. This reaction removes about 90% of the SO2 in the upper respiratory tract. Exposure can trigger allergic-type reactions and asthma in sensitive individuals (as do sulfites used in food preservation). Exposure also aggravates pre-existing respiratory or heart disease. And, as with O_3, low SO_2 concentrations can damage plants and trees.

- **Aerosol effects.** SO_2 itself has a lifetime in the atmosphere of only about a day. Thus, if the problems just noted were the only ones, SO_2 would be more manageable. However, SO_2 is converted into sulfuric acid if moisture is present, or, into sulfate particulates in dry conditions. These tiny particulates, only 0.1 to 1 μm in diameter, are *aerosols*. Aerosols are a gaseous suspension of fine solid or liquid particles. Exposure to sulfate aerosols can affect health because the tiny particles can be deeply inhaled into, and inflame, the lungs.

- **Major environmental effects of aerosols.** The aerosols also have major environmental impacts.
- Aerosols form a haze that affects visibility. Haze also affects the level of sunlight reaching the Earth.
- The acid particulates formed are part of the acid deposition (acid rain) problem.
- The particulates don't directly destroy stratospheric O_3, but do provide surfaces on which O_3-destroying reactions can occur.
- Aerosols have a cooling influence on climate. These impacts are shared with nitrate and nitric acid aerosols.

Food production and sulfur dioxide emissions

China, a nation of 1.5 billion people burns coal to generate most of its electricity, and is the world's largest emitter of SO_2. A study sponsored by the US National Aeronautics and Space Administration, and the Georgia Institute of Technology found disturbing results: haze resulting from sulfuric acid and sulfate aerosols is cutting agricultural production.

This happens because haze partially filters out the solar energy reaching the plants. Researchers estimated that the haze reduces sunlight by 5% to 30%. Moreover, decreased sunlight is affecting up to 70% of China's agricultural areas. This is particularly disturbing. With its huge and stillgrowing population, China already faces the prospect of needing to import food. If crop production is much hindered by pollution, the situation could be even worse.

However, there *are* solutions. One is installing technology to lessen SO_2 emissions. Indeed, China is making major efforts to change how it uses energy. It has closed many small inefficient industrial facilities, started to burn cleaner coals (with less sulfur), and it has switched many residents, who previously burned coal, to gas or electricity.

Sources of sulfur dioxide

In the United States and the industrialized northern hemisphere, human activities produce *five times more SO_2 than do natural sources.*

- Worldwide, the figure is about two times as much. In 1985, electric utilities burning fossil fuels produced about two-thirds of the anthropogenic (produced by human actions) SO_2 in the United States. The worst offenders are utilities that burn high-sulfur coal.

- Metal smelters and other fossil-fuel-burning industrial facilities produce another 15% to 20%. Smelters emit SO_2 because many metal ores contain sulfur.
- Petroleum contains sulfur too, but it can be more readily removed than that in coal, and motor vehicles account for a lesser percentage of SO_2 emissions.

Natural sources

Although most SO_2 comes from human activities, there are many natural sources. These include sea water, marine plankton, bacteria, plants, and geothermal emissions. Erupting volcanoes are a major but periodic source of SO_2. In 1991, the Filipino volcano Mt. Pinatubo ejected about 20 million tons (18.1 million tonnes) of SO_2 into the atmosphere. This huge quantity is believed to have been responsible for cooling the Earth's climate for several years thereafter.

Reducing sulfur dioxide emissions

Over the past 30 years many nations have mandated emission controls on coal-burning electric power plants and other industrial sources of SO_2.

Nitrogen oxides

The gases, nitric oxide (NO) and nitrogen dioxide (NO_2) are the major components of nitrogen oxides (NOx, pronounced ''knocks"). A third gas, *nitrous oxide* (N_2O) is also often grouped into NOx. Nitrogen oxides account for about 6% of US air pollution.

Why care about NOx?

- Human health. Direct exposure to NOx gases irritates the lungs, aggravates asthma, and lowers resistance to infection. Nitrogen dioxide is poisonous to plant life. Converted to the aerosols, nitric acid and nitrate, there can be major deleterious effects. - The tiny aerosols can be deeply inhaled into, and cause inflamation of, the lungs.
- Major environmental effects.
- Formation of haze that affects visibility.
- Acid deposition (acid rain)
- Stratospheric O_3 depletion because the aerosol particles provide surfaces on which O_3-destroying reactions can occur.
- Cooling influence on climate. Notice that the above-mentioned effects are shared with sulfate and sulfuric acid aerosols.
- Remember the following: NOx has two distinctive and very important effects not shared with sulfate and sulfuric acid. These are:
- NOx gases are precursors of ground-level O_3 whereas SO_2 is not.
- Deposited to Earth or water, the nitrogen in nitrate and nitric acid is a major plant nutrient. It can benefit plant life, *but* high concentrations have adverse, even devastating, consequences.

NOx sources

Nitrogen compounds are present in fossil fuels only in very small amounts -- they are not a large source of NOx gases.

- The reason that NOx forms is distinctive. Atmospheric nitrogen, N_2 is a very stable chemical. However, burning fuel at a high temperature promotes a reaction between N_2 and atmospheric oxygen, O_2. It is this reaction that results in NOx formation.

- Human sources. NOx gases are emitted almost anywhere that combustion occurs, especially at high temperatures. This is unfortunate because high temperatures otherwise promote *efficient combustion* and lower the emissions of CO, PAHs and other incomplete products of combustion.

- As is true of CO, the major source of NOx is motor vehicles, including off-road vehicles such as construction equipment. Motor vehicles account for more than 50% of NOx emissions overall, and a greater percentage in urban areas.

- Electric utilities in the United States emit another 25% to 30% of NOx, and industrial combustion about 14%. Smaller amounts result from commercial and residential combustion.

- Natural sources. NOx gases are produced by lightning and volcanoes. Microbes decomposing vegetation in soil produce nitrous oxide (N_2O). If nitrogen fertilizer is added to the soil they produce even larger amounts.

Particulate matter

Any gas such as N_2, O_2, or SO_2 blends into air in a homogeneous manner. *Particulate matter does not.* As the name particulate matter (PM) implies, PM is solid, albeit the particles may be very fine aerosols. PM accounts for about 10% of US air pollution. PM is a confusing pollutant.

- Composition varies. Other criteria pollutants are specific chemicals, CO, O_3, SO_2, NOx , and lead. But, PM has no fixed composition. A particle may contain only one chemical such as sulfate, sulfuric acid, or lead. Another particle may contain a number of pollutants, e.g. sulfate, nitrate, metals, dust, biological matter, etc.

- Size varies greatly. PM may be as large as visible cotton dust, historically found in fabric mills. Or, PM can be tiny, submicroscopic particles, aerosols. Indeed, it is these tiny particles - even at low concentrations - that now pose major problems.

- Other pollutants can become PM. However, a number of other pollutants can be converted to particulates. Gases, SO_2 and NOx, are converted to the aerosols, sulfate and sulfuric acid, and nitrate and nitric acid. Such conversion to PM is not unique. Some organic vapors, VOCs, condense into particulates. Many hazardous air pollutants (see below) are metals, and thus are emitted as particulates. (Elemental mercury, largely emitted as a vapor is an exception.)

Why care about PM?

- Health effects. The idea of a particle seems innocuous -- a tiny piece or speck -- not something to inspire alarm. The reality is different.

- Before the advent of workplace protection laws PM took a terrible toll. Workers with chronic exposure to silica dust developed silicosis. Coal miners, over their years of working, developed blacklung disease from coal dust. Textile workers developed brown-lung disease from cotton dust. Workers inhaling airborne asbestos developed asbestosis, lung cancer, or mesothelioma. All these diseases are disabling or deadly. Many workers inhaled large quantities of all sizes of particles, overwhelming their respiratory systems.

- With less-extreme exposure, larger particles of dirt, dust, or pollen catch in the nose, throat, or windpipe, and can be sneezed, coughed, swallowed, or spat out. _ At the beginning of the twenty-first century, it is the very tiniest of particulates that most trouble us. The US EPA regulates particles with diameters of 10 μm or less (PM_{10}), and an even more dangerous category with diameters of 2.5 μm or less ($PM_{2.5}$). The diameter of $PM_{2.5}$ is barely one-fortieth of the width of a human hair.

Sulfate particles, and some soot and dust particles are as small as 0.01 μm. Deeply inhaled, they reach and can inflame the lung's alveoli (tiny air sacs where oxygen is exchanged with carbon dioxide). The very smallest may be absorbed into the bloodstream and exert systemic effects elsewhere in the body.

- Environmental effects. The health impacts of tiny particulates can be bad, but PM also strongly contributes to the haze or smog seen in many cities and often spreading far into rural areas. Sulfate and sulfuric acid aerosols in US parks illustrate this. Haze has reduced visibility in western national parks by up to 50% compared with earlier years. In eastern parks, which are exposed to a greater number of emission sources, haze has reduced visibility by 80% compared with the 1940s.

Just in the 1980s, the Shenandoah and Great Smoky Mountains parks in Virginia and North Carolina saw a 40% increase in sulfate particles. Instead of blue sky, one often sees pale-white haze or gray fog composed of dilute sulfuric acid. This increase in sulfuric acid and sulfate hazes in the 1980s happened as sulfur dioxide emissions nationwide were decreasing.

There is much that we don't yet understand. _ Beyond haze, individual particulates contribute to a variety of other problems depending on what they contain. The sulfate and sulfuric acid particulates, of course, are a major component of acid deposition.

Sources of PM_{10} and $PM_{2.5}$

- PM10. The major source of PM_{10} is dust from farms, mines, or from roads, unpaved and paved. PM10 also includes pollen. Only about 6% of PM_{10} comes from burning fossil fuels.

- $PM_{2.5}$. Conversely, most $PM_{2.5}$ *does* originate from combustion, especially diesel motor vehicles, electric power plants, and industrial operations such as steel mills emitting SO_2. In the United States electric power plants emit about two-thirds of the SO_2, which becomes sulfuric acid and sulfate aerosols that contribute to haze.

- Power plants and other incinerators also produce *fly ash,* whose very fine particles contain many metal oxides and silicon dioxide. Silicon dioxide is found in a benign form in window glass. But silicon dioxide in fly ash or fine blowing sand is not benign. Fly ash particles also hold on to dioxins formed during combustion.

- When combustion is efficient, almost all organic material is converted to carbon dioxide and water, and little particulate matter and soot is formed. When combustion is less efficient, more particles form.

- Other PM sources. Although fossil-fuel combustion sources dominate PM emissions, especially $PM_{2.5}$, there are many other PM sources. Locales with large numbers of wood-burning stoves, which often burn inefficiently, contribute to particle levels. Rural areas generate airborne particles when burning biomass and from windswept dirt, fertilizer, dried manure, or dried crop residues. Particulates in coastal areas contain high levels of chloride (sea salts), which can corrode local buildings and monuments. Construction sites release large amounts of dust.

Reducing PM emissions

In 1997, the US EPA set new standards for particulates at the same time as its new standard for O_3. And, as is true with O_3, only in 2002 has the new PM standard withstood court appeals to allow its implementation. The old PM10 standard was 50 μg/m3 of air; this standard is retained. However, to this was added a new PM standard specific to $PM_{2.5}$ of 15 $\mu g/m^3$. But $PM_{2.5}$ poses quandaries as we contemplate how to reduce it.

- Despite excellent epidemiological information on the dangers of $PM_{2.5}$, major questions need answers before we can act most effectively to reduce $PM_{2.5}$ emissions: Example: $PM_{2.5}$ contains chemicals that are different on the US east coast compared with the west coast. Moreover, although almost everyone believes that $PM_{2.5}$ is responsible for adverse health effects, we don't know *what* component(s) within $PM_{2.5}$ is most responsible: sulfate, nitrate, elemental carbon, carbon compounds, or metal oxides. If we determine what PM chemicals are most dangerous, we could design more specific controls. This could reduce risk more efficiently and more cost-effectively than attempting to control all fine-particulate sources.

- Combustion. By now it is probably clear that controlling combustion sources producing PM is a major need. And remember that, as was the case for CO, O_3, SO_2, and NOx, we need to control combustion *sources*! For Americans, this would mean committing to lessened dependence on fossil fuels; still an unpopular proposal. The EU countries are focusing attention on means to reduce O_3 and PM too. When the European Union adopted its Clean Air for Europe (CAFE) program, these two were targeted as, ''the air pollutants of greatest concern" in 2001.

Lead

In the 1970s, when the EPA designated it a criteria air pollutant, lead was still added to gasoline in the United States, incinerators were less well controlled than today, and lead emissions in general were less well controlled. Today, most lead emissions have been eliminated or are well controlled. Lead emissions from coal-burning power plants are an exception.

However, a separate set of lead-related problems exist. Lead mobilized into the environment many years ago, remains a significant pollutant today. It is in the paint of houses built before the late 1970s, in the solder of old water pipes, and in roadside soil contaminated with lead from car exhaust. Leaded gasoline is still used in a number of less-developed countries; so is leaded tableware. Recycling of lead-acid batteries in impoverished countries remains an occupational exposure even for children.

Volatile organic chemicals

A great many organic chemicals are volatile; that is, they can evaporate. These fall into the category of ''volatile organic chemicals" (VOCs).

Why care about VOCs?

VOCs contribute to the formation of ground-level ozone. And VOCs all by themselves can have adverse effects: many drivers and pedestrians develop headaches and other symptoms if heavily exposed to volatile hydrocarbons in motor vehicle exhausts. Sensitive individuals may react with attacks of asthma or other respiratory problems. But ozone and other air pollutants also contribute to people feeling ill around heavy traffic.

VOC sources

- Combustion sources. Recall that hydrocarbons are the most common VOCs. Motor vehicles are the largest source of VOCs; they emit up to half of all VOCs in the United States. Hydrocarbons also evaporate as gas tanks are filled, and when vehicles are running, idling, or cooling. Combustion engines are pervasive, used not only in cars, trucks, and buses, but also in airplanes, construction, farm and forestry equipment, gasolinepowered lawn and garden equipment.

- Non-combustion sources. Petroleum refineries, chemical plants and electric power plants can be important *local* sources, but compared with the VOC emissions of motor vehicles, these are small sources.

Other non-combustion sources include a great variety of facilities including gasoline stations, vehicle maintenance shops, paint and print shops, dry-cleaners, wood-drying or wood-painting operations, even freshly painted houses.

- Restaurants and bakeries emit pleasant-smelling VOCs, but large bakeries emit large amounts of ethanol formed by the action of yeast.

- Sewage-treatment plants and composting operations emit VOCs too, often with objectionable odors.
- Natural sources. Trees and plants are a large natural source of hydrocarbons, especially in hot weather. Trees may be a significant source, even in large cities. In Maine, a heavily forested state, trees produce more than 90% of the state's VOCs. Among the many volatile compounds emitted by trees are the terpenes, responsible for the smell of pine trees. Pleasant smelling though they are, they can contribute to ozone formation. However . . . trees do *not* emit the NOx that interacts with VOCs to form ozone. Although natural VOCs are not themselves a problem, they must be included in any strategy to reduce ozone formation.

- Homes and offices. Although we don't produce large amounts of VOCs in homes or offices, they can be a large source of human exposure. Paints, solvents, charcoal broiler starters, aerosol sprays, even deodorants and cosmetics all expose us to VOCs.

Reducing VOC emissions

By 1985, VOC emissions had dropped about 30% compared with 1970 in the United States. This was presumably due to regulations enacted in the 1970 Clean Air Act (CAA) and its 1977 amendments. In 1990, CAA amendments mandated further reductions in motorvehicle emissions. Regions of the United States with the worst air pollution had to commit to a 15% reduction in urban smog by 1996. Southern California has been especially aggressive in reducing not just VOC emissions from motor vehicles, but all vehicle pollutants. This state took measures that included switching to less-polluting fuels, and mandating that motor-vehicle manufacturers sell specified numbers of low-polluting or zero-polluting vehicles.
- Other locales have taken actions such as highway lanes that can be used only by vehicles carrying more than one passenger, or tolls paid only by those who drive in peak-traffic hours. Sometimes employers pay employees to take alternative forms of transportation to work.

2.9 HAZARDOUS AIR POLLUTANTS

As is true of the term criteria air pollutants, legislation was the source of the term ''hazardous air pollutants" (HAPs). Commonly called ''toxic air pollutants," HAPs include 188 chemicals specified as hazardous in the 1990 US Clean Air Act Amendments. Many additional chemicals could qualify as HAPs, but these 188 had a high priority. HAPs are not unique to the United States. They are pollutants that might occur almost anywhere in the world.

About 70% of HAPs are also VOCs; that is, volatile organic chemicals, whereas many others are metals. **Do not conclude** that hazardous air pollutants pose greater problems than do criteria air pollutants. Criteria pollutants are hazardous too, and they are produced in much larger quantities typically than are HAPs.

Sources of HAPs and their adverse effects

The specific adverse effects of a particular HAP depend upon its intrinsic toxicity, and the amount to which a human, animal, or plant is exposed. It is not feasible to consider all HAPs individually, so only examples are given here. *Benzene* is a high-production chemical with many uses.

One use is as an anti-knock agent in gasoline. Benzene can irritate the skin and eyes, cause headaches and dizziness. At the high levels once found in some workplaces, it was associated with leukemia and aplastic anemia. Benzene is found anywhere that gasoline is used: around motor vehicles, lawnmowers, and other equipment. Benzene is also present in cigarette smoke. Of all the HAPs, benzene is one of the most common. Exposure to benzene can be nearly ubiquitous.

Formaldehyde is another high-production HAP. It is released from factories that manufacture furniture or pressed wood products. Formaldehyde can irritate the eyes and lungs. At high doses it is an animal carcinogen. Some people develop severe allergies to formaldehyde.

Chloroform was used as an anesthetic for 100 years before its ability to cause liver damage was fully appreciated. It can also be toxic to the kidney and, at high concentrations, is an animal carcinogen. Chloroform is released from sewage-treatment plants, from pulpbleaching facilities, and other facilities that use chlorine-containing chemicals.

Cadmium is highly toxic and bioaccumulates in plants, shellfish, and animal kidneys and liver. It is a metal HAP emitted by metal-refining facilities and by manufacturing facilities that make cadmium-containing products. Incinerators emit cadmium in small amounts; so do facilities that burn fossil fuels. *Mercury* is a neurotoxin.

It is a metal HAP, and the only metal that is liquid. Mercury is especially toxic after bacteria convert it to methylmercury, which biomagnifies in the food web; see also **Table 1.**

Examples of regulated hazardous air pollutants

Pollutant	Representative sources or uses
Organic	
Benzene	Gasoline, cigarette smoke
Toluene	Gasoline, vehicle exhaust, smoking, paints
Ethylene glycol	Automobile antifreeze, brake fluid
Methanol	Windshield antifreeze, solvent
Chloroform	Formed during water chlorination
Methyl bromide	Fumigant
Formaldehyde	Particle-board and plywood, insulation, cosmetics
Parathion	Insecticide
Styrene	Manufacturing plastics, rubbers, adhesives, and cushions
Vinyl chloride	Manufacture of plastics, new automobile interiors
Inorganic	
Asbestos (fibrous mineral)	Once widely used to fireproof materials
Arsenic (metalloid)	Mining and smelting operations, glass making, petroleum

	refining, metal alloys
Cadmium	Electroplating, NiCad batteries, pigment and plastic stabilizer
Chromium	Electroplating (vehicle parts, bathroom fixtures), chemical catalyst
Mercury	Mercury measuring devices such as thermometers, also lamps, dental amalgams
Nickel	Electroplating, in alloys, chemical catalyst

Other HAP concerns
An individual HAP poses the most concern at or near its point of emission. However, wind currents can carry the HAPs far from their sources.

Chloroform emissions are important only near the facilities that use chlorine-containing chemicals, such as municipal wastewater-treatment or pulp-bleaching facilities. Several HAPs – benzene is a major case - are more widespread. Each benzene source may be local, but sources are ubiquitous. Metal HAPs pose special problems because they are persistent.

Metals have built up in soil and sediment in certain locations. And, because every motor vehicle that burns gasoline emits HAPs, there are many millions of small metalemitting sources. On a larger scale, every fossil-fuel-burning power plant, and many industrial facilities, emit metals. The greatest source of human exposure to HAPs such as benzene, formaldehyde, and a number of others, is often in our own homes.

Reducing HAP emissions
Recall that each criteria air pollutant has an ambient air standard set for it, a standard that is based on its risk. However, setting standards is both time and money consuming, and by 1990 standards had not been set for many of the HAPs considered to be of high risk. As an alternative to ambient air standards, the US 1990 Clean Air Act Amendments mandated ''maximum available control technology" (MACT) to reduce HAP emissions from individual facilities. In the future, as more information on actual risk becomes available, emissions of some HAPs may be further restricted. There are other ways to reduce HAP emissions other than by regulation. Before 1990, more than 1000 companies engaged in a voluntary program to reduce emissions of 17 HAPs produced in large quantities or that posed special risks. HAPs are among the chemical emissions that must be reported on the Toxic Release Inventory. Such public declarations have led many facilities to work harder to reduce their emissions levels.

Ions and Radicals in the Atmosphere
One of the characteristics of the upper atmosphere that is difficult to duplicate under laboratory conditions is the presence of significant levels of electrons and positive ions. Because of the rarefied conditions, these ions may exist in the upper atmosphere for long periods before recombining to form neutral species.

At altitudes of approximately 50 km and up, ions are so prevalent that the region is called the **ionosphere**. The presence of the ionosphere has been known since about 1901, when it was discovered that radio waves could be transmitted over long distances where the curvature of the earth makes line-of-sight transmission impossible.

These radio waves bounce off the ionosphere. Ultraviolet light is the primary producer of ions in the ionosphere. In darkness, the positive ions slowly recombine with free electrons. The process is more rapid in the lower regions of the ionosphere, where the concentration of species is relatively high. Thus, the lower limit of the ionosphere lifts at night and makes possible the transmission of radio waves over much greater distances.

The earth's magnetic field has a strong influence upon the ions in the upper atmosphere. Probably the best-known manifestation of this phenomenon is found in the Van Allen belts, discovered in 1958, consisting of two belts of ionized particles encircling the earth. They can be visualized as two doughnuts with the axis of Earth's magnetic field extending through the holes in the doughnuts. The inner belt consists of protons and the outer belt consists of electrons. A schematic diagram of the Van Allen belts is shown in Figure 133.

Although ions are produced in the upper atmosphere primarily by the action of energetic electromagnetic radiation, they can also be produced in the troposphere by the shearing of water droplets during precipitation. The shearing may be caused by the compression of descending masses of cold air or by strong winds over hot, dry land masses. The last phenomenon is known as the foehn, sharav (in the Near East), or Santa Ana (in southern California). These hot, dry winds cause severe discomfort. The ions they produce consist of electrons and positively charged molecular species.

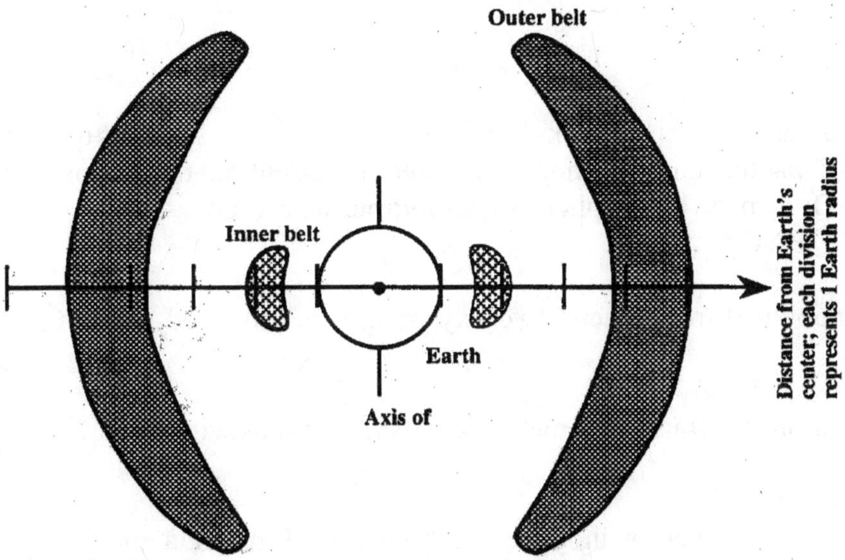

Fig. 133. *Schematic diagram of Van Allen belts*

Free Radicals

In addition to forming ions by photoionization, energetic electromagnetic radiation in the atmosphere may produce atoms or groups of atoms with unpaired electrons called free radicals.

Free radicals are involved with most significant atmospheric chemical phenomena and are of the utmost importance in the atmosphere. Because of their unpaired electrons and the strong pairing tendencies of electrons under most circumstances, free radicals are highly reactive.

The upper atmosphere is so rarefied, however, that at very high altitudes, radicals may have half-lives of several minutes, or even longer. Radicals can take part in chain reactions in which one of the products of each reaction is a radical. Eventually, through processes such as reaction with another radical, one of the radicals in a chain is destroyed and the chain ends, as shown by the reaction of two methyl ($H_3C\bullet$) radicals below:

$H_3C\bullet + H_3C\bullet = C_2H_6$

This process is a chain-terminating reaction. Reactions involving free radicals are responsible for photochemical smog formation.

Free radicals are quite reactive; therefore, they generally have short lifetimes. It is important to distinguish between high reactivity and instability. A totally isolated free radical or atom would be quite stable. Therefore, free radicals and single atoms from diatomic gases tend to persist under the rarefied conditions of very high altitudes because they can travel long distances before colliding with another reactive species. However, electronically excited species have a finite, generally very short lifetime because they can lose energy through radiation without having to react with another species.

Hydroxyl and Hydroperoxyl Radicals in the Atmosphere

The hydroxyl radical, $HO\bullet$, is the single most important reactive intermediate species in atmospheric chemical processes. It is formed by several mechanisms. At higher altitudes, it is produced by photolysis of water:

$H_2O + hn \rightarrow HO\bullet + H$

In the presence of organic matter, hydroxyl radical is produced in abundant quantities as an intermediate in the formation of photochemical smog. To a certain extent in the atmosphere, and for laboratory experimentation, $HO\bullet$ is made by the photolysis of nitrous acid vapor:

$HONO + hn \rightarrow HO\bullet + NO$

In the relatively unpolluted troposphere, hydroxyl radical is produced as the result of the photolysis of ozone:

$O_3 + hn(l < 315\ nm) = O^* + O_2$

followed by the reaction of a fraction of the excited oxygen atoms with water molecules:

$O^* + H_2O = 2HO\bullet$

Hydroxyl radical is a key species in chemical transformations of a number of trace species in the atmosphere. Some aspects of the involvement of hydroxyl radical in such transformations are discussed in the next two chapters. Among the important atmospheric trace species that react with hydroxyl radical are carbon monoxide, sulfur dioxide, hydrogen sulfide, methane, and nitric oxide. Hydroxyl radical is most frequently removed from the troposphere by reaction with methane or carbon monoxide:

$CH_4 + HO\bullet = H_3C\bullet + H_2O$

$CO + HO\bullet = CO_2 + H$

The highly reactive methyl radical, H3C•, reacts with O_2,

$H_3C\bullet + O_2 = H_3COO\bullet$

to form methylperoxyl radical, $H_3COO\bullet$.

$H + O_2 = HOO\bullet$

The hydroperoxyl radical can undergo chain termination reactions, such as

$HOO\bullet + HO\bullet = H_2O + O_2$

$HOO\bullet + HOO\bullet = H_2O_2 + O_2$

or reactions that regenerate hydroxyl radical:

$HOO\bullet + NO = NO_2 + HO\bullet$

$HOO\bullet + O_3 = 2O_2 + HO\bullet$

The global concentration of hydroxyl radical, averaged diurnally and seasonally, is estimated to range from 2 ´ 10 5 to 1 ´ 106 radicals per cm^3 in the troposphere. Because of the higher humidity and higher incident sunlight, which result in elevated O* levels, the concentration of HO• is higher in tropical regions. The Southern Hemisphere probably has about a 20% higher level of HO• than does the Northern Hemisphere because of greater production of anthropogenic, HO•-consuming CO in the Northern Hemisphere.

The hydroperoxyl radical, HOO•, is an intermediate in some important chemical reactions. In addition to its production by the reactions discussed above, in polluted atmospheres, hydroperoxyl radical is made by the following two reactions, starting with photolytic dissociation of formaldehyde to produce a reactive formyl radical.

The hydroperoxyl radical reacts more slowly with other species than does the hydroxyl radical. The kinetics and mechanisms of hydroperoxyl radical reactions are difficult to study because it is hard to retain these radicals free of hydroxyl radicals.

Chemical and Biochemical Processes in Evolution of the Atmosphere

It is now widely believed that the earth's atmosphere originally was very different from its present state and that the changes were brought about by biological activity and accompanying chemical changes. Approximately 3.5 billion years ago, when the first primitive life molecules were formed, the atmosphere was probably free of oxygen and consisted of a variety of gases such as carbon dioxide, water vapor, and perhaps even methane, ammonia, and hydrogen.

The atmosphere was bombarded by intense, bond-breaking ultraviolet light which, along with lightning and radiation from radionuclides, provided the energy to bring about chemical reactions that resulted in the production of relatively complicated molecules, including even amino acids and sugars, which are produced and used by living organisms.

From the rich chemical mixture in the sea, life molecules evolved. Initially, these very primitive life forms derived their energy from fermentation of organic matter formed by chemical and photochemical processes, but eventually they gained the capability to produce organic matter, "$\{CH_2O\}$," by photosynthesis:

$CO_2 + H_2O + hn = \{CH_2O\} + O_2(g)$

Photosynthesis released oxygen, thereby setting the stage for the massive biochemical transformation that resulted in the production of almost all the atmosphere's O2.

The elemental oxygen initially produced by photosynthesis was probably quite toxic to primitive life forms. However, much of this oxygen was converted to iron oxides by reaction with soluble iron(II):

$4Fe^{2+} + O_2 + 4H_2O = 2Fe_2O_3 + 8H^+$

The existence of enormous deposits of iron oxides provides major evidence for the liberation of free O_2 in the primitive atmosphere.

Eventually, enzyme systems developed that enabled organisms to mediate the reaction of waste-product oxygen with oxidizable organic matter in the sea. Later, this mode of waste-product disposal was utilized by organisms to produce energy by respiration, which is now the mechanism by which nonphotosynthetic organisms obtain energy.

In time, O_2 accumulated in the atmosphere, providing an abundant source of oxygen for respiration. It had an additional benefit in that it enabled the formation of an ozone shield. The ozone shield absorbs bond-rupturing ultraviolet light. With the ozone shield protecting tissue from destruction by highenergy ultraviolet radiation, the earth became a much more hospitable environment for life, and life forms were enabled to move from the sea to land.

Acid-base reactions in the atmosphere

Acid-base reactions occur between acidic and basic species in the atmosphere. The atmosphere is normally at least slightly acidic because of the presence of a low level of carbon dioxide, which dissolves in atmospheric water droplets and dissociates slightly:

$CO_2(g) + \text{ water} = CO_2(aq)$

$CO_2(aq) + H_2O = H^+ + HCO_3^-$

Atmospheric sulfur dioxide forms a somewhat stronger acid when it dissolves in water:

$SO_2(g) + H_2O = H^+ + HSO_3^-$

In terms of pollution, however, strongly acidic HNO_3 and H_2SO_4 formed by the atmospheric oxidation of N oxides, SO_2, and H_2S are much more important because they lead to the formation of damaging acid rain.

As reflected by the generally acidic pH of rainwater, basic species are relatively less common in the atmosphere. Particulate calcium oxide, hydroxide, and carbonate can get into the atmosphere from ash and ground rock, and can react with acids such as in the following reaction:

$Ca(OH)_2(s) + H_2SO_4(aq) = CaSO_4(s) + 2H_2O$

The most important basic species in the atmosphere is gas-phase ammonia, NH_3.
The major source of atmospheric ammonia is from biodegradation of nitrogencontaining biological matter and from bacterial reduction of nitrate:

$NO_3^- (aq) + 2\{CH_2O\}(\text{biomass}) + H^+ \rightarrow NH_3(g) + 2CO_2 + H_2O$

Ammonia is particularly important as a base in the air because it is the only watersoluble base present at significant levels in the atmosphere. When it is dissolved in atmospheric water droplets, ammonia plays a strong role in neutralizing atmospheric acids, as shown by the following reactions:

$NH_3(aq) + HNO_3(aq) = NH_4NO_3(aq)$

$NH_3(aq) + H_2SO_4(aq) = NH_4HSO_4(aq)$

These reactions have three effects: (1) They result in the presence of NH_4^+ ion in the atmosphere as dissolved or solid salts, (2) they serve in part to neutralize acidic consituents of the atmosphere, and (3) they produce relatively corrosive ammonium salts.

2.10 WATER QUALITY AND QUANTITY

Throughout history, the quality and quantity of water available to humans have been vital factors in determining their well-being. Whole civilizations have disappeared because of water shortages resulting from changes in climate. Even in temperate climates, fluctuations in precipitation cause problems. Devastating droughts in Africa during the 1980s resulted in catastrophic crop failures and starvation. Clean water - and enough of it - is essential to any and all life, animals, plants, and microbes.

Fish are vulnerable to polluted water. Indeed, there are places in the world where the water is so polluted that fish have disappeared. In many other places fish or shellfish survive, but are not safe to eat because their flesh is contaminated. Humans enjoy being around water, but contamination with infectious organisms makes swimming unsafe; or if water has obnoxious odors or scum, being near it is not pleasant. Clean water is vital.

In developed countries, many water bodies are cleaner than 30 years ago. When Congress passed the Clean Water Act in 1972, only 30% of US waters were judged fishable and swimmable. By 1994, it was greater than 60%. ''Fishable" means that fish from the water are safe to eat; ''swimmable," that it can be used for swimming without fear of infectious organisms or other unhealthy contaminants. Critics point out that the 60% figure may be deceiving. Although citizens may see improvements in specific water bodies, they cannot know the overall quality of the country's water bodies because large information gaps exist. In the United States, only 19% of rivers and streams are monitored and only 6% of ocean and shoreline waters.

Methods used to monitor water also vary widely, and many consider the statistics unreliable.

Water quality and quantity

Waterborne diseases such as cholera and typhoid killed millions of people in the past and still cause great misery in less developed countries. Ambitious programs of dam and dike construction have reduced flood damage, but they have had a number of undesirable side effects in some areas, such as inundation of farmland by reservoirs and failure of unsafe dams. Globally, problems with quantity and quality of water supply remain and in some respects are becoming more serious. These problems include increased water use due to population growth, contamination of drinking water by improperly discarded hazardous wastes, and destruction of wildlife by water pollution.

Aquatic chemistry must consider water in rivers, lakes, estuaries, oceans, and underground, as well as the phenomena that determine the distribution and circulation of chemical species in natural waters. Its study requires some understanding of the sources, transport, characteristics, and composition of water. The chemical reactions that occur in water and the chemical species found in it are strongly influenced by the environment in which the water is found.

The chemistry of water exposed to the atmosphere is quite different from that of water at the bottom of a lake. Microorganisms play an essential role in determining the chemical composition of water. Thus, in discussing water chemistry, it is necessary to consider the many general factors that influence this chemistry.

The study of water is known as **hydrology** and is divided into a number of subcategories. **Limnology** is the branch of the science dealing with the characteristics of fresh water including biological properties, as well as chemical and physical properties. **Oceanography** is the science of the ocean and its physical and chemical characteristics. The chemistry and biology of the Earth's vast oceans are unique because of the ocean's high salt content, great depth, and other factors.

Sources and Uses of Water: The Hydrologic Cycle

The world's water supply is found in the five parts of the **hydrologic cycle** (Fig. 134).

About 97% of Earth's water is found in the oceans. Another fraction is present as water vapor in the atmosphere (clouds). Some water is contained in the solid state as ice and snow in snowpacks, glaciers, and the polar ice caps. Surface water is found in lakes, streams, and reservoirs. Groundwater is located in aquifers underground.

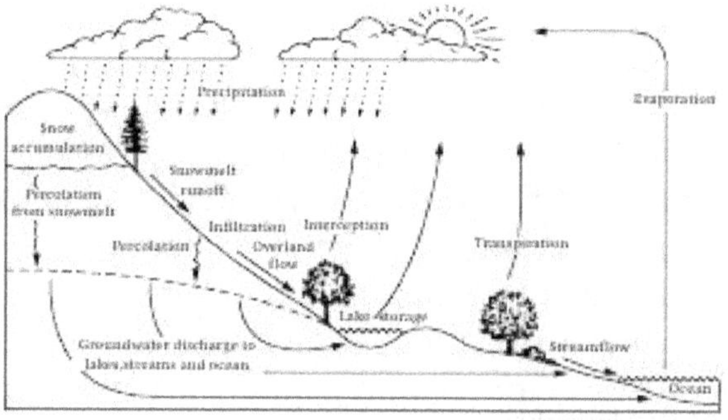

Source: ANZECC & AWRC

Fig. 134. *Hydrologic cycle*

There is a strong connection between the *hydrosphere,* where water is found, and the *lithosphere*, which is that part of the geosphere accessible to water. Human activities affect both. For example, disturbance of land by conversion of grasslands or forests to agricultural land or intensification of agricultural production may reduce vegetation cover, decreasing **transpiration** (loss of water vapor by plants) and affecting the microclimate. The result is increased rain runoff, erosion, and accumulation of silt in bodies of water. The nutrient cycles may be accelerated, leading to nutrient enrichment of surface waters. This, in turn, can profoundly affect the chemical and biological characteristics of bodies of water.

The water that humans use is primarily fresh surface water and groundwater, the sources of which may differ from each other significantly. In arid regions, a small fraction of the water supply comes from the ocean, a source that is likely to become more important as the world's supply of fresh water dwindles relative to demand. Saline or brackish groundwaters may also be utilized in some areas.

A major problem with water supply is its nonuniform distribution with location and time. This causes difficulties because people in areas with low precipitation often consume more water than people in regions with more rainfall. These problems are minor, however, in comparison to those in some parts of Africa where water shortages are contributing to real famine conditions.

The properties of water, a unique substance

Water has a number of unique properties that are essential to life. Some of the special characteristics of water include its polar character, tendency to form hydrogen bonds, and ability to hydrate metal ions.

Important Properties of Water

Property	Effects and Significance
Excellent solvent	Transport of nutrients and waste products, making biological processes possible in an aqueous medium
Highest dielectric constant of any common liquid	High solubility of ionic substances and their ionization in solution
Higher surface tension than any other liquid	Controlling factor in physiology; governs drop and surface phenomena
Transparent to visible and longer-wavelength fraction of ultraviolet light	Colorless, allowing light required for photosynthesis to reach considerable depths in bodies of water
Maximum density as a liquid at 4°C	Ice floats; vertical circulation restricted in stratified bodies of water
Higher heat of evaporation than any other material	Determines transfer of heat and water molecules between the atmosphere and bodies of water
Higher latent heat of fusion than any other liquid except ammonia	Temperature stabilized at the freezing point of water
Higher heat capacity than any other liquid except ammonia	Stabilization of temperatures of organisms and geographical regions

The characteristics of bodies of water

The physical condition of a body of water strongly influences the chemical and biological processes that occur in water. **Surface water** occurs primarily in streams, lakes, and reservoirs. **Wetlands** are flooded areas in which the water is shallow enough to enable growth of bottom-rooted plants. **Estuaries** are arms of the ocean into which streams flow. The mixing of fresh and salt water gives estuaries unique chemical and biological properties. Estuaries are the breeding grounds of much marine life, which makes their preservation very important.

Water's unique temperature-density relationship results in the formation of distinct layers within nonflowing bodies of water, as shown in Figure 21. During the summer a surface layer (**epilimnion**) is heated by solar radiation and, because of its lower density, floats upon the bottom layer, or **hypolimnion**. This phenomenon is called **thermal stratification (Fig. 135)**. When an appreciable temperature difference exists between the two layers, they do not mix but behave independently and have very different chemical and biological properties.

The epilimnion, which is exposed to light, may have a heavy growth of algae. As a result of exposure to the atmosphere and (during daylight hours) because of the photosynthetic activity of algae, the epilimnion contains relatively higher levels of dissolved oxygen and generally is aerobic. In the hypolimnion, bacterial action on biodegradable organic material may cause the water to become anaerobic (lacking dissolved oxygen). As a consequence, chemical species in a relatively reduced form tend to predominate in the hypolimnion.

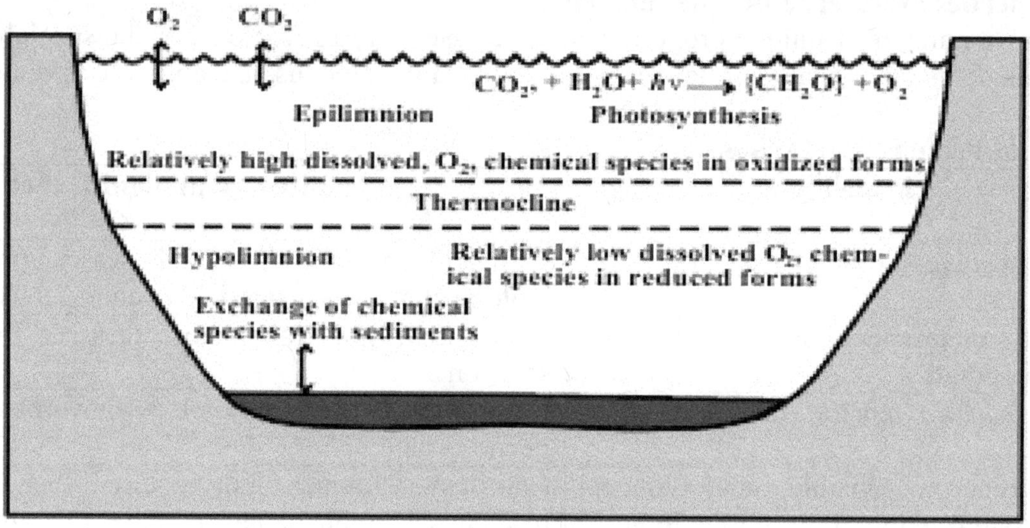

Fig. 135. *Stratification of a lake*

The shear-plane, or layer between epilimnion and hypolimnion, is called the **thermocline**. During the autumn, when the epilimnion cools, a point is reached at which the temperatures of the epilimnion and hypolimnion are equal.

This disappearance of thermal stratification causes the entire body of water to behave as a hydrological unit, and the resultant mixing is known as **overturn**. An overturn also generally occurs in the spring. During the overturn, the chemical and physical characteristics of the body of water become much more uniform, and a number of chemical, physical, and biological changes may result. Biological activity may increase from the mixing of nutrients. Changes in water composition during overturn may cause disruption in water-treatment processes.

Aquatic life

The living organisms (**biota**) in an aquatic ecosystem may be classified as either autotrophic or heterotrophic. **Autotrophic** organisms utilize solar or chemical energy to fix elements from simple, nonliving inorganic material into complex life molecules that compose living organisms. Algae are the most important autotrophic aquatic organisms because they are **producers** that utilize solar energy to generate biomass from CO_2 and other simple inorganic species.

Heterotrophic organisms utilize the organic substances produced by autotrophic organisms as energy sources and as the raw materials for the synthesis of their own biomass. **Decomposers** (or **reducers**) are a subclass of the heterotrophic organisms and consist of chiefly bacteria and fungi, which ultimately break down material of biological origin to the simple compounds originally fixed by the autotrophic organisms.

The ability of a body of water to produce living material is known as its **productivity.** Productivity results from a combination of physical and chemical factors. High productivity requires an adequate supply of carbon (CO_2), nitrogen (nitrate), phosphorus (orthophosphate), and trace elements such as iron. Water of low productivity generally is desirable for water supply or for swimming. Relatively high productivity is required for the support of fish and to serve as the basis of the food chain in an aquatic ecosystem. Excessive productivity results in decay of the biomass produced, consumption of dissolved oxygen, and odor production, a condition called **eutrophication**.

Life forms higher than algae and bacteria—fish, for example—comprise a comparatively small fraction of the biomass in most aquatic systems. The influence of these higher life forms upon aquatic chemistry is minimal. However, aquatic life is strongly influenced by the physical and chemical properties of the body of water in which it lives. *Temperature, transparency,* and *turbulence* are the three main physical properties affecting aquatic life.

Very low water temperatures result in very slow biological processes, whereas very high temperatures are fatal to most organisms. The transparency of water is particularly important in determining the growth of algae. Turbulence is an important factor in mixing processes and transport of nutrients and waste products in water. Some small organisms (**plankton**) depend upon water currents for their own mobility.

Dissolved oxygen (**DO**) frequently is the key substance in determining the extent and kinds of life in a body of water. Oxygen deficiency is fatal to many aquatic animals such as fish. The presence of oxygen can be equally fatal to many kinds of anaerobic bacteria. **Biochemical oxygen demand, BOD,** refers to the amount of oxygen utilized when the organic matter in a given volume of water is degraded biologically.

Carbon dioxide is produced by respiratory processes in water and sediments and can also enter water from the atmosphere. Carbon dioxide is required for the photosynthetic production of biomass by algae and in some cases is a limiting factor. High levels of carbon dioxide produced by the degradation of organic matter in water can cause excessive algal growth and productivity.

The salinity of water also determines the kinds of life forms present. Irrigation waters may pick up harmful levels of salt. Marine life obviously requires or tolerates salt water, whereas many freshwater organisms are intolerant of salt.

Oxygen in Water

Without an appreciable level of dissolved oxygen, many kinds of aquatic organisms cannot exist in water. Dissolved oxygen is consumed by the degradation of organic matter in water. Many fish kills are caused not from the direct toxicity of pollutants but from a deficiency of oxygen because of its consumption in the biodegradation of pollutants.

Most elemental oxygen comes from the atmosphere, which is 20.95% oxygen by volume of dry air. Therefore, the ability of a body of water to reoxygenate itself by contact with the atmosphere is an important characteristic. Oxygen is produced by the photosynthetic action of algae, but this process is really not an efficient means of oxygenating water because some of the oxygen formed by photosynthesis during the daylight hours is lost at night when the algae consume oxygen as part of their metabolic processes. When the algae die, the degradation of their biomass also consumes oxygen.

The solubility of oxygen in water depends upon water temperature, the partial pressure of oxygen in the atmosphere, and the salt content of the water. It is important to distinguish between oxygen *solubility*, which is the maximum dissolved oxygen concentration at equilibrium, and dissolved oxygen *concentration*, which is generally not the equilibrium concentration and is limited by the rate at which oxygen dissolves.

The concentration of oxygen in water at 25°C in equilibrium with air at atmospheric pressure is only 8.32 mg/L. Thus, water in equilibrium with air cannot contain a high level of dissolved oxygen compared to many other solute species. If oxygen-consuming processes are occurring in the water, the dissolved oxygen level may rapidly approach zero unless some efficient mechanism for the reaeration of water is operative, such as turbulent flow in a shallow stream or air pumped into the aeration tank of an activated sludge secondary waste treatment facility.

The problem becomes largely one of kinetics, in which there is a limit to the rate at which oxygen is transferred across the air-water interface. This rate depends upon turbulence, air bubble size, temperature, and other factors.

If organic matter of biological origin is represented by the formula $\{CH_2O\}$, the consumption of oxygen in water by the degradation of organic matter may be expressed by the following biochemical reaction:

$$\{CH_2O\} + O_2 \rightarrow CO_2 + H_2O$$

The weight of organic material required to consume the 8.3 mg of O_2 in a liter of water in equilibrium with the atmosphere at 25°C is given by a simple stoichiometric calculation which yields a value of 7.8 mg of $\{CH_2O\}$. Thus, the microorganism-mediated degradation of only 7 or 8 mg of organic material can completely consume the oxygen in one liter of water initially saturated with air at 25°C.

The depletion of oxygen to levels below those that will sustain aerobic organisms requires the degradation of even less organic matter at higher temperatures (where the solubility of oxygen is less) or in water not initially saturated with atmospheric oxygen. Furthermore, there are no common aquatic chemical reactions that replenish dissolved oxygen; except for oxygen provided by photosynthesis, it must come from the atmosphere.

The temperature effect on the solubility of gases in water is especially important in the case of oxygen. The solubility of oxygen in water decreases from 14.74 mg/L at 0°C to 7.03 mg/L at 35°C. At higher temperatures, the decreased solubility of oxygen, combined with the increased respiration rate of aquatic organisms, frequently causes a condition in which a higher demand for oxygen accompanied by lower solubility of the gas in water results in severe oxygen depletion.

Water acidity and carbon dioxide in water

Acid-base phenomena in water involve loss and acceptance of H^+ ion. Many species act as **acids** in water by releasing H^+ ion, others act as **bases** by accepting H+, and the water molecule itself does both. An important species in the acid-base chemistry of water is bicarbonate ion, HCO_3^-, which may act as either an acid or a base:

$$HCO_3^- \rightarrow CO_3^{2-} + H^+$$

$$HCO_3^- + H^+ \rightarrow CO_2(aq) + H_2O$$

Acidity as applied to natural water and wastewater is the capacity of the water to neutralize OH^-; it is analogous to alkalinity, the capacity to neutralize H^+. Although virtually all water has some alkalinity acidic water is not frequently encountered, except in cases of severe pollution. Acidity generally results from the presence of weak acids, particularly CO_2, but sometimes includes others such as $H_2PO_4^-$, H_2S, proteins, and fatty acids. Acidic metal ions, particularly Fe3+, may also contribute to acidity.

From the pollution standpoint, strong acids are the most important contributors to acidity. The term *free mineral acid* is applied to strong acids such as H_2SO_4 and HCl in water. Acid mine water is a common water pollutant that contains an appreciable concentration of free mineral acid. Whereas total acidity is determined by titration with base to the phenolphthalein endpoint (pH 8.2), free mineral acid is determined by titration with base to the methyl orange endpoint (pH 4.3). The acidic character of some hydrated metal ions may contribute to acidity:

$$Al(H_2O)_6^{3+} \rightarrow Al(H_2O)_5OH^{2+} + H^+ \quad (4)$$

Some industrial wastes, such as spent steel pickling liquor, contain acidic metal ions and often some excess strong acid. The acidity of such wastes must be measured in calculating the amount of lime or other chemicals required to neutralize the acid.

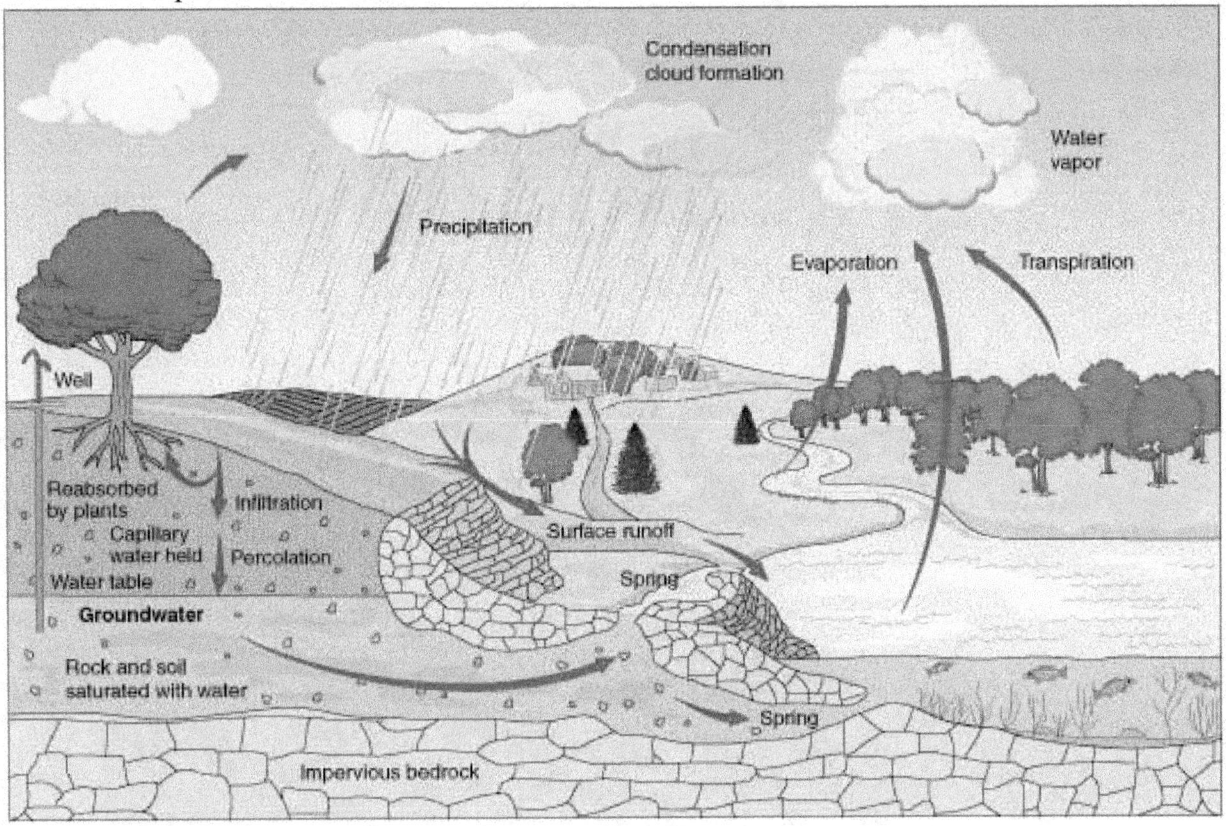

Fig. 136. *Water cycle*

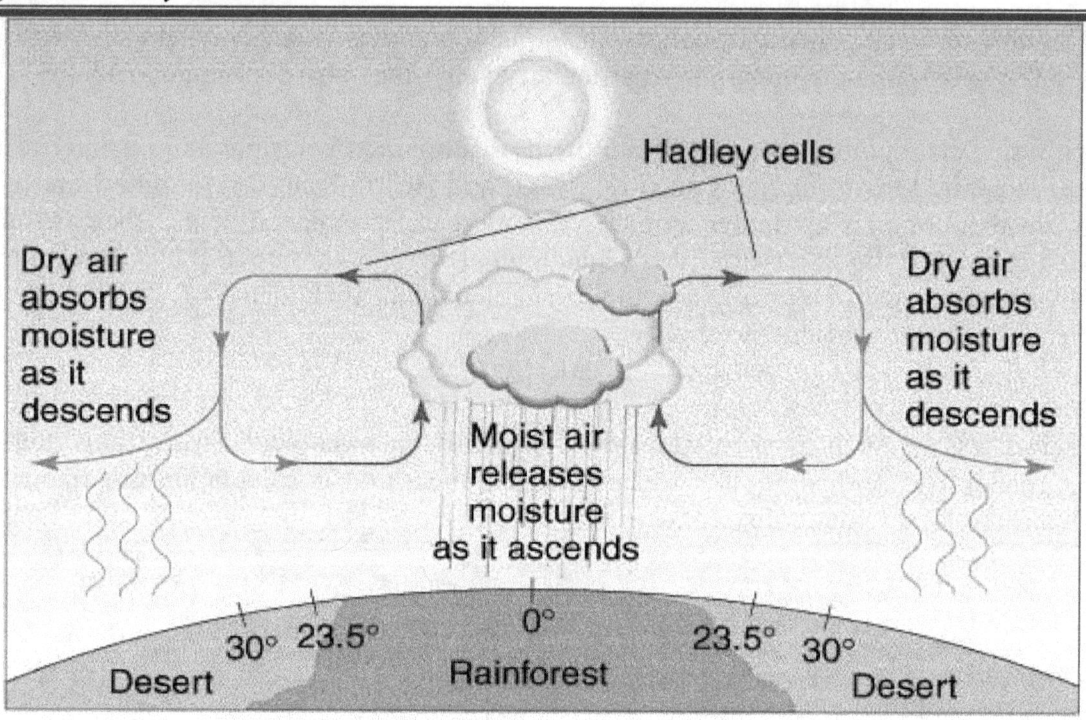

(a) Hadley cells at the equator

Fig. 137. *Haedley cells at the equator*

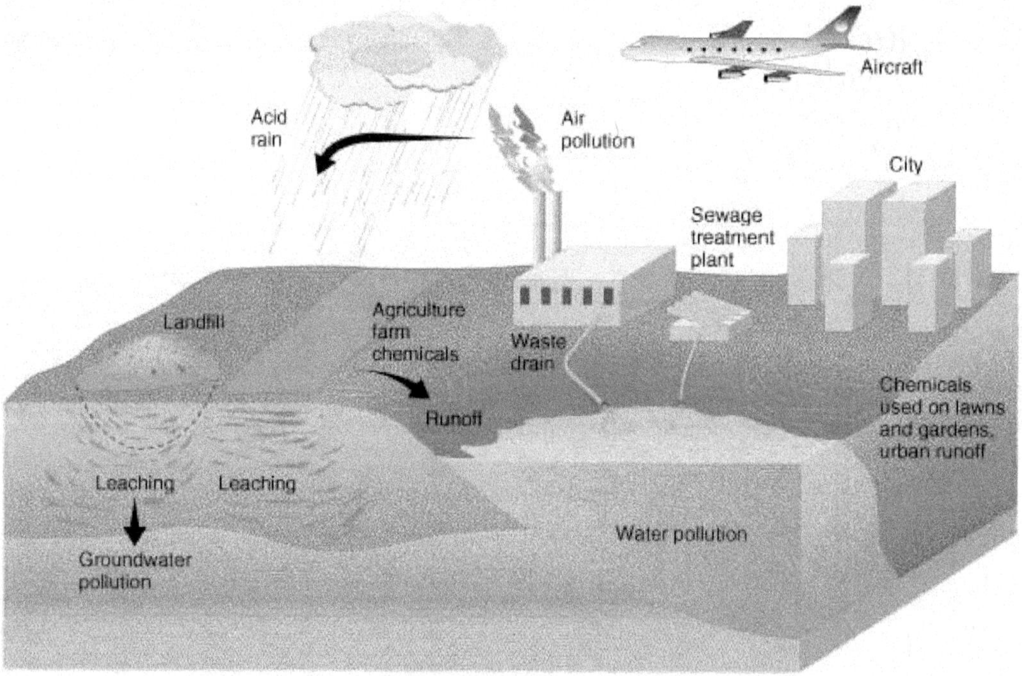

Fig. 138. *Water pollution.*

Water pollution is the contamination of water bodies (e.g. lakes, rivers, oceans and groundwater).

Water pollution affects plants and organisms living in these bodies of water; and, in almost all cases the effect is damaging not only to individual species and populations, but also to the natural biological communities.

Water pollution occurs when pollutants are discharged directly or indirectly into water bodies without adequate treatment to remove harmful compounds.

Water pollution is a major problem in the global context. It has been suggested that it is the leading worldwide cause of deaths and diseases, and that it accounts for the deaths of more than 14,000 people daily. An estimated 700 million Indians have no access to a proper toilet, and 1,000 Indian children die of diarrheal sickness every day. Some 90% of China's cities suffer from some degree of water pollution, and nearly 500 million people lack access to safe drinking water.

In addition to the acute problems of water pollution in developing countries, industrialized countries continue to struggle with pollution problems as well. In the most recent national report on water quality in the United States, 45 percent of assessed stream miles, 47 percent of assessed lake acres, and 32 percent of assessed bay and estuarine square miles were classified as polluted.

Water is typically referred to as polluted when it is impaired by anthropogenic contaminants and either does not support a human use, like serving as drinking water, and/or undergoes a marked shift in its ability to support its constituent biotic communities, such as fish. Natural phenomena such as volcanoes, algae blooms, storms, and earthquakes also cause major changes in water quality and the ecological status of water.

Pollutant movement in the environment

Conventional pollutants

These are the ''conventional,'' ''non-conventional,'' and ''toxic'' pollutants. Just as the term ''criteria air pollutants'' did not reveal the seriousness of those six pollutants, the term ''conventional water pollutants'' does not tell us that these can have serious, even devastating, effects. However, ''conventional'' does correctly imply that these are common pollutants produced in large amounts. These conventional pollutants are as follows: *biochemical oxygen demand, nutrients, pH, suspended solids, oil and grease*, and *pathogenic microorganisms*. Note that none of these is an individual chemical. Indeed, one, microorganisms, refers to whole living organisms. These conventional pollutants will now be examined in turn.

Biochemical oxygen demand

Microorganisms decompose organic matter discharged to a water body. They require oxygen to do so. The amount of oxygen required to decompose a given amount of organic pollutant is the ''biochemical oxygen demand'' (BOD). Natural BOD, such as plant debris and wildlife feces, is almost always present. However, a high BOD often indicates human activity, such as sewage or industrial discharge. Human activities that lead to a discharge of BOD include municipal wastewatertreatment plants, food-processing operations, chemical plants, pulp and paper operations, tanneries, and slaughterhouses.

A high BOD can reduce or deplete the oxygen in water. In a large water body, fish can swim away from low-oxygen (''hypoxic'') conditions, but crabs and snails and sedentary organisms may die. Professor R. Diaz of the College of William and Mary's School of Marine Sciences has noted, ''Low oxygen now causes more mass fish deaths than any other single agent, including oil spills, and it ranks as a leading threat to commercial fisheries and the marine environment in general.''Hypoxic water is a problem in the United States and worldwide. Hypoxia can be an irregular or seasonal occurrence in a water body, or an ongoing problem. An inorganic nutrient is not BOD, but notice that a nutrient can generate BOD.

Nutrients

A nutrient is a substance required for life, but has a more sinister face at high concentrations. Man-made fertilizers, containing concentrated *reactive nitrogen*2 and phosphorus are a major source of nutrients to water bodies. When fertilizer is added to agricultural fields, the excess runs off with rainwater into water bodies. There water plants, especially algae, ingest the nutrients. This has a cascade of effects. Although synthetic fertilizer is the major offender, any organic matter has nutrient value. Human activities that lead to the discharge of organic matter include municipal wastewater-treatment plants, food-processing operations, chemical plants, pulp and paper operations, tanneries, and slaughterhouses. Natural sources of nutrients likewise are similar to those noted for BOD, i.e., plant and animal debris and wildlife feces.

Continuing input of excess nutrients can lead to *eutrophication*, a process ''during which a lake, estuary, or bay evolves into a bog or marsh and eventually disappears.'' A water body naturally becomes eutrophic, but over many years as it slowly accumulates nutrients. During later stages of eutrophication, the water is choked with plant life, in particular algal ''blooms.'' Blooms may form a scum on the water surface, produce offensive smells, give the water a bad taste, and make it unfit for swimming. Human activities that put excess amounts of nutrients into water accelerate eutrophication.
Nitrate and ammonia, as well as many organic chemicals, contain nitrogen in a form bioavailable to plants and algae. Excess nutrients – the ''nitrogen glut'' -- have become a global problem.

Nitrate and phosphorus are two nutrients found in commercial fertilizer. As inorganic chemicals, they themselves do not exert BOD. However, they are fertilizers, stimulating plant growth. Nutrients stimulate algal growth in water, sometimes an algal bloom. Zooplankton eat the proliferating algae. Bacteria, using the oxygen in water to do so, digest the fecal pellets of the zooplankton as well as dead plankton and dead vegetation. Even if oxygen is only partially depleted, aquatic organisms suffer. If most of the oxygen is depleted, a dead zone can result.

In estuaries and coastal areas, excessive nitrate is the major culprit although phosphorus, another important nutrient, also contributes to the problem. In freshwater lakes, phosphorus is often the major culprit stimulating excessive growth of vegetation.

Suspended solids

This physical pollutant is found naturally in water to varying extents. As usual, it is an excess that is deleterious. Also recall that in air, it is the very fine particles which cause the greatest health problems. Similarly, fine particles in soil runoff become fine suspended solids in water, which can cause serious problems. Increased suspendedsolids content makes water more turbid or cloudy. This limits the sunlight reaching aquatic plants and stunts their growth. Fine suspended solids can also clog fish gills and harm the respiration of other water animals. Suspended solids can interfere with efficient water disinfection by shielding microorganisms from the disinfectant.

Surviving microorganisms can then contaminate drinking water. A major *source* of suspended solids is soil runoff from agricultural fields, especially in row crops. Forestry and construction activities contribute too. Point sources of suspended solids are facilities that discharge various solids including those that create BOD.

Oil and grease

Oil spills are a major problem in some near-coastal waters, killing or adversely affecting fish, other aquatic organisms, birds, and mammals. These spills can also kill or reduce organisms living in coastal sands and rocks, and may kill the worms and insects that are food to birds and wildlife. When the spills intrude into coastal marshes, the oil can damage or kill fish, shrimp, birds, and other animals. Oil spills can also foul beaches used for swimming and recreation.

Despite the sometimes horrendous damage caused by oil spills, they are seen as a relatively minor problem for fish and the marine environment in comparison to chronic nutrient pollution. Depending upon the amount and type of oil spilled, where it is spilled, and weather conditions, ecosystem recovery can be quick or painfully slow.

Spills are not the only source of oil in water: oil leaking from vehicles, or released during accidents, washes off roads with rainwater and then reaches water bodies. A portion also percolates down to groundwater. Used oil from motor vehicles is often improperly disposed of too. Direct releases of oil into water bodies also occur.

Motor and other recreational boats release up to 30% of their fuel, unburned, into water. These individually small, but ongoing events add up to much more oil than is spilled in a dramatic event such as the *Exxon Valdez*. However, the effects of a major spill are obvious whereas the environmental impact of ongoing small events is harder to assess.

Pathogenic microorganisms

Most microorganisms are not pathogens, and do not cause disease. Most perform useful, often vital, functions for humans including assisting the digestion in our intestines.We depend on microbes to degrade organic wastes in the environment and to biodegrade the organic material in landfills. We use microbes in fermentations to make food products, pharmaceuticals, and other chemicals. Microbes are almost ubiquitous in our environment, found almost anywhere that one looks. Professors Bruce Levin and Rustom Antia state it well, ‘‘Almost every time we eat, brush our teeth, scrape our skin, we are confronted with populations of microbes that are capable of colonizing the mucosa lining our orifices and alimentary tract and proliferating in fluids and cells within us. Nevertheless, we rarely get sick much less succumb to these infections.” It is specific microorganisms, pathogens that are capable of causing infectious disease, that concern us. A pathogen can be a bacterium, virus, fungus, protozoan, or toxic algal species.

Dangers of pathogens

If infectious microbes or their toxins are found in shellfish, their harvesting for food use is halted. Pathogenic viruses and bacteria in coastal water can infect swimmers and others. Viruses are abundant in marine waters, often surviving in salt water longer than bacteria. Infections can result not just by ingesting water containing pathogens, but through the skin. In the United States, as many as 19 out of

1000 swimmers each year are reported to suffer gastroenteritis caused by swimming in water containing infectious microbes.

Sources of pathogens

Pathogenic microbes in a water body are often anthropogenic, generated by human activities.
Runoff. They may arrive in water bodies in runoff of storm water, and from improperly operating septic systems, or runoff from livestock operations.

Point sources. Pathogens sometimes come from point sources, especially poorly performing municipal sewage-treatment plants.

All these sources exist in developed countries. The situation is worse in less-developed nations where most sewage remains untreated, and is often dumped into rivers and oceans.

Priority (toxic) pollutants

If you remember that a specific pollutant, such as lead or chloroform, often contaminates more than one environmental medium, you won't be surprised to learn that many common air pollutants are common water pollutants too. Recall too that hazardous air pollutants (HAPs, Chapter 5) are often called toxic air pollutants. Likewise, priority water pollutants are often called toxic water pollutants. The US EPA, under the Clean Water Act, regulates 126 priority pollutants including metals such as arsenic, cadmium, lead, mercury, nickel, copper, and zinc. These metals are not only priority water pollutants.

They are also HAPs. Among the priority pollutants that are organic chemicals, are the widely used industrial chemicals such as benzene, toluene, and many pesticides. Many of these too are HAPs.

A high concentration of a priority water pollutant such as a pesticide may cause acute illnesses or death in aquatic life. In the United States, hundreds of fish kills are still reported each year that result from runoff of spilled pesticides or other chemicals. In smaller quantities, many priority pollutants present a chronic health risk, e.g., some pesticides may act as environmental hormones.

If the priority pollutant comes from a point source, e.g., a wastewater-treatment facility, it can usually be well controlled. Control is more difficult for priority pollutants found in non-point-source runoff, as with pesticides from agricultural fields, organic chemicals in runoff from city streets, or polycyclic aromatic hydrocarbons (PAHs) deposited from air.

Banned discharges

The Clean Water Act totally forbids the discharge of some substances, including radioactive chemicals, and chemical and biological warfare agents. Other chemicals whose discharge is prohibited are PCBs, which in the United States are regulated by the Toxic Substances Control Act (TSCA). In industrialized countries, the manufacture of PCBs was banned in the 1970s, but they still occur in runoff from old spills, waste sites, or leaks in old electrical equipment. As a result of discharges into water when PCB manufacture was still legal, these long-lived chemicals are still to be found in sediments. PCB concentration is lower in water because of their poor water solubility.

Non-conventional and non-toxic pollutants

A third group of water pollutants is the non-conventional and non-toxic pollutants. Here we find ammonia, chloride (as from sodium chloride, salt), iron, aluminum, total phenols, and color. Many facilities -- textile factories are an example -- discharge colored effluents. The intensity of the discharged color is regulated by law. Heat is a regulated pollutant too. Electric power plants especially, but also many industrial facilities discharge heated effluents. Thermal pollution can cause problems but is not ordinarily as serious as many other pollutants.

Reducing point sources

One point source of water pollutant is municipal sewage-treatment plants and industrial wastewater-treatment plants. Wastewater is tap water after it has been used in homes, businesses, and institutions for drinking, bathing, flushing toilets, and other purposes.

Before a municipal or industrial facility can discharge wastewater into receiving water in the United States, it must first treat it to remove pollutants to levels in compliance with its particular permit.

Because elimination was not immediately possible, laws allowed permits to be issued to municipal and industrial facilities. These permits allowed discharges of specified amounts of particular pollutants. A facility must comply with its permit and regularly monitor its discharges to assure its compliance. The goal of permits is to limit discharges to a point that is protective of human health and aquatic life.

Over the years wastewater treatment has become increasingly effective at removing contaminants, but pollutant release has not been eliminated. Unfortunately, in many less-developed countries pollutant release is poorly controlled or uncontrolled.

2.11 WATER POLLUTION

Water pollution is any chemical, physical or biological change in the quality of water that has a harmful effect on any living thing that drinks or uses or lives (in) it. When humans drink polluted water it often has serious effects on their health. Water pollution can also make water unsuited for the desired use.

Major water pollutants

There are several classes of water pollutants. The first are disease-causing agents. These are bacteria, viruses, protozoa and parasitic worms that enter sewage systems and untreated waste.

A second category of water pollutants is oxygen-demanding wastes; wastes that can be decomposed by oxygen-requiring bacteria. When large populations of decomposing bacteria are converting these wastes it can deplete oxygen levels in the water. This causes other organisms in the water, such as fish, to die. A third class of water pollutants is water-soluble inorganic pollutants, such as acids, salts and toxic metals. Large quantities of these compounds will make water unfit to drink and will cause the death of aquatic life.

Another class of water pollutants are nutrients; they are water-soluble nitrates and phosphates that cause excessive growth of algae and other water plants, which deplete the water's oxygen supply. This kills fish and, when found in drinking water, can kill young children.

Water can also be polluted by a number of organic compounds such as oil, plastics and pesticides, which are harmful to humans and all plants and animals in the water. A very dangerous category is suspended sediment, because it causes depletion in the water's light absorption and the particles spread dangerous compounds such as pesticides through the water.

Finally, water-soluble radioactive compounds can cause cancer, birth defects and genetic damage and are thus very dangerous **water pollutants.**

Environmental Science 10
Water pollutant and water treatment
General Types of Water Pollutants (1)

Class of pollutant	Significance
Trace Elements	Health, aquatic biota, toxicity
Heavy metals	Health, aquatic biota, toxicity
Organically-bound metals	Metal transport
Radionuclides	Toxicity
Inorganic pollutants	Toxicity, aquatic biota
Asbestos	Human health
Algal nutrients	Eutrophication
Acidity, alkalinity, salinity (in excess)	Water quality, aquatic life
Trace organic pollutants	Toxicity
Polychlorinated biphenyls	Possible biological effects
Pesticides	Toxicity, aquatic biota, wildlife
Petroleum wastes	Effect on wildlife, esthetics
Sewage, human and animal wastes	Water quality, oxygen levels
Biochemical oxygen demand	Water quality, oxygen levels
Pathogens	Health effects
Detergents	Eutrophication, wildlife, esthetics
Chemical carcinogens	Incidence of cancer
Sediments	Water quality, aquatic biota, wildlife
Taste, odor, and color	Esthetics

Elemental pollutants

Trace element is a term that refers to those elements that occur at very low levels of a few parts per million or less in a given system. The term **trace substance** is a more general one applied to both elements and chemical compounds. Table 2 summarizes the more important trace elements encountered in natural waters. Some of these are recognized as nutrients required for animal and plant life, including some that are essential at low levels but toxic at higher levels. This is typical behavior for many substances in the aquatic environment, a point that must be kept in mind in judging whether a particular element is beneficial or detrimental. Some of these elements, such as lead or mercury, have such toxicological and environmental significance that they are discussed in detail in separate sections.

Important Trace Elements in Natural Waters (2)

Element	Sources	Effects and Significance
Arsenic	Mining byproduct, chemical waste	Toxic1, possibly carcinogenic
Beryllium	Coal, industrial wastes	Toxic
Boron	Coal, detergents, wastes	Toxic
Chromium	Metal plating	Essential as Cr(III), toxic as Cr(VI)
Copper	Metal plating, mining, indus trial waste and algae at higher levels	Essential trace element, toxic to plants
Fluorine (F-)	Natural geological sources, wastes, water additive	Prevents tooth decay at around 1 mg/L, toxic at higher levels
Iodine (I-)	Industrial wastes, brines, seawater intrusion	Prevents goiter
Iron	Industrial wastes, corrosion, acid mine water, microbial action	Essential nutrient, damages fixtures by staining
Lead	Industrial waste, mining, fuels	Toxic, harmful to wildlife
Manganese	Industrial wastes, acid mine, water, microbial action	Toxic to plants, damages fixtures by staining
Mercury	Industrial waste, mining, coal	Toxic, mobilized as methyl mercury compounds by anaerobic bacteria
Molybdenum	Industrial wastes, natural sources	Essential to plants, toxic to animals
Selenium	Natural sources, coal	Essential at lower levels, toxic at higher levels
Zinc	Industrial waste, metal plating, plumbing	Essential element, toxic to plants at higher levels

Some of the **heavy metals** are among the most harmful of the elemental pollutants and are of particular concern because of their toxicities to humans. These elements are in general the transition metals, and some of the representative elements, such as lead and tin, in the lower right-hand corner of the periodic table.

Heavy metals include essential elements like iron as well as toxic metals like cadmium and mercury. Most of them have a tremendous affinity for sulfur and disrupt enzyme function by forming bonds with sulfur groups in enzymes. Protein carboxylic acid (–COOH) and amino (–NH_2) groups are also chemically bound by heavy metals. Cadmium, copper, lead, and mercury ions bind to cell membranes, hindering transport processes through the cell wall. Heavy metals may also precipitate phosphate biocompounds or catalyze their decomposition.

Some of the **metalloids**, elements on the borderline between metals and nonmetals, are significant water pollutants. Arsenic, selenium, and antimony are of particular interest.

Inorganic chemicals manufacture has the potential to contaminate water with trace elements. Among the industries regulated for potential trace element pollution of water are those producing chlor-alkali, hydrofluoric acid, sodium dichromate (sulfate process and chloride ilmenite process), aluminum fluoride, chrome pigments, copper sulfate, nickel sulfate, sodium bisulfate, sodium hydrosulfate, sodium bisulfite, titanium dioxide, and hydrogen cyanide.

Cadmium

Pollutant **cadmium** in water may arise from industrial discharges and mining wastes. Cadmium is widely used in metal plating. Chemically, cadmium is very similar to zinc, and these two metals frequently undergo geochemical processes together. Both metals are found in water in the +2 oxidation state.

The effects of acute cadmium poisoning in humans are very serious. Among them are high blood pressure, kidney damage, destruction of testicular tissue, and destruction of red blood cells. It is believed that much of the physiological action of cadmium arises from its chemical similarity to zinc. Specifically, cadmium may replace zinc in some enzymes, thereby altering the stereostructure of the enzyme and impairing its catalytic activity. Disease symptoms ultimately result.

Cadmium and zinc are common water and sediment pollutants in harbors surrounded by industrial installations. Concentrations of more than 100 ppm dry weight sediment have been found in harbor sediments. Typically, during periods of calm in the summer when the water stagnates, the anaerobic bottom layer of harbor water has a low soluble Cd concentration because microbial reduction of sulfate produces sulfide,

$2\{CH_2O\} + SO_4^{2-} + H^+ \rightarrow 2CO_2 + HS- + 2H_2O$

which precipitates cadmium as insoluble cadmium sulfide:

$CdCl^+$(*chloro complex in seawater*) + $HS^- \rightarrow CdS(s) + H+ + Cl-$

Mixing of bay water from outside the harbor and harbor water by high winds during the winter results in desorption of cadmium from harbor sediments by aerobic bay water. This dissolved cadmium is carried out into the bay where it is absorbed by suspended solid materials, which then become incorporated with the bay sediments. This is an example of the sort of complicated interaction of hydraulic, chemical solution-solid, and microbiological factors involved in the transport and distribution of a pollutant in an aquatic system.

Lead

Inorganic **lead** arising from a number of industrial and mining sources occurs in water in the +2 oxidation state. Lead from leaded gasoline used to be a major source of atmospheric and terrestrial lead, much of which eventually entered natural water systems. In addition to pollutant sources, lead-bearing limestone and galena (PbS) contribute lead to natural waters in some locations.

Despite greatly increased total use of lead by industry, evidence from hair samples and other sources indicates that body burdens of this toxic metal have decreased during recent decades. This may be the result of less lead used in plumbing and other products that come in contact with food or drink.

Acute lead poisoning in humans causes severe dysfunction in the kidneys, reproductive system, liver, and the brain and central nervous system. Sickness or death results. Lead poisoning from environmental exposure is thought to have caused mental retardation in many children. Mild lead poisoning causes anemia. The victim may have headaches and sore muscles, and may feel generally fatigued and irritable.

Except in isolated cases, lead is probably not a major problem in drinking water, although the potential exists in cases where old lead pipe is still in use. Lead used to be a constituent of solder and some pipe-joint formulations, so that household water does have some contact with lead. Water that has stood in household plumbing for some time may accumulate significant levels of lead (along with zinc, cadmium, and copper) and should be drained for a while before use.

Mercury

Because of its toxicity, mobilization as methylated forms by anaerobic bacteria, and other pollution factors, **mercury** generates a great deal of concern as a heavymetal pollutant. Mercury is found as a trace component of many minerals, with continental rocks containing an average of around 80 parts per billion, or slightly less, of this element. Cinnabar, red mercuric sulfide, is the chief commercial mercury ore. Fossil fuel coal and lignite contain mercury, often at levels of 100 parts per billion or even higher, a matter of some concern with increased use of these fuels for energy resources.

Metallic mercury is used as an electrode in the electrolytic generation of chlorine gas, in laboratory vacuum apparatus, and in other applications. Significant quantities of inorganic mercury(I) and mercury(II) compounds are used annually. Organic mercury compounds used to be widely applied as pesticides, particularly fungicides. These mercury compounds include aryl mercurials such as phenyl mercuric dimethyldithiocarbamate (formerly used in paper mills as a slimicide and as a mold retardant for paper), and alkyl-mercurials such as ethylmercuric chloride, C_2H_5HgCl, that was used as a seed fungicide. Because of their resistance to degradation and their mobility, the alkyl mercury compounds are generally considered to be more of an environmental threat than either the aryl or inorganic compounds.

Mercury enters the environment from a large number of miscellaneous sources related to human use of the element. These include discarded laboratory chemicals, batteries, broken thermometers, amalgam tooth fillings, and formerly lawn fungicides and pharmaceutical products. Taken individually, each of these sources may not contribute much of the toxic metal, but the total effect can be substantial. Sewage effluent sometimes contains up to 10 times the level of mercury found in typical natural waters.

The toxicity of mercury was tragically illustrated in the Minamata Bay area of Japan during the period 1953-1960. A total of 111 cases of mercury poisoning and 43 deaths were reported among people who had consumed seafood from the bay that had been contaminated with mercury waste from a chemical plant that drained into Minamata Bay. Congenital defects were observed in 19 babies whose mothers had consumed seafood contaminated with mercury. The level of metal in the contaminated seafood was 5-20 parts per million.

Among the toxicological effects of mercury are neurological damage, including irritability, paralysis, blindness, or insanity; chromosome breakage; and birth defects. The milder symptoms of mercury poisoning such as depression and irritability have a psychopathological character. Because of the resemblance of these symptoms to common human behavior, mild mercury poisoning may escape detection.

Some forms of mercury are relatively nontoxic and were formerly used as medicines, for example, in the treatment of syphilis. Other forms of mercury, particularly organic compounds, are highly toxic.

Because there are few major natural sources of mercury, and since most inorganic compounds of this element are relatively insoluble, it was assumed for some time that mercury was not a serious water pollutant. However, in 1970, alarming mercury levels were discovered in fish in Lake Saint Clair located between Michigan and Ontario, Canada.

A subsequent survey by the U.S. Federal Water Quality Administration revealed a number of other waters contaminated with mercury. It was found that several chemical plants, particularly caustic chemical manufacturing operations, were each releasing up to 14 or more kilograms of mercury in wastewaters each day.

The unexpectedly high concentrations of mercury found in water and in fish tissues result from the formation of soluble monomethylmercury ion, CH_3Hg^+, and volatile dimethylmercury, $(CH_3)_2Hg$, by anaerobic bacteria in sediments. Mercury from these compounds becomes concentrated in fish lipid (fat) tissue and the concentration factor from water to fish may exceed 103. The methylating agent by which inorganic mercury is converted to methylmercury compounds is methylcobalamin, a vitamin B_{12} analog.

It is believed that the bacteria that synthesize methane produce methylcobalamin as an intermediate in the synthesis. Thus, waters and sediments in which anaerobic decay is occurring provide the conditions under which methylmercury production occurs. In neutral or alkaline waters, the formation of dimethylmercury, $(CH_3)_2Hg$, is favored. This volatile compound can escape to the atmosphere.

2.12 WATER TREATMENTS

Water treatment and water use

The treatment of water may be divided into three major categories:

- Purification for domestic use
- Treatment for specialized industrial applications
- Treatment of wastewater to make it acceptable for release or reuse.

The type and degree of treatment are strongly dependent upon the source and intended use of the water. Water for domestic use must be thoroughly disinfected to eliminate disease-causing microorganisms, but may contain appreciable levels of dissolved calcium and magnesium (hardness). Water to be used in boilers may contain bacteria but must be quite soft to prevent scale formation. Wastewater being discharged into a large river may require less rigorous treatment than water to be reused in an arid region. As world demand for limited water resources grows, more sophisticated and extensive means will have to be employed to treat water.

Most physical and chemical processes used to treat water involve similar phenomena, regardless of their application to the three main categories of water treatment listed above. Therefore, after introductions to water treatment for municipal use, industrial use, and disposal, each major kind of treatment process is discussed as it applies to all of these applications.

Municipal water treatment

The modern water treatment plant (Fig. 139) is often called upon to perform wonders with the water fed to it. The clear, safe, even tasteful water that comes from a faucet may have started as a murky liquid pumped from a polluted river laden with mud and swarming with bacteria. Or, its source may have been well water, much too hard for domestic use and containing high levels of stain-producing dissolved iron and manganese. The water treatment plant operator's job is to make sure that the water plant product presents no hazards to the consumer.

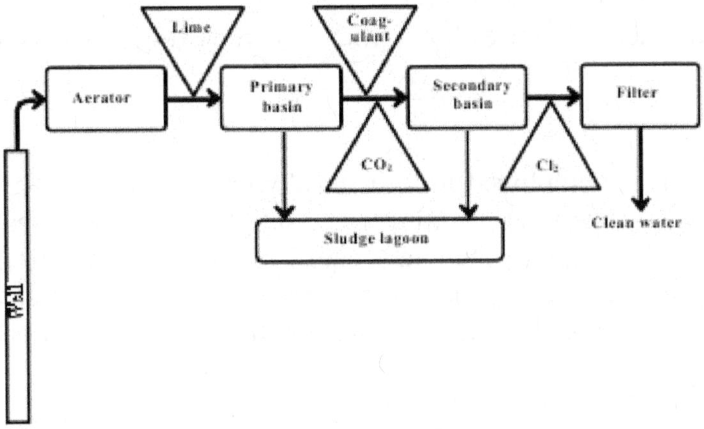

Fig 139. *Treatment of drinking water*

This particular facility treats water containing excessive hardness and a high level of iron. The raw water taken from wells first goes to an aerator. Contact of the water with air removes volatile solutes such as hydrogen sulfide, carbon dioxide, methane, and volatile odorous substances such as methane thiol (CH_3SH) and bacterial metabolites. Contact with oxygen also aids iron removal by oxidizing soluble iron(II) to insoluble iron(III). The addition of lime as CaO or $Ca(OH)_2$ after aeration raises the pH and results in the formation of precipitates containing the hardness ions Ca^{2+} and Mg^{2+}. These precipitates settle from the water in a primary basin. Much of the solid material remains in suspension and requires the addition of coagulants (such as iron(III) and aluminum sulfates, which form gelatinous metal hydroxides) to settle the colloidal particles. Activated silica or synthetic polyelectrolytes may also be added to stimulate coagulation or flocculation.

The settling occurs in a secondary basin after the addition of carbon dioxide to lower the pH. Sludge from both the primary and secondary basins is pumped to a sludge lagoon. The water is finally chlorinated, filtered, and pumped to the city water mains.

Treatment of water for industrial use

Water is widely used in various process applications in industry. Other major industrial uses are boiler feedwater and cooling water. The kind and degree of treatment of water in these applications depends upon the end use. As examples, cooling water may require only minimal treatment, removal of corrosive substances and scale-forming solutes is essential for boiler feedwater, and water used in food processing must be free of pathogens and toxic substances. Improper treatment of water for industrial use can cause problems such as corrosion, scale formation, reduced heat transfer in heat exchangers, reduced water flow, and product contamination.

These effects may cause reduced equipment performance or equipment failure, increased energy costs due to inefficient heat utilization or cooling, increased costs for pumping water, and product deterioration. Obviously, the effective treatment of water at minimum cost for industrial use is a very important area of water treatment.

Numerous factors must be taken into consideration in designing and operating an industrial water treatment facility. These include the following:
- Water requirement
- Quantity and quality of available water sources
- Sequential use of water (successive uses for applications requiring progressively lower water quality)
- Water recycle
- Discharge standards

The various specific processes employed to treat water for industrial use are discussed later. **External treatment**, usually applied to the plant's entire water supply, uses processes such as aeration, filtration, and clarification to remove material that may cause problems from water. Such substances include suspended or dissolved solids, hardness, and dissolved gases. Following this basic treatment, the water may be divided into different streams, some to be used without further treatment, and the rest to be treated for specific applications.

Internal treatment is designed to modify the properties of water for specific applications. Examples of internal treatment include the following:
- Reaction of dissolved oxygen with hydrazine or sulfite
- Addition of chelating agents to react with dissolved Ca^{2+} and prevent formation of calcium deposits
- Addition of precipitants, such as phosphate used for calcium removal
- Treatment with dispersants to inhibit scale
- Addition of inhibitors to prevent corrosion
- Adjustment of pH
- Disinfection for food processing uses or to prevent bacterial growth in cooling water

Sewage treatment

Typical municipal sewage contains oxygen-demanding materials, sediments, grease, oil, scum, pathogenic bacteria, viruses, salts, algal nutrients, pesticides, refractory organic compounds, heavy metals, and an astonishing variety of flotsam ranging from children's socks to sponges. It is the job of the waste treatment plant to remove as much of this material as possible.

Several characteristics are used to describe sewage. These include turbidity (international turbidity units), suspended solids (ppm), total dissolved solids (ppm), acidity (H+ ion concentration or pH), and dissolved oxygen (in ppm O_2). Biochemical oxygen demand is used as a measure of oxygen-demanding substances.

Current processes for the treatment of wastewater may be divided into three main categories of primary treatment, secondary treatment, and tertiary treatment, each of which is discussed separately. Also discussed are total wastewater treatment systems, based largely upon physical and chemical processes Waste from a municipal water system is normally treated in a **publicly owned treatment works, POTW**.

Primary Waste Treatment

Primary treatment of wastewater consists of the removal of insoluble matter such as grit, grease, and scum from water. The first step in primary treatment normally is screening. Screening removes or reduces the size of trash and large solids that get into the sewage system. These solids are collected on screens and scraped off for subsequent disposal. Most screens are cleaned with power rakes.

Comminuting devices shred and grind solids in the sewage. Particle size may be reduced to the extent that the particles can be returned to the sewage flow. Grit in wastewater consists of such materials as sand and coffee grounds which do not biodegrade well and generally have a high settling velocity. **Grit removal** is practiced to prevent its accumulation in other parts of the treatment system, to reduce clogging of pipes and other parts, and to protect moving parts from abrasion and wear. Grit normally is allowed to settle in a tank under conditions of low flow velocity, and it is then scraped mechanically from the bottom of the tank.

Primary sedimentation removes both settleable and floatable solids. During primary sedimentation there is a tendency for flocculent particles to aggregate for better settling, a process that may be aided by the addition of chemicals. The material that floats in the primary settling basin is known collectively as grease. In addition to fatty substances, the grease consists of oils, waxes, free fatty acids, and insoluble soaps containing calcium and magnesium. Normally, some of the grease settles with the sludge and some floats to the surface, where it may be removed by a skimming device.

Secondary Waste Treatment by Biological Processes

The most obvious harmful effect of biodegradable organic matter in wastewater is BOD, consisting of a biochemical oxygen demand for dissolved oxygen by microorganism-mediated degradation of the organic matter. **Secondary wastewater treatment** is designed to remove BOD, usually by taking advantage of the same kind of biological processes that would otherwise consume oxygen in water receiving the wastewater. Secondary treatment by biological processes takes many forms but consists basically of the action of microorganisms provided with added oxygen degrading organic material in solution or in suspension until the BOD of the waste has been reduced to acceptable levels. The waste is oxidized biologically under conditions controlled for optimum bacterial growth, and at a site where this growth does not influence the environment.

One of the simplest biological waste treatment processes is the **trickling filter** in which wastewater is sprayed over rocks or other solid support material covered with microorganisms. The structure of the trickling filter is such that contact of the wastewater with air is allowed and degradation of organic matter occurs by the action of the microorganisms.

Rotating biological reactors (**contactors**), another type of treatment system, consist of groups of large plastic discs mounted close together on a rotating shaft. The device is positioned such that at any particular instant half of each disc is immersed in wastewater and half exposed to air. The shaft rotates constantly, so that the submerged portion of the discs is always changing. The discs, usually made of high-density polyethylene or polystyrene, accumulate thin layers of attached biomass, which degrades organic matter in the sewage. Oxygen is absorbed by the biomass and by the layer of wastewater adhering to it during the time that the biomass is exposed to air.

Both trickling filters and rotating biological reactors are examples of fixed-film biological (FFB) or attached growth processes. The greatest advantage of these processes is their low energy consumption. The energy consumption is minimal because it is not necessary to pump air or oxygen into the water, as is the case with the popular activated sludge process described below. The trickling filter has long been a standard means of wastewater treatment, and a number of wastewater treatment plants use trickling filters at present.

The **activated sludge process**, is probably the most versatile and effective of all wastewater treatment processes. Microorganisms in the aeration tank convert organic material in wastewater to microbial biomass and CO_2. Organic nitrogen is converted to ammonium ion or nitrate. Organic phosphorus is converted to orthophosphate. The microbial cell matter formed as part of the waste degradation processes is normally kept in the aeration tank until the microorganisms are past the log phase of growth, at which point the cells flocculate relatively well to form settleable solids.

These solids settle out in a settler and a fraction of them is discarded. Part of the solids, the return sludge, is recycled to the head of the aeration tank and comes into contact with fresh sewage. The combination of a high concentration of "hungry" cells in the return sludge and a rich food source in the influent sewage provides optimum conditions for the rapid degradation of organic matter.

The degradation of organic matter that occurs in an activated sludge facility also occurs in streams and other aquatic environments. However, in general, when a degradable waste is put into a stream, it encounters only a relatively small population of microorganisms capable of carrying out the degradation process. Thus, several days may be required for the buildup of a sufficient population of organisms to degrade the waste. In the activated sludge process, continual recycling of active organisms provides the optimum conditions for waste degradation, and a waste may be degraded within the very few hours that it is present in the aeration tank.
The activated sludge process provides two pathways for the removal of BOD. BOD may be removed by (1) oxidation of organic matter to provide energy for the metabolic processes of the microorganisms, and (2) synthesis, incorporation of the organic matter into cell mass. In the first pathway, carbon is removed in the gaseous form as CO_2.

The second pathway provides for removal of carbon as a solid in biomass. That portion of the carbon converted to CO2 is vented to the atmosphere and does not present a disposal problem. The disposal of waste sludge, however, is a problem, primarily because it is only about 1% solids and contains many undesirable components. Normally, partial water removal is accomplished by drying on sand filters, vacuum filtration, or centrifugation.

The dewatered sludge may be incinerated or used as landfill. To a certain extent, sewage sludge may be digested in the absence of oxygen by methane-producing anaerobic
bacteria to produce methane and carbon dioxide, $2\{CH_2O\} \rightarrow CH_4 + CO_2$
a process that reduces both the volatile-matter content and the volume of the sludge by about 60%. A carefully designed plant may produce enough methane to provide for all of its power needs.

One of the most desirable means of sludge disposal is to use it to fertilize and condition soil. However, care has to be taken that excessive levels of heavy metals are not applied to the soil as sludge contaminants.

Activated sludge wastewater treatment is the most common example of an aerobic suspended culture process. Many factors must be considered in the design and operation of an activated sludge wastewater treatment system.1 These include parameters involved with the process modeling and kinetics. The microbiology of the system must be considered. In addition to BOD removal, phosphorus and nitrogen removal must also be taken into account. Oxygen transfer and solids separation are important. Industrial wastes and the fates and effects of industrial chemicals (xenobiotics) must also be considered.

Nitrification (the microbially mediated conversion of ammonium nitrogen to nitrate; is a significant process that occurs during biological waste treatment. Ammonium ion is normally the first inorganic nitrogen species produced in the biodegradation of nitrogenous organic compounds. It is oxidized, under the appropriate conditions, first to nitrite by *Nitrosomonas* bacteria, then to nitrate by *Nitrobacter*:

$2NH_4^+ + 3O_2 \rightarrow 4H^+ + 2NO_2^- + 2H_2O$

$2NO_2^- + O_2 \rightarrow 2NO_3^-$

These reactions occur in the aeration tank of the activated sludge plant and are favored in general by long retention times, low organic loadings, large amounts of suspended solids, and high temperatures. Nitrification can reduce sludge settling efficiency because the denitrification reaction $4NO3^- + 5\{CH_2O\} + 4H^+ \rightarrow 2N_2(g) + 5CO_2(g) + 7H_2O$ occurring in the oxygen-deficient settler causes bubbles of N_2 to form on the sludge floc (aggregated sludge particles), making it so buoyant that it floats to the top. This prevents settling of the sludge and increases the organic load in the receiving waters. Under the appropriate conditions, however, advantage can be taken of this phenomenon to remove nutrient nitrogen from water.

Tertiary Waste Treatment

Unpleasant as the thought may be, many people drink used water—water that has been discharged from a municipal sewage treatment plant or from some industrial process. This raises serious questions about the presence of pathogenic organisms or toxic substances in such water. Because of high population density and heavy industrial development, the problem is especially acute in Europe where some municipalities process 50% or more of their water from “used” sources. Obviously, there is a great need to treat wastewater in a manner that makes it amenable to reuse. This requires treatment beyond the secondary processes.

Tertiary waste treatment (sometimes called **advanced waste treatment**) is a term used to describe a variety of processes performed on the effluent from secondary waste treatment.2 The contaminants removed by tertiary waste treatment fall into the general categories of (1) suspended solids, (2) dissolved organic compounds, and (3) dissolved inorganic materials, including the important class of algal nutrients. Each of these categories presents its own problems with regard to water quality.

Suspended solids are primarily responsible for residual biological oxygen demand in secondary sewage effluent waters. The dissolved organics are the most hazardous from the standpoint of potential toxicity. The major problem with dissolved inorganic materials is that presented by algal nutrients, primarily nitrates and phosphates. In addition, potentially hazardous toxic metals may be found among the dissolved inorganics.

In addition to these chemical contaminants, secondary sewage effluent often contains a number of disease-causing microorganisms, requiring disinfection in cases where humans may later come into contact with the water. Among the bacteria that may be found in secondary sewage effluent are organisms causing tuberculosis, dysenteric bacteria (*Bacillus dysenteriae*, *Shigella dysenteriae*, *Shigella paradysenteriae*, *Proteus vulgaris*), cholera bacteria (*Vibrio cholerae*), bacteria causing mud fever (*Leptospira icterohemorrhagiae*), and bacteria causing typhoid fever (*Salmonella typhosa*, *Salmonella paratyphi*). In addition, viruses causing diarrhea, eye infections, infectious hepatitis, and polio may be encountered. Ingestion of sewage still causes disease, even in more developed nations.

Physical-Chemical Treatment of Municipal Wastewater

Complete physical-chemical wastewater treatment systems offer both advantages and disadvantages relative to biological treatment systems. The capital costs of physical-chemical facilities can be less than those of biological treatment facilities, and they usually require less land. They are better able to cope with toxic materials and overloads. However, they require careful operator control and consume relatively large amounts of energy.

Basically, a physical-chemical treatment process involves:

- Removal of scum and solid objects
- Clarification, generally with addition of a coagulant, and frequently with the addition of other chemicals (such as lime for phosphorus removal)
- Filtration to remove filterable solids
- Activated carbon adsorption
- Disinfection

During the early 1970s, it appeared likely that physical-chemical treatment would largely replace biological treatment. However, higher chemical and energy costs since then have slowed the development of physical-chemical facilities.

GROUNDWATER IN THE GEOSPHERE

Groundwater is a vital resource in its own right that plays a crucial role in geochemical processes, such as the formation of secondary minerals. The nature, quality, and mobility of groundwater are all strongly dependent upon the rock formations in which the water is held. Physically, an important characteristic of such formations is their **porosity**, which determines the percentage of rock volume available to contain water. A second important physical characteristic is **permeability**, which describes the ease of flow of the water through the rock. High permeability is usually associated with high porosity. However, clays tend to have low permeability even when a large percentage of the volume is filled with water.

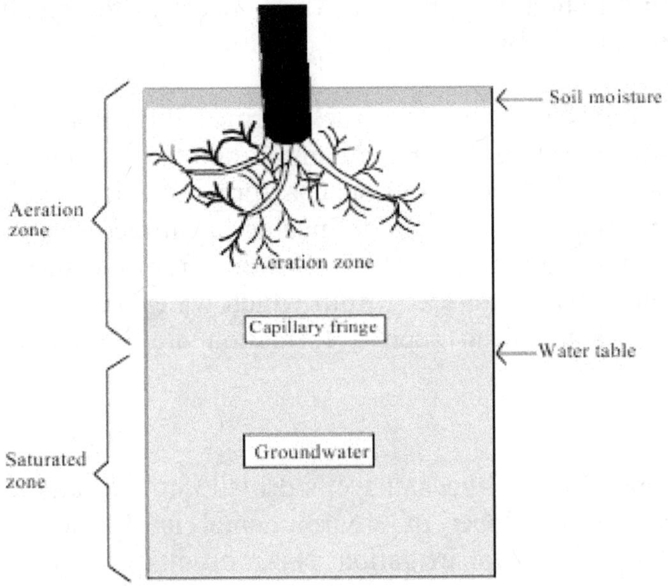

Fig. 140. *Ground water in the geosphere*

Most groundwater originates as **meteoric** water from precipitation in the form of rain or snow. If water from this source is not lost by evaporation, transpiration, or to stream runoff, it may infiltrate into the ground. Initial amounts of water from precipitation onto dry soil are held very tightly as a film on the surfaces and in the micropores of soil particles in a **belt of soil moisture**. At intermediate levels, the soil particles are covered with films of water, but air is still present in larger voids in the soil.

The region in which such water is held is called the **unsaturated zone** or **zone of aeration** and the water present in it is **vadose water**. At lower depths in the presence of adequate amounts of water, all voids are filled to produce a **zone of saturation**, the upper level of which is the **water table**. Water present in a zone of saturation is called **groundwater**. Because of its surface tension, water is drawn somewhat above the water table by capillary-sized passages in soil in a region called the capillary fringe.

The **water table** is crucial in explaining and predicting the flow of wells and springs and the levels of streams and lakes. It is also an important factor in determining the extent to which pollutant and hazardous chemicals underground are likely to be transported by water. The water table can be mapped by observing the equilibrium level of water in wells, which is essentially the same as the top of the saturated zone. The water table is usually not level, but tends to follow the general contours of the surface topography. It also varies with differences in permeability and water infiltration. The water table is at surface level in the vicinity of swamps and frequently above the surface where lakes and streams are encountered. The water level in such bodies may be maintained by the water table. **Influent** streams or reservoirs are located above the water table; they lose water to the underlying aquifer and cause an upward bulge in the water table beneath the surface water.

Groundwater **flow** is an important consideration in determining the accessibility of the water for use and transport of pollutants from underground waste sites. Various parts of a body of groundwater are in hydraulic contact so that a change in pressure at one point will tend to affect the pressure and level at another point. For example, infiltration from a heavy, localized rainfall may affect the water table at a point remote from the infiltration. Groundwater flow occurs as the result of the natural tendency of the water table to assume even levels by the action of gravity.

Groundwater flow is strongly influenced by rock permeability. Porous or extensively fractured rock is relatively highly **pervious**, meaning that water can migrate through the holes, fissures, and pores in such rock. Because water can be extracted from such a formation, it is called an **aquifer**. By contrast, an **aquiclude** is a rock formation that is too impermeable or unfractured to yield groundwater. Impervious rock in the unsaturated zone may retain water infiltrating from the surface to produce a **perched water table** that is above the main water table and from which water may be extracted. However, the amounts of water that can be extracted from such a formation are limited and the water is vulnerable to contamination.

Water Wells

Most groundwater is tapped for use by water wells drilled into the saturated zone. The use and misuse of water from this source has a number of environmental implications. In the U.S. about 2/3 of the groundwater pumped is consumed for irrigation; lesser amounts of ground water are used for industrial and municipal applications.

As water is withdrawn, the water table in the vicinity of the well is lowered. This **drawdown** of water creates a **zone of depression**. In extreme cases the groundwater is severely depleted and surface land levels can even subside (which is one reason that Venice, Italy is now very vulnerable to flooding). Heavy drawdown can result in infiltration of pollutants from sources such as septic tanks, municipal refuse sites, and hazardous waste dumps. When soluble iron(II) or manganese(II) are present in groundwater, exposure to air at the well wall can result in the formation of deposits of insoluble iron(III) and manganese(IV) oxides produced by bacterially catalyzed processes:

$$4Fe^{2+}(aq) + O_2(aq) + 10H_2O \rightarrow 4Fe(OH)_3(s) + 8H^+$$

$$2Mn^{2+}(aq) + O_2(aq) + (2x + 2)H_2O \rightarrow 2MnO_2{\bullet}xH_2O(s) + 4H^+$$

Deposits of iron(III) and manganese(IV) that result from the processes outlined above line the surfaces from which water flows into the well with a coating that is relatively impermeable to water. The deposits fill the spaces that water must traverse to enter the well. As a result, they can seriously impede the flow of water into the well from the water-bearing aquifer. This creates major water source problems for municipalities using groundwater for water supply. As a result of this problem, chemical or mechanical cleaning, drilling of new wells, or even acquisition of new water sources may be required.

2.13 ENVIRONMENTAL ASPECTS OF THE GEOSPHERE

Most of the remainder of this chapter deals specifically with the environmental aspects of geology and human interactions with the geosphere. It discusses how natural geological phenomena affect the environment through occurrences such as volcanic eruptions that may blast so much particulate matter and acid gas into the atmosphere that it may have a temporary effect on global climate, or massive earthquakes that disrupt surface topography and disturb the flow and distribution of groundwater and surface water. Also discussed are human influences on the geosphere and the strong connection between the geosphere and the anthrosphere.

Going back several billion years to its formation as a ball of dust particles collected from the universe and held together by gravitational forces, earth has witnessed constant environmental change and disruption.

During its earlier eons earth was a most inhospitable place for humans and, indeed, for any form of life. Heat generated by gravitational compression of primitive earth and by radioactive elements in its interior caused much of the mass of the planet to liquify. Relatively high density iron sank into the core, and lighter minerals, primarily silicates, solidified and floated to the surface.

Although in the scale of a human lifetime earth changes almost imperceptibly, the planet is in fact in a state of constant change and turmoil. It is known that continents have formed, broken apart, and moved around. Rock formations produced in ancient oceans have been thrust up onto continental land and huge masses of volcanic rock exist where volcanic activity is now unknown. Today the angry bowels of earth unleash enormous forces that push molten rock to the surface and move continents continuously as evidenced from volcanic activity, and from earthquakes resulting from the movement of great land masses relative to each other.

Earth's surface is constantly changing as new mountain ranges are heaved up and old ones are worn down. Humans have learned to work with, against, and around natural earth processes and phenomena to exploit earth's resources and to make these processes and phenomena work for the benefit of humankind. Human efforts have been moderately successful in mitigating some of the major hazards posed by natural geospheric phenomena, although such endeavors often have had unforeseen detrimental consequences, sometimes many years after they were first applied.

The survival of modern civilization and, indeed, of humankind will depend upon how intelligently humans work with the earth. That is why it is so important for humans to have a fundamental understanding of the geospheric environment.

An important consideration in human interaction with the geosphere is the application of engineering to geology. Engineering geology takes account of the geological characteristics of soil and rock in designing buildings, dams, highways, and other structures in a manner compatible with the geological strata on which they rest. Engineering geology must consider a large number of geological factors including type, strength, and fracture characteristics of rock, tendency for landslides to occur, susceptibility to settling, and likelihood of erosion. Engineering geology is an important consideration in land-use planning.

Natural Hazards

Earth presents a variety of natural hazards to the creatures that dwell on it. Some of these are the result of internal processes that arise from the movement of land masses relative to each other and from heat and intrusions of molten rock from below the surface. The most common such hazards are **earthquakes and volcanoes**.

Whereas internal processes tend to force matter upward, often with detrimental effects, surface processes are those that generally result from the tendency of matter to seek lower levels. Such processes include erosion, landslides, avalanches, mudflows, and subsidence.

A number of natural hazards result from the interaction and conflict between solid Earth and liquid and solid water. Perhaps the most obvious such hazard consists of floods when too much water falls as precipitation and seeks lower levels through streamflow. Wind can team with water to increase destructive effects, such as beach erosion and destruction of beachfront property resulting from wind-driven seawater. Ice, too, can have some major effects on solid earth. Evidence of such effects of the Ice

Age times include massive glacial moraines left over from deposition of till from melting glaciers, and landscape features carved by advancing ice sheets.

Anthropogenic Hazards

All too often, attempts to control and reshape the geosphere to human demands have been detrimental to the geosphere and dangerous to human life and well-being. Such attempts may exacerbate damaging natural phenomena. A prime example of this interaction occurs when efforts are made to control the flow of rivers by straightening them and building levees. The initial results can be deceptively favorable in that a modified stream may exist for decades, flowing smoothly and staying within the confines imposed by humans. But eventually, the forces of nature are likely to overwhelm the efforts of humans to control them, such as when a record flood breaks levees and destroys structures constructed in flood-prone areas. Landslides of mounds of earthen material piled up from mining can be very destructive. Destruction of wetlands in an effort to provide additional farmland can have some detrimental effects upon wildlife and upon the overall health of ecosystems.

Earthquake

The loss of life and destruction of property by earthquakes makes them some of nature's more damaging natural phenomena. The destructive effects of an earthquake are due to the release of energy. The released energy moves from the quake's focus as seismic waves, discussed in Section 15.4. Literally millions of lives have been lost in past earthquakes, and damage from an earthquake in a developed urban area can easily run into billions of dollars. A tragic August 1999 earthquake in Turkey killed several thousand people and caused massive damage. Earthquakes may cause catastrophic secondary effects, especially large, destructive ocean waves called tsunamis.

Adding to the terror of earthquakes is their lack of predictability. An earthquake can strike at any time—during the calm of late night hours or in the middle of busy rush hour traffic. Although the exact prediction of earthquakes has so far eluded investigators, the locations where earthquakes are most likely to occur are much more well-known. These are located in lines corresponding to boundaries along which tectonic plates collide and move relative to each other, building up stresses that are suddenly released when earthquakes occur. Such interplate boundaries are locations of preexisting faults and breaks. Occasionally, however, an earthquake will occur within a plate, made more massive and destructive because for it to occur the thick lithosphere has to be ruptured. The scale of earthquakes can be estimated by the degree of motion that they cause and by their destructiveness. The former is termed the **magnitude** of an earthquake and is commonly expressed by the **Richter scale**. The Richter scale is open-ended, and each unit increase in the scale reflects a ten-fold increase in magnitude.

Several hundred thousand earthquakes with magnitudes from two to three occur each year; they are detected by seismographs, but are not felt by humans. Minor earthquakes range from four to five on the Richter scale, and earthquakes cause damage at a magnitude greater than about five. Great earthquakes, which occur about once or twice a year, register over eight on the Richter scale. The **intensity** of an earthquake is a subjective estimate of its potential destructive effect. On the Mercalli intensity scale, an intensity III earthquake feels like the passage of heavy vehicles; one with an intensity of VII causes difficulty in standing, damage to plaster, and dislodging of loose brick, whereas a quake with an intensity of XII causes virtually total destruction, throws objects upward, and shifts huge masses of earthen material. Intensity does not correlate exactly with magnitude.

Distance from the epicenter, the nature of underlying strata, and the types of structures affected may all result in variations in intensity from the same earthquake. In general, structures built on bedrock will survive with much less damage than those constructed on poorly consolidated material. Displacement of ground along a fault can be substantial, for example, up to six meters along the San Andreas fault during the 1906 San Francisco earthquake. Such shifts can break pipelines and destroy roadways. Highly destructive surface waves can shake vulnerable structures apart.

The shaking and movement of ground are the most obvious means by which earthquakes cause damage. In addition to shaking it, earthquakes can cause the ground to rupture, subside, or rise. **Liquefaction** is an important phenomenon that occurs during earthquakes with ground that is poorly consolidated and in which the water table may be high. Liquefaction results from separation of soil particles accompanied by water infiltration such that the ground behaves like a fluid.

Another devastating phenomenon consists of **tsunamis**, large ocean waves resulting from earthquake-induced movement of ocean floor. Tsunamis sweeping onshore at speeds up to 1000 km/hr have destroyed many homes and taken many lives, often large distances from the epicenter of the earthquake, itself. This effect occurs when a tsunami approaches land and forms huge breakers, some as high as 10-15 meters, or even higher. On April 1, 1946, an earthquake off the coast of Alaska generated a Tsunami estimated to be more than 30 meters high that killed 5 people on a nearby lighthouse.

About 5 hours later a Tsunami generated by the same earthquake reached Hilo, Hawaii, and killed 159 people with a wave exceeding 15 meters high. The March 27, 1964, Alaska earthquake generated a tsunami over 10 meters high that hit a freighter docked at Valdez, tossing it around like matchwood.

Miraculously, nobody on the freighter was killed, but 28 people on the dock died. Literally millions of lives have been lost in past earthquakes, and damage from an earthquake in a developed urban area can easily run into billions of dollars. As examples, a massive earthquake in Egypt and Syria in 1201 A.D. took over 1 million lives, one in Tangshan, China, in 1976 killed about 650,000, and the 1989 Loma Prieta earthquake in California cost about 7 billion dollars. Significant progress has been made in designing structures that are earthquake resistant. As evidence of that, during a 1964 earthquake in Niigata, Japan, some buildings tipped over on their sides due to liquefaction of the underlying soil, but remained structurally intact! Other areas of endeavor that can lessen the impact of earthquakes is the identification of areas susceptible to earthquakes, discouraging development in such areas, and educating the public about earthquake hazards.

Accurate prediction would be a tremendous help in lessening the effects of earthquakes, but so far has been generally unsuccessful. Most challenging of all is the possibility of preventing major earthquakes. One unlikely possibility would be to detonate nuclear explosives deep underground along a fault line to release stress before it builds up to an excessive level. Fluid injection to facilitate slippage along a fault has also been considered.

Volkanoes

Many volcanic disasters have been recorded throughout history. Perhaps the best known of these is the 79 A.D. eruption of Mount Vesuvius, which buried the Roman city of Pompei with volcanic ash.

Temperatures of **lava**, molten rock flowing from a volcano, typically exceed 500°C and may get as high as 1400°C or more. Lava flows destroy everything in their paths, causing buildings and forests to burn and burying them under rock that cools and becomes solid. Often more dangerous than a lava flow are the **pyroclastics** produced by volcanoes and consisting of fragments of rock and lava.

Some of these particles are large and potentially very damaging, but they tend to fall quite close to the vent. Ash and dust may be carried for large distances and, in extreme cases, as was the case in ancient Pompei, may bury large areas to some depth with devastating effects. The explosion of Tambora volcano in Indonesia in 1815 blew out about 30 cubic kilometers of solid material. The ejection of so much solid into the atmosphere had such a devastating effect on global climate that the following year was known as "the year without a summer," causing widespread hardship and hunger because of global crop failures.

A special kind of particularly dangerous pyroclastic consists of **nuée ardente**. This term, French for "glowing cloud," refers to a dense mixture of hot toxic gases and fine ash particles reaching temperatures of 1000°C that can flow down the slopes of a volcano at speeds of up to 100 km/hr. In 1902 a nuée ardente was produced by the eruption of Mont Pelée on Martinique in the Caribbean. Of as many as 40,000 people in the town of St. Pierre, the only survivor was a terrified prisoner shielded from the intense heat by the dungeon in which he was imprisoned.

One of the more spectacular and potentially damaging volcanic phenomena is a **phreatic eruption** that occurs when infiltrating water is superheated by hot magma and causes a volcano to literally explode. This happened in 1883 when uninhabited Krakatoa in Indonesia blew up with an energy release of the order of 100 megatons of TNT. Dust was blown 80 kilometers into the stratosphere, and a perceptible climatic cooling was noted for the next 10 years. As is the case with earthquakes, volcanic eruptions may cause the devastating tsunamis. Krakatoa produced a tsunami 40 meters high that killed 30 to 40 thousand people on surrounding islands.

Some of the most damaging health and environmental effects of volcanic eruptions are caused by gases released to the atmosphere. Huge quantities of water vapor are often evolved. Dense carbon dioxide gas can suffocate people near the point of release. Highly toxic H2S and CO gases may be released by volcanoes.

Volcanoes tend to give off acid gases such as hydrogen chloride produced by the subduction and heating of sodium chloride entrained in ocean sediment. Sulfur oxides released by volcanoes may affect the atmosphere. In 1982 El Chichón erupted in Mexico, producing comparatively little dust but huge quantities of sulfur oxides.

These gases were converted to sulfuric acid droplets in the atmosphere, which reflected enough sunlight to cause a perceptible cooling in climate. Eventually the sulfuric acid released fell as acidic precipitation, "acid rain."

Volcanic activity could change the global environment dramatically. Massive volcanic eruptions many millions of years ago were probably responsible for widespread extinctions of organisms on earth's surface. These effects occur primarily by the ejection of particles and sulfuric acid precursors into the atmosphere causing global cooling and potential harm to the protective stratospheric ozone layer.

Although such an extinction event is unlikely in modern times, a volcanic eruption such as that of the Tambora volcano described above could certainly happen. With humankind "living on the edge" as far as grain supplies are concerned, widespread starvation resulting from a year or two of crop failures would almost certainly occur.

The Threat of Rising Sea Levels

Large numbers of people live at a level near, or in some cases below, sea level. As a result, any significant temporary or permanent rise in sea level poses significant risks to lives and property. Such an event occurred on February 1, 1953, when high tides and strong winds combined to breach the system of dikes protecting much of the Netherlands from seawater. About one sixth of the country was flooded as far inland as 64 kilometers from the coast, killing about 2000 people and leaving approximately 100,000 without homes.

Although isolated instances of flooding by seawater caused by combinations of tidal and weather phenomena will continue to occur, a much more long-lasting threat is posed by long-term increases in sea level. These could result from global warming due to the greenhouse gas emissions. Several factors could raise ocean levels to destructive highs, also a result of greenhouse warming. Simple expansion of warmed oceanic water could raise sea levels by about 1/3 meter over the next century. The melting of glaciers, such as those in the Alps, has probably raised ocean levels about 5 cm during the last century, and the process is continuing.

The greatest concern, however, is that global warming could cause the great West Antarctic ice sheet to melt, which would raise sea levels by as much as 6 meters. A great deal of uncertainty exists regarding the possibility of the West Antarctic ice sheet melting with consequent rises in sea level. Current computer models show a compensating effect in that hotter air produced by greenhouse warming could carry much more atmospheric moisture to the Antarctic regions where the moisture would be deposited as snow.

The net result could well be an *increase* in solid snow and ice in the Antarctic, and an accompanying *decrease* in sea levels. Some of the uncertainty regarding the status of the West Antarctic ice sheet may be alleviated in the future by highly accurate space satellite measurements. The measurement of sea levels has proven to be a difficult task because the levels of the surface of land keep changing. Land most recently covered with Ice Age glaciers in areas such as Scandinavia is still "springing back" from the immense mass of the glaciers, so that sea levels measured by gauges fixed on land actually appear to be dropping by several millimeters per year in such locations. An opposite situation exists on the east coast of North America where land was pushed outward and raised around the edge of the enormous sheet of ice that covered Canada and the northern U.S. about 20,000 years ago and is now settling back. Factors such as these illustrate the advantages of remarkably accurate satellite technology now used in the determination of sea levels.

Effects of ice

The power of ice to alter the geosphere is amply demonstrated by the remains of past glacial activity from the Ice Age. Those large areas of the earth's surface that were once covered with layers of glacial ice 1 or 2 kilometers in thickness show evidence of how the ice carved the surface, left massive piles of rock and gravel, and left rich deposits of fresh water. The enormous weight of glaciers on earth's surface compressed it, and in places it is still springing back 10,000 or so years after the glaciers retreated. Today the influence of ice on earth's surface is minimal, and there is substantial concern that melting of glaciers by greenhouse warming will raise sea levels so high that coastal areas will be inundated.

Glaciers form at sufficiently high latitudes and altitudes such that snow does not melt completely each summer. This occurs when snow becomes compacted over several to several thousand years such that the frozen water turns to crystals of true ice. Huge masses of ice with areas of several thousand square kilometers or more, and often around 1 kilometer thick, occur in polar regions and are called **continental glaciers**. Both Greenland and the Antarctic are covered by continental glaciers.

Alpine glaciers occupy mountain valleys. Glaciers on a slope flow as a consequence of their mass. This rate of flow is usually only a few meters per year, but may reach several kilometers per year. If a glacier flows into the sea, it may lose masses of ice as icebergs, a process called **calving**. Ice may also be lost by melting along the edges. The processes by which ice is lost is termed **ablation**.

Glacial ice affects the surface of the geosphere by both erosion and deposition. It is easy to imagine that a flowing mass of glacial ice is very efficient in scraping away the surface over which it flows, a process called **abrasion**. Adding to the erosive effect is the presence of rocks frozen into the glaciers which can act like tools to carve the surface of the underlying rock and soil. Whereas abrasion tends to wear rock surfaces away producing a fine rock powder, larger bits of rock can be dislodged from the surface over which the glacier flows and be carried along with the glacial ice.

When glacial ice melts, the rock that has been incorporated into it is left behind. This material is called **till**, or if it has been carried for some distance by water running off the melting glacier it is called **outwash**. Piles of rock left by melting glaciers produce unique structures called **moraines**.

Although the effects of glaciers described above are the most spectacular manifestations of the action of ice on the geosphere, at a much smaller level ice can have some very substantial effects. Freezing and expansion of water in pores and small crevices in rock is a major contributor to physical weathering processes. Freeze/thaw cycles are also very destructive to some kinds of structures, such as stone buildings.

2.14 SOLID WASTE

Municipal solid waste (MSW) is but one of many wastes. There are also construction and demolition debris, municipal sludge, combustion ash, mining and drilling debris, agricultural wastes, industrialprocess wastes including some sludge, hazardous waste, and others.

Altogether the United States alone produces about 10.1 *billion* tons of waste each year. MSW is barely 3% of that gigantic stream, but it is important: each of us is intimately associated with it. Its composition and amount is the result of our behavior. Moreover, if we move back one step, the businesses producing the products going into MSW generate much of those billions of pounds of waste.

Why do we generate so much waste? Why does society generate so much waste? Think about a metal. To obtain it, an ore is mined. However, the metal is but a small portion of even an enriched ore, so most of the mined material becomes waste. And the *overburden* -- soil disturbed during mining - becomes waste too, as does the disturbed rock. If these are not handled properly, they are damaging materials. Refining and smelting of the ore produce more waste; so does manufacturing a product from the refined metal, and the product's disposal or recycling. Or, consider the book in front of you. Starting in the forest, it produced waste when harvesting the wood to make it, more waste when the wood was transformed into paper, and yet more when the book was printed.

Think perhaps too about a head of lettuce. You discard its outer leaves. Before you bought it, other lettuce waste was produced by the farmer and the grocery store. A food waste such as lettuce is organic and can be composted. But often this does not happen. For a sense of the amount of waste that producing a product generates, look at chemical products. Recovering oil may pollute the surroundings of the well from which it was recovered and the oil-refining process also produces pollution. However, oil refining produces little waste -- as compared with the mining mentioned above - because almost all petroleum is refined into products, albeit mostly burnable products. Only a few per cent of the oil is used to manufacture chemicals. ''Bulk chemicals" are those produced in large quantities, such as benzene, and typically require only a few steps to make so, although the tons of waste produced per ton of product is much higher than in oil refining, it is still relatively low. Notice that fine chemicals and pharmaceuticals have a higher waste/product ratio.

This is because it takes more steps to make them, and each step produces waste. And what happens to a product, whether a chemical or a book, at the end of its useful life? Very often it becomes waste too. If recycled, that produces waste too. Wealthy societies produce much more waste per person, but less-developed countries also produce staggering amounts. Even though a poor person produces less waste per person, there are so many people and typically poor waste management, so discarded wastes can be a major problem.

Municipal solid waste

You may not notice your contributions to air pollution. You may likewise casually send the contaminated water you produce down drains and toilets. It's somewhat more difficult to ignore the solid waste that you produce. At a minimum, you must take it to a curbside. Historically, trash was the bane of many cities, as people dumped anything, including bodily wastes, onto streets. Solid waste still poses many problems.

What is in municipal solid waste and who produces it? Household waste contains anything that we choose to discard. That includes still-usable and recyclable items: waste food, papers and newspapers, packaging, bottles, metal cans, batteries, grass clippings and other yard waste, clothing, furniture and appliances, paint and other discarded household chemicals. The ''Other" category includes appliances (white goods), furniture, batteries, and household hazardous waste (HHW). HHW includes residual paints, oily products such as automotive-maintenance products, and pesticides. For many years paper has been the largest proportion of MSW in the United States, currently about 37% by weight. Some states consider construction and demolition debris to be MSW too, where it equals about 12% by weight. Some communities consider other wastes to be MSW too, sludge from drinkingwater-treatment and wastewater-treatment plants, septic-tank sludge, medical and slaughterhouse waste, and fast-food grease.

Households are only one contributor to MSW. Others generating MSW are institutions such as hospitals, government offices, schools, and prisons. Commercial businesses including restaurants, grocery stores, and offices are large generators too. And industries also generate MSW although industrial-*process* wastes are *not* MSW, and are handled separately.

Product	Time needed to degrade in the open environment
Glass bottle	1 000 000 years
Plastic six-pack holder	450 years
Aluminum can	200–400 years
Steel (tin) can	80–100 years
Rubber sole	50–80 years
Leather belt	40–50 years
Nylon fabric	30–40 years
Hard plastic container	20–30 years
Disposable diaper	10–20 years
Wooden stick	10–15 years
Cigarette butt	2–5 years
Hemp rope	1–14 months
Woolen cap	12 months
Cotton rag	1–5 months
Banana or orange peel	3–5 weeks
Paper	2–4 weeks

Why does MSW concern us? Almost any garbage is at best unpleasant. It has unpleasant odors. Diapers and other sanitary items contain microorganisms, sometimes infectious; so do rotting food and yard wastes. HHW is only a small part of MSW, but enough to sometimes catch on fire, or injure workers whose skin or eyes come into contact with it, or who breathe in its fumes. However, a material need not be hazardous to present problems. Consider packaging, generated in large quantities. The composition of packaging may make it very difficult to recycle, or prevent it from biodegrading. And if it contains even trace amounts of hazardous metals, expensive controls are necessary to prevent metal emissions if it is incinerated. Or, if disposed of in a landfill, much MSW can last indefinitely.

Even dumped in the open environment where there is sunlight, heat, and water, many items are long lived. Even if MSW had no other problematic characteristics, the quantity generated creates difficulties for communities. The United States alone, with its 100 million households, generated 232 million tons (210 million tonnes) of MSW in 2000. In earlier times most families lived in rural locations, and household waste was not a major problem.

But if every household today had to manage its own garbage, a crisis would quickly develop. The average American produces 4.3 lbs (2 kg) of MSW each day. For a household of four, this is 17.2 lbs (7.8 kg) per day or 3.2 tons (2.9 tonnes) per year. Handling waste is expensive whether it is incinerated or landfilled -- or recycled. Americans pay about $30 billion per year for waste management, second or third only to the amount communities spend on education and police protection.

Plastics

The amount of plastic discarded in the United States has increased a striking 10% every year for 20 years. Plastics are used for containers and many other products, including component parts of motor vehicles, airplanes, and computers. In 1960, a motor vehicle used only 40 lbs (18.1 kg) of plastic; this had increased to 300 lbs (136 kg) by 1992.

Recycling plastics is environmentally valuable, as demonstrated by the polyethylene terephthalate (PET) in soft-drink bottles. Recycling PET uses 47% less energy than making it from raw materials. Most common plastics are thermoplastics; that is, they can be melted down and reshaped many times into new plastic products.

Recycling plastics poses challenges. One is the difficulty of collecting and separating the many types of plastics. It's even more difficult to recover plastic from complicated products such as a computer's plastic casing. Another is that individual polymers are often incompatible when being processed into new products.

In 2002, New York City suspended glass and plastic recycling because recycling them didn't pay for itself. Used plastics are valuable, but businesses say they cannot be assured of a dependable supply. Moreover, the price that plastic recyclers receive varies with the price of oil, so there is no dependable price. Ten US states have bottle bills, laws mandating that deposits be paid on plastic, glass, and metal containers. Companies making containers vigorously resist extending bottle bills to other states although this would increase the dependability of a recycling supply. Meanwhile in the United States, about 4 billion PET soda bottles a year go into landfills or become litter.

The situation for recovering plastic casings from computers is worse and is deteriorating. About 315 million computers will have been discarded by 2004 containing about 2 million tons (1.8 million tonnes) of plastic.

Discarded plastic – once collected and sorted – is valuable and goes into products such as TyvekTM (disposable medical suits and insulation), fleece garments, plastic "lumber," and water-repellent coatings for shipping boxes, and asphalt too. But products made from recycled plastic are often more expensive. Plastic box coating costs \$2.50/lb (0.454 kg) compared with \$0.50/lb for wax coating. However, compared with other products, plastics may be more durable, perform better, or the product weighs less than its counterparts. For example, plastic "lumber" can last much longer than real wood. However, people resist the higher prices of recycled plastics.

Composting and ''biotreatment". Organic wastes such as grass, leaves, and other yard wastes can be composted. So can most food wastes. Composting is considered to be recycling. It is also a form of *biotreatment* because microorganisms degrade the organic materials to yield a soil-like product. Composting food wastes with worms is another form of biotreatment. In addition to composting, other examples of beneficially using biotreatment are seen in this text: microorganisms degrade organic matter in landfills, in secondary wastewater treatment, and can treat some hazardous wastes. More broadly, the degradation of wastes by microorganisms, worms, and other creatures is a major natural service.

Promoting recycling

Plastics contain a number of different resins, which, if mixed together, are useful in manufacturing only a few products. As technology develops to collect and sort plastics more efficiently and economically, each can be recycled to specific higher-value uses. New technologies may also find more uses for mixed plastics. *Economic incentives* can promote recycling, for example when government subsidizes businesses that use secondary (used) materials rather than virgin ones. *Promoting social and behavioral changes* can improve recycling.

Governments can: support recycling and waste-reduction education in schools; sponsor television, radio, or newspaper advertisements promoting recycling; and emphasize the important role of citizens in sorting trash from recyclables or buying products with recycled content. *Governmental mandates* may help to enhance recycling. They can: require government offices to give preference to environmentally preferable products when making purchases; and require citizens to pay deposits on beverage containers, tires, car batteries, furniture, or appliances at the time of purchase. A bottle bill to promote container return is an example of this; so is a law charging a tax on each new tire sold, the revenue is used to subsidize recycling.

Businesses making voluntary changes. In cooperation with the US EPA WasteWi$e program, L. L. Bean of Maine formed teams to reduce waste. Bean's set waste-reduction goals and an action schedule to meet them. As the name WasteWi$e implies, waste reduction can provide financial rewards.

The EPA has urged businesses to preferentially buy recycled products when making purchasing decisions, and manufacturers to include recycled content in their products.

Another voluntary action is to use DfE. Designers work to develop products that are easily disassembled into reusable or recyclable parts. They use fewer types of plastics in products, label plastics for easy identification, and use plastic in ways that allow their easy separation from other components at the end of the product's life.

It is harder to persuade businesses to design for durability and easy repair of a product, because they want to sell more not less. Another voluntary program in the United States is **take-back**. Sony and several other corporations have begun pilot take-back programs for consumer electronics including television and computer monitors. In Europe, there are mandatory take-back programs.

Recycling is not just about paper and cans

Organic wastes	Yard and kitchen waste can be composted in backyards or community composting programs. Properly treated sewage, paper-mill sludge, and animal waste can be used as a fertilizer or soil amendment.
Old clothing and textiles	In the nineteenth century, paper was made from cotton rags, and new clothing from discarded woolens. Today, clothes are recycled by selling them at yard sales or donating them to others. Old clothing is often exported to less-developed countries. Clothing that cannot be reused may be processed into rags or mixed into asphalt to make roof shingles.
Disposable diapers	A child may use 5000 diapers by the age of 30 months. A number of cities worldwide, including Toronto, recycle them. Wood pulp is extracted as are the super-absorbent polymers and plastics.
Grease	Fast-food restaurants discard billions of lbs per year of used cooking grease. This can be processed into ingredients of poultry and cattle feed, or used as lubricating oil. Another growing use is as a biodiesel fuel for motor vehicles.
Television sets and telephones	European television and glass manufacturers formed a consortium to recycle picture tubes from discarded television sets. Germany's largest telephone company has formed a consortium to recycle telephones.
Building materials	Wood and other materials from old buildings can be recovered, and often reused. Builders can incorporate materials into new construction – discarded tires, plastic

	bottles, or metal cans, but it may be difficult to obtain building permits. Incinerator ash can go into concrete, and old newspapers into insulation.
Toilets	California had a program to crush discarded toilets into an aggregate, which was added to concrete for road-building projects.
Telephone poles	Old telephone poles contain creosote and pentachlorophenol, wood preservatives that are now banned. These made recycling impossible until a *bioremediation* process was developed. Poles are chipped into small pieces and composted with microorganisms that can degrade creosote and pentachlorophenol. Clean wood fiber is then sold to paper mills.
Fluorescent lamps	Fluorescent lamps are more energy efficient than incandescent bulbs, and the United States uses at least a half a billion lamps yearly. They contain a glass tube, aluminum end caps, inert gases, phosphor powders, and 20 to 50 mg of mercury. Because of the mercury, many states ban them from MSW landfills. Businesses have developed techniques to recover and recycle the mercury, aluminum, and other components. Newer lamps contain much less mercury.
Radioactive materials	Without a long-term storage site, the nuclear power industry recycles metal from used radioactive pumps into new pumps used within the industry.
Old subway cars	These can be remanufactured into new cars. Those that cannot be reused can be submerged along coastal areas for use as artificial reefs.
Recycling as an automatic way of thinking	- San Francisco converted a former Air Force base, the Presidio, into a park; 70 acres (28.3 ha) of asphalt and concrete from its air field were crushed and reused beneath pathways and parking lots. - Curitiba, Brazil converted an 11-acre (4.5 ha) garbage dump into a botanical garden. It converted a derelict quarry into a Free University of the Environment, which holds courses for everyone on land-use and environmental issues.

2.15 SOIL AND AGRICULTURE

Soil and agriculture

Soil and agricultural practices are strongly tied with the environment. Some of these considerations are addressed later in this chapter along with a discussion of soil erosion and conservation. Cultivation of land and agricultural practices can influence both the atmosphere and the hydrosphere. Although this chapter deals primarily with soil, the topic of agriculture in general is introduced for perspective.

Agriculture, the production of food by growing crops and livestock, provides for the most basic of human needs. No other industry impacts as much as agriculture does on the environment. Agriculture is absolutely essential to the maintenance of the huge human populations now on earth. The displacement of native plants, destruction of wildlife habitat, erosion, pesticide pollution, and other environmental aspects of agriculture have enormous potential for environmental damage. Survival of humankind on earth demands that agricultural practice become as environmentally friendly as possible. On the other hand, growth of domestic crops removes (at least temporarily) greenhouse gas carbon dioxide from the atmosphere and provides potential sources of renewable resources of energy and fiber that can substitute for petroleum-derived fuels and materials.

Agriculture can be divided into the two main categories of **crop farming**, in which plant photosynthesis is used to produce grain, fruit, and fiber, and **livestock farming**, in which domesticated animals are grown for meat, milk, and other animal products. The major divisions of crop farming include production of cereals, such as wheat, corn or rice; animal fodder, such as hay; fruit; vegetables; and specialty crops, such as sugarcane, pineapple, sugar beets, tea, coffee, tobacco, cotton, and cacao. Livestock farming involves the raising of cattle, sheep, goats, swine, asses, mules, camels, buffalo, and various kinds of poultry. In addition to meat, livestock produce dairy products, eggs, wool, and hides. Freshwater fish and even crayfish are raised on "fish farms." Beekeeping provides honey.

Agriculture is based on domestic plants engineered by early farmers from their wild plant ancestors. Without perhaps much of an awareness of what they were doing, early farmers selected plants with desired characteristics for the production of food. This selection of plants for domestic use brought about a very rapid evolutionary change, so profound that the products often barely resemble their wild ancestors.

Plant breeding based on scientific principles of heredity is a very recent development dating from early in the present century. One of the major objectives of plant breeding has been to increase yield. An example of success in this area is the selection of dwarf varieties of rice, which yield much better and mature faster than the varieties that they replaced. Such rice was largely responsible for the "green revolution" dating from about the 1950s.

Yields of crops can also be increased by selecting for resistance to insects, drought, and cold. In some cases the goal is to increase nutritional value, such as in the development of corn high in lysine, an amino acid essential for human nutrition, such that corn becomes a more complete food. The development of hybrids has vastly increased yields and other desired characteristics of a number of important crops. Basically, **hybrids** are the offspring of crosses between two different **true-breeding** strains.

Often quite different from either parent strain, hybrids tend to exhibit "hybrid vigor" and to have significantly higher yields. The most success with hybrid crops has been obtained with corn (maize). Corn is one of the easiest plants to hybridize because of the physical separation of the male flowers, which grow as tassels on top of the corn plant, from female flowers, which are attached to incipient ears on the side of the plant. Despite past successes by more conventional means and some early disappointments with "genetic engineering," application of recombinant DNA technology will probably eventually overshadow all the advances ever made in plant breeding.

In addition to plant strains and varieties, numerous other factors are involved in crop production. Weather is an obvious factor, and shortages of water, chronic in many areas of the world, are mitigated by irrigation. Here, automated techniques and computer control are beginning to play an important, more environmentally friendly role by minimizing the quantities of water required. The application of chemical fertilizer has vastly increased crop yields. The judicious application of pesticides, especially herbicides, but including insecticides and fungicides as well, has increased crop yields and reduced losses greatly. Use of herbicides has had an environmental benefit in reducing the degree of mechanical cultivation of soil required. Indeed, "no-till" and "low-till" agriculture are now widely practiced on a large scale.

The crops that provide for most of human caloric food intake, as well as much food for animals, are **cereals**, which are harvested for their starch-rich seeds. In addition to corn, mentioned above, wheat used

for making bread and related foods, and rice consumed directly, other major cereal crops include barley, oats, rye, sorghum, and millet. As applied to agriculture and food, **vegetables** are plants or their products that can be eaten directly by humans. A large variety of different parts of plants are consumed as vegetables. These include leaves (lettuce), stems (asparagus), roots (carrots), tubers (potato), bulb (onion), immature flower (broccoli), immature fruit (cucumber), mature fruit (tomato), and seeds (pea). Fruits, which are bodies of plant tissue containing the seed, may be viewed as a subclassification of vegetables. Common fruits include apple, peach, apricot, citrus (orange, lemon, lime, grapefruit), banana, cherry, and various kinds of berries.
The rearing of domestic animals may have significant environmental effects. The Netherlands' pork industry has been so productive that hog manure and its byproducts have caused serious problems. Overflow from waste lagoons associated with hog production caused very damaging water pollution in Eastern North Carolina following the record rainfall from Hurricane Floyd in September 1999.

Goats and sheep have destroyed pastureland in the Near East, Northern Africa, Portugal, and Spain. Of particular concern are the environmental effects of raising cattle. Significant amounts of forest land have been converted to marginal pasture land to raise beef. Production of one pound of beef requires about four times as much water and four times as much feed as does production of one pound of chicken. An interesting aspect of the problem is emission of greenhouse-gas methane by anaerobic bacteria in the digestive systems of cattle and other ruminant animals; cattle rank right behind wetlands and rice paddies as producers of atmospheric methane. However, because of the action of specialized bacteria in their stomachs, cattle and other ruminant animals are capable of converting otherwise unusable cellulose to food.

Pesticides and Agriculture
Pesticides, particularly insecticides and herbicides, are an integral part of modern agricultural production. The problem has been especially severe for **minor-use pesticides** for which the market is not very large. In contrast to pesticides used on approximately 220 million acres of major crops in the U.S.—corn, soybeans, wheat, and cotton— minor-use pesticides are applied on only about 8 million acres of orchards, trees, ornamental plants, turf grass, fruits, nuts, and vegetables. Despite their limited use, about 40% of the monetary value of agricultural pesticides resides with minor-use pesticides. Complicating the reregistration process for many years was the wellintentioned but unrealistic 1958 Delaney Amendment to the Food, Drug, and Cosmetic Act which prohibited in food any chemical that had been shown to cause cancer in animals and humans. Many chemicals that are almost certainly safe as trace-level contaminants in food will, indeed, cause cancer when fed in massive doses to experimental animals. This effect has been attributed in part to the fact that huge doses of some chemicals destroy tissue, which the organism attempts to replace by growing new cells. Cells that are reproducing rapidly in an attempt to make up for tissue loss are more likely to become cancerous.

An interesting development regarding the use of herbicides in the late 1990s was the production of transgenic crops resistant to specific herbicides. The Monsanto company pioneered this approach with the development of "Roundup Ready" crops that resist the herbicidal effects of Monsanto's flagship Roundup® herbicide, a glyphosate compound with an annual market exceeding $2 billion. The seedlings of crops resistant to the herbicide are not harmed by exposure to it, whereas competing weeds are killed. In an interesting marketing strategy, Monsanto has licensed manufacturing and distribution rights for Roundup to other companies for application onto "Roundup Ready" crops sold by Monsanto.

Soil

Soil, consisting of a finely divided layer of weathered minerals and organic matter upon which plants grow, is the most fundamental requirement for agriculture. To humans and most terrestrial organsms, soil is the most important part of the geosphere. Though only a tissue-thin layer compared to the earth's total diameter, soil is the medium that produces most of the food required by most living things.

Good soil—and a climate conducive to its productivity—is the most valuable asset a nation can have. In addition to being the site of most food production, soil is the receptor of large quantities of pollutants, such as particulate matter from power plant smokestacks. Fertilizers, pesticides, and some other materials applied to soil often contribute to water and air pollution. Therefore, soil is a key component of environmental chemical cycles.

Soils are formed by the weathering of parent rocks as the result of interactive geological, hydrological, and biological processes. Soils are porous and are vertically stratified into horizons as the result of downward-percolating water and biological processes, including the production and decay of biomass. Soils are open systems that undergo continual exchange of matter and energy with the atmosphere, hydrosphere, and biosphere.

Nature and composition of soil

Soil is a variable mixture of minerals, organic matter, and water capable of supporting plant life on the earth's surface. 3 It is the final product of the weathering action of physical, chemical, and biological processes on rocks, which largely produces clay minerals. The organic portion of soil consists of plant biomass in various stages of decay. High populations of bacteria, fungi, and animals such as earthworms may be found in soil. Soil contains air spaces and generally has a loose texture.

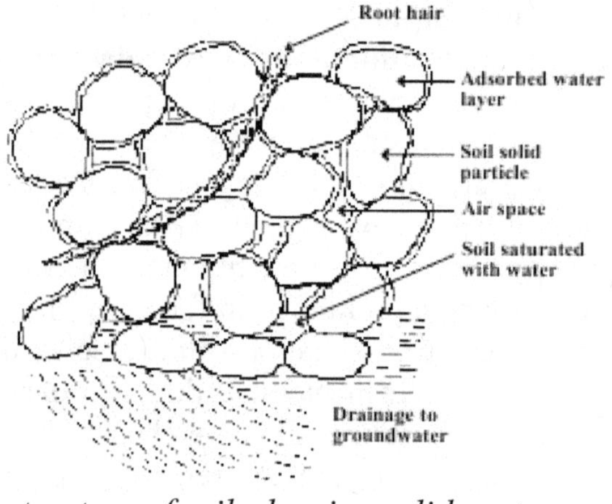

Fig. 141. *Fine structure of soil, showing solid, water, and air phases*

The solid fraction of typical productive soil is approximately 5% organic matter and 95% inorganic matter. Some soils, such as peat soils, may contain as much as 95% organic material. Other soils contain as little as 1% organic matter.

Typical soils exhibit distinctive layers with increasing depth. These layers are called **horizons**. Horizons form as the result of complex interactions among processes that occur during weathering. Rainwater percolating through soil carries dissolved and colloidal solids to lower horizons where they are deposited.

Biological processes, such as bacterial decay of residual plant biomass, produces slightly acidic CO_2, organic acids, and complexing compounds that are carried by rainwater to lower horizons where they interact with clays and other minerals, altering the properties of the minerals. The top layer of soil, typically several inches in thickness, is known as the A horizon, or **topsoil**. This is the layer of maximum biological activity in the soil and contains most of the soil organic matter. Metal ions and clay particles in the A horizon are subject to considerable leaching. The next layer is the B horizon, or **subsoil**. It receives material such as organic matter, salts, and clay particles leached from the topsoil. The C horizon is composed of weathered parent rocks from which the soil originated.

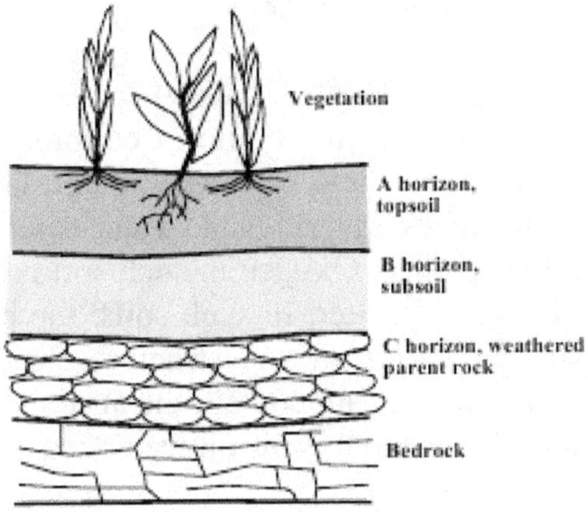

Fig. 142. *Soil profile showing soil horizons.*

Soils exhibit a large variety of characteristics that are used for their classification for various purposes, including crop production, road construction, and waste disposal. Soil profiles are discussed above. The parent rocks from which soils are formed obviously play a strong role in determining the composition of soils. Other soil characteristics include strength, workability, soil particle size, permeability, and degree of maturity. One of the more important classes of productive soils is the **podzol** type of soil formed under relatively high rainfall conditions in temperate zones of the world. These generally rich soils tend to be acidic (pH 3.5–4.5) such that alkali and alkaline earth metals and, to a lesser extent aluminum and iron, are leached from their A horizons, leaving kaolinite as the predominant clay mineral. At somewhat higher pH in the B horizons, hydrated iron oxides and clays are redeposited.

From the engineering standpoint, especially, the mechanical properties of soil are emphasized. These properties, which may have important environmental implications in areas such as waste disposal, are largely determined by particle size. According to the United Classification System (UCS), the four major categories of soil particle sizes are the following: Gravels (2–60 mm) > sands (0.06–2 mm) > silts (0.06-0.006 mm) > clays (less than 0.002 mm). In the UCS classification scheme clays represent a size fraction rather than a specific class of mineral matter.

Water and Air in Soil

Large quantities of water are required for the production of most plant materials. For example, several hundred kg of water are required to produce one kg of dry hay. Water is part of the three-phase, solid-liquid-gas system making up soil. It is the basic transport medium for carrying essential plant nutrients from solid soil particles into plant roots and to the farthest reaches of the plant's leaf structure.

The water enters the atmosphere from the plant's leaves, a process called **transpiration**. Normally, because of the small size of soil particles and the presence of small capillaries and pores in the soil, the water phase is not totally independent of soil solid matter. The availability of water to plants is governed by gradients arising from capillary and gravitational forces. The availability of nutrient solutes in water depends upon concentration gradients and electrical potential gradients. Water present in larger spaces in soil is relatively more available to plants and readily drains away. Water held in smaller pores or between the unit layers of clay particles is held much more strongly. Soils high in organic matter may hold appreciably more water than other soils, but it is relatively less available to plants because of physical and chemical sorption of the water by the organic matter.

There is a very strong interaction between clays and water in soil. Water is absorbed on the surfaces of clay particles. Because of the high surface/volume ratio of colloidal clay particles, a great deal of water may be bound in this manner. Water is also held between the unit layers of the expanding clays, such as the montmorillonite clays. As soil becomes waterlogged (water-saturated) it undergoes drastic changes in physical, chemical, and biological properties. Oxygen in such soil is rapidly used up by the respiration of microorganisms that degrade soil organic matter. In such soils, the bonds holding soil colloidal particles together are broken, which causes disruption of soil structure. Thus, the excess water in such soils is detrimental to plant growth, and the soil does not contain the air required by most plant roots. Most useful crops, with the notable exception of rice, cannot grow on waterlogged soils.

One of the most marked chemical effects of waterlogging is a reduction of pE by the action of organic reducing agents acting through bacterial catalysts. Thus, the redox condition of the soil becomes much more reducing, and the soil pE may drop from that of water in equilibrium with air (+13.6 at pH 7) to 1 or less. One of the more significant results of this change is the mobilization of iron and manganese as soluble iron(II) and manganese(II) through reduction of their insoluble higher oxides:

$MnO_2 + 4H^+ + 2e^- \rightarrow Mn^{2+} + 2H_2O$

$Fe_2O_3 + 6H^+ + 2e- \rightarrow 2Fe^{2+} + 3H_2O$

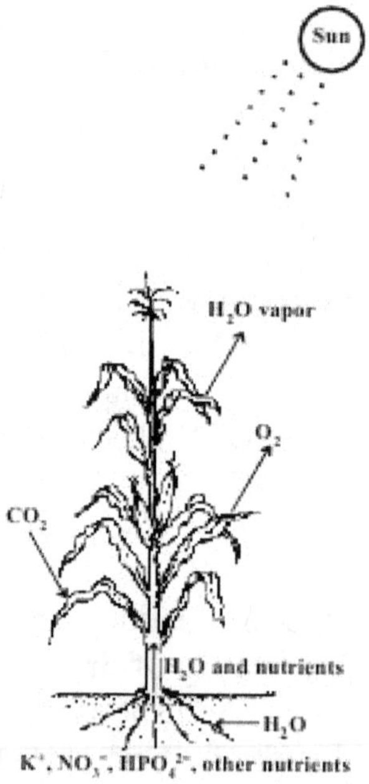

Fig. 143. *Plants transport water from the soil to the atmosphere by transpiration.*

Nutrients are also carried from the soil to the plant extremities by this process. Plants remove CO_2 from the atmosphere and add O_2 by photosynthesis. The reverse occurs during plant respiration.

Although soluble manganese generally is found in soil as Mn2+ ion, soluble iron(II) frequently occurs as negatively charged iron-organic chelates. Strong chelation of iron(II) by soil fulvic acids apparently enables reduction of iron(III) oxides at more positive pE values than would otherwise be possible.

Some soluble metal ions such as Fe^{2+} and Mn^{2+} are toxic to plants at high levels. Their oxidation to insoluble oxides may cause formation of deposits of Fe_2O_3 and MnO_2, which clog tile drains in fields.

Roughly 35% of the volume of typical soil is composed of air-filled pores. Whereas the normal dry atmosphere at sea level contains 21% O_2 and 0.03% CO_2 by volume, these percentages may be quite different in soil air because of the decay of organic matter:

$$\{CH_2O\} + O_2 = CO_2 + H_2O$$

This process consumes oxygen and produces CO_2. As a result, the oxygen content of air in soil may be as low as 15%, and the carbon dioxide content may be several percent. Thus, the decay of organic matter in soil increases the equilibrium level of dissolved CO_2 in groundwater. This lowers the pH and contributes to weathering of carbonate minerals, particularly calcium. CO_2 also shifts the equilibrium of the process by which roots absorb metal ions from soil.

The Inorganic Components of Soil

The weathering of parent rocks and minerals to form the inorganic soil components results ultimately in the formation of inorganic colloids. These colloids are repositories of water and plant nutrients, which may be made available to plants as needed. Inorganic soil colloids often absorb toxic substances in soil, thus playing a role in detoxification of substances that otherwise would harm plants. The abundance and nature of inorganic colloidal material in soil are obviously important factors in determining soil productivity.

The uptake of plant nutrients by roots often involves complex interactions with the water and inorganic phases. For example, a nutrient held by inorganic colloidal material has to traverse the mineral/water interface and then the water/root interface. This process is often strongly influenced by the ionic structure of soil inorganic matter.

The most common elements in the earth's crust are oxygen, silicon, aluminum, iron, calcium, sodium, potassium, and magnesium. Therefore, minerals composed of these elements — particularly silicon and oxygen — constitute most of the mineral fraction of the soil. Common soil mineral constituents are finely divided quartz (SiO2), orthoclase ($KAlSi_3O_8$), albite ($NaAlSi_3O_8$), epidote ($4CaO{\bullet}3(AlFe)_2O_3{\bullet}6SiO2{\bullet}H_2O$), geothite ($FeO(OH)$), magnetite ($Fe_3O_4$), calcium and magnesium carbonates ($CaCO_3$, $CaCO_3{\bullet}MgCO_3$), and oxides of manganese and titanium.

Organic Matter in Soil

Though typically comprising less than 5% of a productive soil, organic matter largely determines soil productivity. It serves as a source of food for microorganisms, undergoes chemical reactions such as ion exchange, and influences the physical properties of soil. Some organic compounds even contribute to the weathering of mineral matter, the process by which soil is formed.

For example, $C_2O_4^{2-}$, oxalate ion, produced as a soil fungi metabolite, occurs in soil as the calcium salts whewellite and weddelite. Oxalate in soil water dissolves minerals, thus speeding the weathering process and increasing the availability of nutrient ion species. This weathering process involves oxalate complexation of iron or aluminum in minerals, represented by the reaction

$3H^+ + M(OH)_3(s) + 2CaC_2O_4(s) \rightarrow M(C_2O4)^{2-}(aq) + 2Ca^{2+}(aq) + 3H_2O$

in which M is Al or Fe. Some soil fungi produce citric acid and other chelating organic acids which react with silicate minerals and release potassium and other nutrient metal ions held by these minerals.

The strong chelating agent 2-ketogluconic acid is produced by some soil bacteria. By solubilizing metal ions, it may contribute to the weathering of minerals. It may also be involved in the release of phosphate from insoluble phosphate compounds.

Biologically active components of the organic soil fraction include polysaccharides, amino sugars, nucleotides, and organic sulfur and phosphorus compounds. Humus, a water-insoluble material that biodegrades very slowly, makes up the bulk of soil organic matter.

The accumulation of organic matter in soil is strongly influenced by temperature and by the availability of oxygen. Since the rate of biodegradation decreases with decreasing temperature, organic matter does not degrade rapidly in colder climates and tends to build up in soil. In water and in waterlogged soils, decaying vegetation does not have easy access to oxygen, and organic matter accumulates. The organic content may reach 90% in areas where plants grow and decay in soil saturated with water.

The presence of naturally occurring polynuclear aromatic (PAH) compounds is an interesting feature of soil organic matter. These compounds, some of which are carcinogenic, are discussed as air pollutants in lections 3-4. PAH compounds found in soil include fluoranthene, pyrene, and chrysene. PAH compounds in soil result in part from combustion from both natural sources (grass fires) or pollutant sources. Terpenes also occur in soil organic matter. Extraction of soil with ether and alcohol yields the pigments b-carotene, chlorophyll, and xanthophyll.

Major Classes of Organic Compounds in Soil

Compound type	Composition	Significance
Humus	Degradation-resistant residue from plant decay, largely C, H, and O	Most abundant organic component, improves soil physical properties, exchanges nutrients, reservoir of fixed N
Fats, resins, and waxes	Lipids extractable by organic solvents	Generally, only several percent of soil organic matter, may adversely affect soil physical properties by repelling water, perhaps phytotoxic
Saccharides	Cellulose, starches, hemicellulose, gums	Major food source for soil microorganisms, help stabilize soil aggregates
N-containing organics	Nitrogen bound to humus, amino acids, amino sugars, other compounds	Provide nitrogen for soil fertility
Phosphorus compounds	Phosphate esters, inositol phosphates (phytic acid), phospholipids	Sources of plant phosphate

Soil Humus

Soil humus is by far the most significant. Humus, composed of a base-soluble fraction called humic and fulvic acids, and an insoluble fraction called humin, is the residue left when bacteria and fungi biodegrade plant material. The bulk of plant biomass consists of relatively degradable cellulose and degradation-resistant lignin, which is a polymeric substance with a higher carbon content than cellulose. Among lignin's prominent chemical components are aromatic rings connected by alkyl chains, methoxyl groups, and hydroxyl groups. These structural artifacts occur in soil humus and give it many of its characteristic properties.

The process by which humus is formed is called **humification**. Soil humus is similar to its lignin precursors, but has more carboxylic acid groups. Part of each molecule of humic substance is nonpolar and hydrophobic, and part is polar and hydrophilic. Such molecules are called **amphiphiles**, and they form micelles in which the nonpolar parts compose the inside of small colloidal particles and the polar functional groups are on the outside. Amphiphilic humic substances probably also form bilayer surface coatings on mineral grains in soil.
An increase in nitrogen/carbon ratio is a significant feature of the transformation of plant biomass to humus through the humification process. This ratio starts at approximately 1/100 in fresh plant biomass. During humification, microorganisms convert organic carbon to CO_2 to obtain energy. Simultaneously, the bacterial action incorporates bound nitrogen with the compounds produced by the decay processes.

The result is a nitrogen/carbon ratio of about 1/10 upon completion of humification. As a general rule, therefore, humus is relatively rich in organically bound nitrogen. Humic substances influence soil properties to a degree out of proportion to their small percentage in soil. They strongly bind metals, and serve to hold micronutrient metal ions in soil. Because of their acid-base character, humic substances serve as buffers in soil. The water-holding capacity of soil is significantly increased by humic substances. These materials also stabilize aggregates of soil particles, and increase the sorption of organic compounds by soil.

Humic materials in soil strongly sorb many solutes in soil water and have a particular affinity for heavy polyvalent cations. Soil humic substances may contain levels of uranium more than 104 times that of the water with which they are in equilibrium. Thus, water becomes depleted of its cations (or purified) in passing through humic-rich soils. Humic substances in soils also have a strong affinity for organic compounds with low water-solubility such as DDT or Atrazine, a herbicide widely used to kill weeds in corn fields. In some cases, there is a strong interaction between the organic and inorganic portions of soil. This is especially true of the strong complexes formed between clays and humic (fulvic) acid compounds. In many soils, 50-100% of soil carbon is complexed with clay.

These complexes play a role in determining the physical properties of soil, soil fertility, and stabilization of soil organic matter. One of the mechanisms for the chemical binding between clay colloidal particles and humic organic particles is probably of the flocculation type in which anionic organic molecules with carboxylic acid functional groups serve as bridges in combination with cations to bind clay colloidal particles together as a floc. Support is given to this hypothesis by the known ability of NH_4^+, Al^{3+}, Ca^{2+}, and Fe^{3+} cations to stimulate clay-organic complex formation. The synthesis, chemical reactions, and biodegradation of humic materials are affected by interaction with clays. The lower-molecular-weight fulvic acids may be bound to clay, occupying spaces in layers in the clay.

Macronutrients in soil

One of the most important functions of soil in supporting plant growth is to provide essential plant nutrients—macronutrients and micronutrients. Macronutrients are those elements that occur in substantial levels in plant materials or in fluids in the plant. Micronutrients are elements that are essential only at very low levels and generally are required for the functioning of essential enzymes.

The elements generally recognized as essential macronutrients for plants are carbon, hydrogen, oxygen, nitrogen, phosphorus, potassium, calcium, magnesium, and sulfur. Carbon, hydrogen, and oxygen are obtained from the atmosphere. The other essential macronutrients must be obtained from soil. Of these, nitrogen, phosphorus, and potassium are the most likely to be lacking and are commonly added to soil as fertilizers.

Calcium-deficient soils are relatively uncommon. Application of lime, a process used to treat acid soils, provides a more than adequate calcium supply for plants. However, calcium uptake by plants and leaching by carbonic acid may produce a calcium deficiency in soil. Acid soils may still contain an appreciable level of calcium which, because of competition by hydrogen ion, is not available to plants. Treatment of acid soil to restore the pH to nearneutrality generally remedies the calcium deficiency. In alkaline soils, the presence of high levels of sodium, magnesium, and potassium sometimes produces calcium deficiency because these ions compete with calcium for availability to plants.

Most of the 2.1% of magnesium in earth's crust is rather strongly bound in minerals. Exchangeable magnesium held by ion-exchanging organic matter or clays is considered available to plants. The availability of magnesium to plants depends upon the calcium/magnesium ratio. If this ratio is too high, magnesium may not be available to plants and magnesium deficiency results. Similarly, excessive levels of potassium or sodium may cause magnesium deficiency.

Sulfur is assimilated by plants as the sulfate ion, SO_4^{2-}. In addition, in areas where the atmosphere is contaminated with SO_2, sulfur may be absorbed as sulfur dioxide by plant leaves. Atmospheric sulfur dioxide levels have been high enough to kill vegetation in some areas. However, some experiments designed to show SO2 toxicity to plants have resulted in increased plant growth where there was an unexpected sulfur deficiency in the soil used for the experiment.

Soils deficient in sulfur do not support plant growth well, largely because sulfur is a component of some essential amino acids and of thiamin and biotin. Sulfate ion is generally present in the soil as immobilized insoluble sulfate minerals, or as soluble salts which are readily leached from the soil and lost as soil water runoff.

Unlike the case of nutrient cations such as K^+, little sulfate is adsorbed to the soil (that is, bound by ion exchange binding) where it is resistant to leaching while still available for assimilation by plant roots.

Soil sulfur deficiencies have been found in a number of regions of the world. Whereas most fertilizers used to contain sulfur, its use in commercial fertilizers has declined. With continued use of sulfur-deficient fertilizers, it is possible that sulfur will become a limiting nutrient in more cases.

The reaction of FeS_2 with acid in acid-sulfate soils may release H_2S, which is very toxic to plants and which also kills many beneficial microorganisms. Toxic hydrogen sulfide can also be produced by reduction of sulfate ion through microorganism-mediated reactions with organic matter. Production of hydrogen sulfide in flooded soils may be inhibited by treatment with oxidizing compounds, one of the most effective of which is KNO_3.

Nitrogen, phosphorus and potassium in soil

Nitrogen, phosphorus, and potassium are plant nutrients that are obtained from soil. They are so important for crop productivity that they are commonly added to soil as fertilizers.

Nitrogen

Figure below summarizes the primary sinks and pathways of nitrogen in soil. In most soils, over 90% of the nitrogen content is organic. This organic nitrogen is primarily the product of the biodegradation of dead plants and animals. It is eventually hydrolyzed to NH4+, which can be oxidized to NO_3^- by the action of bacteria in the soil.

Nitrogen bound to soil humus is especially important in maintaining soil fertility. Unlike potassium or phosphate, nitrogen is not a significant product of mineral weathering. Nitrogen-fixing organisms ordinarily cannot supply sufficient nitrogen to meet peak demand. Inorganic nitrogen from fertilizers and rainwater is often largely lost by leaching. Soil humus, however, serves as a reservoir of nitrogen required by plants. It has the additional advantage that its rate of decay, hence its rate of nitrogen release to plants, roughly parallels plant growth—rapid during the warm growing season, slow during the winter months.

Nitrogen is an essential component of proteins and other constituents of living matter. Plants and cereals grown on nitrogen-rich soils not only provide higher yields, but are often substantially richer in protein and, therefore, more nutritious. Nitrogen is most generally available to plants as nitrate ion, NO_3^-. Some plants such as rice may utilize ammonium nitrogen; however, other plants are poisoned by this form of nitrogen. When nitrogen is applied to soils in the ammonium form, nitrifying bacteria perform an essential function in converting it to available nitrate ion.

Plants may absorb excessive amounts of nitrate nitrogen from soil. This phenomenon occurs particularly in heavily fertilized soils under drought conditions. Forage crops containing excessive amounts of nitrate can poison ruminant animals such as cattle or sheep. Plants having excessive levels of nitrate can endanger people when used for ensilage, an animal food consisting of finely chopped plant material such as partially matured whole corn plants, fermented in a structure called a silo. Under the reducing conditions of fermentation, nitrate in ensilage may be reduced to toxic NO_2 gas, which can accumulate to high levels in enclosed silos. There have been many cases reported of persons being killed by accumulated NO_2 in silos.

Nitrogen fixation is the process by which atmospheric N_2 is converted to nitrogen compounds available to plants. Human activities are resulting in the fixation of a great deal more nitrogen than would otherwise be the case. Artificial sources now account for 30-40% of all nitrogen fixed. These include chemical fertilizer manufacture, nitrogen fixed during fuel combustion, combustion of nitrogencontaining fuels, and the increased cultivation of nitrogen-fixing. A major concern with this increased fixation of nitrogen is the possible effect upon the atmospheric ozone layer by N_2O released during denitrification of fixed nitrogen.

Before the widespread introduction of nitrogen fertilizers, soil nitrogen was provided primarily by legumes. These are plants such as soybeans, alfalfa, and clover, which contain on their root structures bacteria capable of fixing atmospheric nitrogen. Leguminous plants have a symbiotic (mutually advantageous) relationship with the bacteria that provide their nitrogen. Legumes may add significant quantities of nitrogen to soil, up to 10 pounds per acre per year, which is comparable to amounts commonly added as synthetic fertilizers. Soil fertility with respect to nitrogen may be maintained by rotating plantings of nitrogen-consuming plants with plantings of legumes, a fact recognized by agriculturists as far back as the Roman era.

The nitrogen-fixing bacteria in legumes exist in special structures on the roots called root nodules. The rod-shaped bacteria that fix nitrogen are members of a special genus, *Rhizobium*. These bacteria may exist independently, but cannot fix nitrogen except in symbiotic combination with plants. Although all species of *Rhizobium* appear to be very similar, they exhibit a great deal of specificity in their choice of host plants. Curiously, legume root nodules also contain a form of hemoglobin, which must somehow be involved in the nitrogen-fixation process.

Nitrate pollution of some surface waters and groundwater has become a major problem in some agricultural areas. Although fertilizers have been implicated in such pollution, there is evidence that feedlots are a major source of nitrate pollution. The growth of livestock populations and the concentration of livestock in feedlots have aggravated the problem. Such concentrations of cattle, coupled with the fact that a steer produces approximately 18 times as much waste material as a human, have resulted in high levels of water pollution in rural areas with small human populations. Streams and reservoirs in such areas frequently are just as polluted as those in densely populated and highly industrialized areas.

Nitrate in farm wells is a common and especially damaging manifestation of nitrogen pollution from feedlots because of the susceptibility of ruminant animals to nitrate poisoning. The stomach contents of ruminant animals such as cattle and sheep constitute a reducing medium (low pE) and contain bacteria capable of reducing nitrate ion to toxic nitrite ion:

$$NO_3^- + 2H^+ + 2e^- \rightarrow NO_2^- + H_2O$$

The origin of most nitrate produced from feedlot wastes is amino nitrogen present in nitrogen-containing waste products. Approximately one-half of the nitrogen excreted by cattle is contained in the urine. Part of this nitrogen is proteinaceous and the other part is in the form of urea, NH_2CONH_2. As a first step in the degradation process, the amino nitrogen is probably hydrolyzed to ammonia, or ammonium ion:
$RNH_2 + H_2O = R\text{-}OH + NH_3\ (NH_4^+)$

This product is then oxidized through microorganism-catalyzed reactions to nitrate
ion: $NH_3 + 2O_2 \rightarrow H^+ + NO_3^- + H_2O$

Under some conditions, an appreciable amount of the nitrogen originating from the degradation of feedlot wastes is present as ammonium ion. Ammonium ion is rather strongly bound to soil (recall that soil is a generally good cation exchanger), and a small fraction is fixed as nonexchangeable ammonium ion in the crystal lattice of clay minerals. Because nitrate ion is not strongly bound to soil, it is readily carried through soil formations by water. Many factors, including soil type, moisture, and level of organic matter, affect the production of ammonia and nitrate ion originating from feedlot wastes, and a marked variation is found in the levels and distributions of these materials in feedlot areas.

Phosphorus

Although the percentage of phosphorus in plant material is relatively low, it is an essential component of plants. Phosphorus, like nitrogen, must be present in a simple inorganic form before it can be taken up by plants. In the case of phosphorus, the utilizable species is some form of orthophosphate ion. In the pH range that is present in most soils, H2PO4- and HPO42- are the predominant orthophosphate species.

Orthophosphate is most available to plants at pH values near neutrality. It is believed that in relatively acidic soils, orthophosphate ions are precipitated or sorbed by species of Al(III) and Fe(III). In alkaline soils, orthophosphate may react with calcium carbonate to form relatively insoluble hydroxyapatite:
$3HPO_4^{2-} + 5CaCO_3(s) + 2H_2O = Ca_5(PO_4)_3(OH)(s) + 5HCO_3^- + OH^-$

In general, because of these reactions, little phosphorus applied as fertilizer leaches from the soil. This is important from the standpoint of both water pollution and utilization of phosphate fertilizers.

Potassium

Relatively high levels of potassium are utilized by growing plants. Potassium activates some enzymes and plays a role in the water balance in plants. It is also essential for some carbohydrate transformations. Crop yields are generally greatly reduced in potassium-deficient soils. The higher the productivity of the crop, the more potassium is removed from soil. When nitrogen fertilizers are added to soils to increase productivity, removal of potassium is enhanced. Therefore, potassium may become a limiting nutrient in soils heavily fertilized with other nutrients.

Potassium is one of the most abundant elements in the earth's crust, of which it makes up 2.6%; however, much of this potassium is not easily available to plants. For example, some silicate minerals such as leucite, $K2O \cdot Al2O3 \cdot 4SiO_2$, contain strongly bound potassium. Exchangeable potassium held by clay minerals is relatively more available to plants.

Micronutrients in soil

Boron, chlorine, copper, iron, manganese, molybdenum (for N-fixation), and zinc are considered essential plant **micronutrients**. These elements are needed by plants only at very low levels and frequently are toxic at higher levels. There is some chance that other elements will be added to this list as techniques for growing plants in environments free of specific elements improve. Most of these elements function as components of essential enzymes. Manganese, iron, chlorine, and zinc may be involved in photosynthesis.Though not established for all plants, it is possible that sodium, silicon, and cobalt may also be essential plant nutrients.

Iron and manganese occur in a number of soil minerals. Sodium and chlorine (as chloride) occur naturally in soil and are transported as atmospheric particulate matter from marine sprays. Some of the other micronutrients and trace elements are found in primary (unweathered) minerals that occur in soil. Boron is substituted isomorphically for Si in some micas and is present in tourmaline, a mineral with the formula $NaMg_3Al_6B_3Si_6O_27(OH,F)_4$. Copper is isomorphically substituted for other elements in feldspars, amphiboles, olivines, pyroxenes, and micas; it also occurs as trace levels of copper sulfides in silicate minerals.

Molybdenum occurs as molybdenite (MoS_2). Vanadium is isomorphically substituted for Fe or Al in oxides, pyroxenes, amphiboles, and micas. Zinc is present as the result of isomorphic substitution for Mg, Fe, and Mn in oxides, amphiboles, olivines, and pyroxenes and as trace zinc sulfide in silicates. Other trace elements that occur as specific minerals, sulfide inclusions, or by isomorphic substitution for other elements in minerals are chromium, cobalt, arsenic, selenium, nickel, lead, and cadmium.

The trace elements listed above may be coprecipitated with secondary minerals that are involved in soil formation. Such secondary minerals include oxides of aluminum, iron, and manganese (precipitation of hydrated oxides of iron and manganese very efficiently removes many trace metal ions from solution); calcium and magnesium carbonates; smectites; vermiculites; and illites.

Some plants accumulate extremely high levels of specific trace metals. Those accumulating more than 1.00 mg/g of dry weight are called **hyperaccumulators**. Nickel and copper both undergo hyperaccumulation in some plant species. As an example of a metal hyperaccumulater, *Aeolanthus biformifolius DeWild* growing in copper-rich regions of Shaba Province, Zaire, contains up to 1.3% copper (dry weight) and is known as a "copper flower".

The hyperaccumulation of metals by some plants has led to the idea of **phytoremediation** in which plants growing on contaminated ground accumulate metals, which are then removed with the plant biomass. *Brassica juncea* and *Brassica chinensis* (Chinese cabbage) have been shown to hyperaccumulate as much as 5 grams of uranium per kg plant dry weight when grown on uranium-contaminated soil. Uranium accumulation in the plants was enhanced by the addition of citrate, which complexes uranium and makes it more soluble.

Fertilizers

Crop fertilizers contain nitrogen, phosphorus, and potassium as major components. Magnesium, sulfate, and micronutrients may also be added. Fertilizers are designated by numbers, such as 6-12-8, showing the respective percentages of nitrogen expressed as N (in this case 6%), phosphorus as P_2O_5 (12%), and potassium as K_2O (8%). Farm manure corresponds to an approximately 0.5-0.24-0.5 fertilizer.

The organic fertilizers such as manure must undergo biodegradation to release the simple inorganic species (NO_3^-, $HxPO_{4x}{}^{-3}$, K^+) assimilable by plants.

Most modern nitrogen fertilizers are made by the Haber process, in which N2 and H2 are combined over a catalyst at temperatures of approximately 500°C and pressures up to 1000 atm: $N_2 + 3H_2 = 2NH_3$

The anhydrous ammonia product has a very high nitrogen content of 82%. It may be
added directly to the soil, for which it has a strong affinity because of its water solubility and formation of ammonium ion:

$NH_3(g)$ (water) = $NH_3(aq)$

$NH_3(aq) + H_2O = NH_4^+ + OH-$

Special equipment is required, however, because of the toxicity of ammonia gas. Aqua ammonia, a 30% solution of NH_3 in water, may be used with much greater safety. It is sometimes added directly to irrigation water. It should be pointed out that ammonia vapor is toxic and NH3 is reactive with some substances. Improperly discarded or stored ammonia can be a hazardous waste.

Ammonium nitrate, NH_4NO_3, is a common solid nitrogen fertilizer. It is made by oxidizing ammonia over a platinum catalyst, converting the nitric oxide product to nitric acid, and reacting the nitric acid with ammonia. The molten ammonium nitrate product is forced through nozzles at the top of a *prilling tower* and solidifies to form small pellets while falling through the tower. The particles are coated with a water repellent. Ammonium nitrate contains 33.5% nitrogen.

Although convenient to apply to soil, it requires considerable care during manufacture and storage because it is explosive. Ammonium nitrate also poses some hazards. It is mixed with fuel oil to form an explosive that serves as a substitute for dynamite in quarry blasting and construction. This mixture was used to devastating effect in the dastardly bombing of the Oklahoma City Federal Building in 1995.
Urea is easier to manufacture and handle than ammonium nitrate. It is now the favored solid nitrogen-containing fertilizer. The overall reaction for urea synthesis is

$CO_2 + 2NH_3 = CO(NH_2)_2 + H_2O$ involving a rather complicated process in which ammonium carbamate, chemical formula $NH_2CO_2NH_4$, is an intermediate.

Other compounds used as nitrogen fertilizers include sodium nitrate (obtained largely from Chilean deposits), calcium nitrate, potassium nitrate, and ammonium phosphates. Ammonium sulfate, a by-product of coke ovens, used to be widely applied as fertilizer. The alkali metal nitrates tend to make soil alkaline, whereas ammonium sulfate leaves an acidic residue.

The principal mineral is fluorapatite, $Ca_5(PO_4)_3F$. The phosphate from fluorapatite is relatively unavailable to plants, and fluorapatite is frequently treated with phosphoric or
sulfuric acids to produce superphosphates:

$2Ca_5(PO_4)_3F(s) + 14H_3PO_4 + 10H_2O = 2HF(g) + 10Ca(H_2PO_4)_2{\bullet}H2O$
$2Ca_5(PO_4)_3F(s) + 7H_2SO_4 + 3H_2O = 2HF(g) + 3Ca(H_2PO_4)_2{\bullet}H_2O + 7CaSO_4$

The superphosphate products are much more soluble than the parent phosphate minerals. The HF produced as a byproduct of superphosphate production can create air pollution problems.

Phosphate minerals are rich in trace elements required for plant growth, such as boron, copper, manganese, molybdenum, and zinc. Ironically, these elements are lost to a large extent when the phosphate minerals are processed to make fertilizer.

Ammonium phosphates are excellent, highly soluble phosphate fertilizers. Liquid ammonium polyphosphate fertilizers consisting of ammonium salts of pyrophosphate, triphosphate, and small quantities of higher polymeric phosphate anions in aqueous solution work very well as phosphate fertilizers. The polyphosphates are believed to have the additional advantage of chelating iron and other micronutrient metal ions, thus making the metals more available to plants.

Potassium fertilizer components consist of potassium salts, generally KCl. Such salts are found as deposits in the ground or may be obtained from some brines. Very large deposits are found in Saskatchewan, Canada. These salts are all quite soluble in water. One problem encountered with potassium fertilizers is the luxury uptake of potassium by some crops, which absorb more potassium than is really needed for their maximum growth. In a crop where only the grain is harvested, leaving the rest of the plant in the field, luxury uptake does not create much of a problem because most of the potassium is returned to the soil with the dead plant. However, when hay or forage is harvested, potassium contained in the plant as a consequence of luxury uptake is lost from the soil.

Wastes and pollutants in soil

Soil receives large quantities of waste products. Much of the sulfur dioxide emitted in the burning of sulfur-containing fuels ends up as soil sulfate. Atmospheric nitrogen oxides are converted to nitrates in the atmosphere, and the nitrates eventually are deposited on soil. Soil sorbs NO and NO_2, and these gases are oxidized to nitrate in the soil. Carbon monoxide is converted to CO_2 and possibly to biomass by soil bacteria and fungi. Particulate lead from automobile exhausts is found at elevated levels in soil along heavily traveled highways. Elevated levels of lead from lead mines and smelters are found on soil near such facilities.

Soil is the receptor of many hazardous wastes from landfill leachate, lagoons, and other sources. In some cases, land farming of degradable hazardous organic wastes is practiced as a means of disposal and degradation. The degradable material is worked into the soil, and soil microbial processes bring about its degradation.

Volatile organic compounds (VOC) such as benzene, toluene, xylenes, dichloromethane, trichloroethane, and trichloroethylene, may contaminate soil in industrialized and commercialized areas, particularly in countries in which enforcement of regulations is not very stringent. One of the more common sources of these contaminants is leaking underground storage tanks. Landfills built before current stringent regulations were enforced and improperly discarded solvents are also significant sources of soil VOCs.

Measurements of levels of polychlorinated biphenyls (PCBs) in soils that have been archived for several decades provide interesting insight into the contamination of soil by pollutant chemicals and subsequent loss of these substances from soil. Analyses of soils from the United Kingdom dating from the early 1940s to 1992 showed that the PCB levels increased sharply from the 1940s, reaching peak levels around 1970. Subsequently, levels fell sharply and now are back to early 1940s concentrations.

This fall was accompanied by a shift in distribution to the more highly chlorinated PCBs, which was attributed by those doing the study to volatilization and long range transport of the lighter PCBs away from the soil. These trends parallel levels of PCB manufacture and use in the United Kingdom from the early 1940s to the present. This is consistent with the observation that relatively high concentrations of PCBs have been observed in remote Arctic and sub-Arctic regions, attributed to condensation in colder climates of PCBs volatilized in warmer regions.

Some pollutant organic compounds are believed to become bound with humus during the humification process that occurs in soil. This largely immobilizes and detoxifies the compounds. Binding of pollutant compounds by humus is particularly likely to occur with compounds that have structural similarities to humic substances, such as phenolic and anilinic compounds.

Such compounds can become covalently bonded to humic substance molecules, largely through the action of microbial enzymes. After binding they are known as **bound residues** and are highly resistant to extraction with solvents by procedures that would remove unbound parent compounds. Compounds in the bound residues are resistant to biological and chemical attack. Soil receives enormous quantities of pesticides as an inevitable result of their application to crops.

The degradation and eventual fate of these pesticides on soil largely determines their ultimate environmental effects. Detailed knowledge of these effects are now required for licensing of a new pesticide. Among the factors to be considered are the sorption of the pesticide by soil; leaching of the pesticide into water, as related to its potential for water pollution; effects of the pesticide on microorganisms and animal life in the soil; and possible production of relatively more toxic degradation products.

Adsorption by soil is a key aspect of pesticide degradation and plays a strong role in the speed and degree of degradation. The degree of adsorption and the speed and extent of ultimate degradation are influenced by a number of other factors. Some of these, including solubility, volatility, charge, polarity, and molecular structure and size, are properties of the medium. Adsorption of a pesticide by soil components may have several effects. Under some circumstances, it retards degradation by separating the pesticide from the microbial enzymes that degrade it, whereas under other circumstances the reverse is true. Purely chemical degradation reactions may be catalyzed by adsorption. Loss of the pesticide by volatilization or leaching is diminished. The toxicity of a herbicide to plants may be reduced by sorption on soil.

The forces holding a pesticide to soil particles may be of several types. Physical adsorption involves van der Waals forces arising from dipole-dipole interactions between the pesticide molecule and charged soil particles. Ion exchange is especially effective in holding cationic organic compounds, such as the herbicide paraquat, to anionic soil particles. Some neutral pesticides become cationic by protonation and are bound as the protonated positive form. Hydrogen bonding is another mechanism by which some pesticides are held to soil. In some cases a pesticide may act as a ligand coordinating to metals in soil mineral matter.

Soil loss and degradation

Soil is a fragile resource that can be lost by erosion or become so degraded that it is no longer useful to support crops. The physical properties of soil and, hence, its susceptibility to erosion, are strongly affected by the cultivation practices to which the soil is subjected. **Desertification** refers to the process associated with drought and loss of fertility by which soil becomes unable to grow significant amounts of plant life. Desertification caused by human activities is a common problem globally, occurring in diverse locations such as Argentina, the Sahara, Uzbekistan, the U.S. Southwest, Syria, and Mali.

It is a very old problem dating back many centuries to the introduction of domesticated grazing animals to areas where rainfall and groundcover were marginal. The most notable example is desertification aggravated by domesticated goats in the Sahara region. Desertification involves a number of interrelated factors, including erosion, climate variations, water availability, loss of fertility, loss of soil humus, and deterioration of soil chemical properties.

A related problem is **deforestation**, loss of forests. The problem is particularly acute in tropical regions, where the forests contain most of the existing plant and animal species. In addition to extinction of these species, deforestation can cause devastating deterioration of soil through erosion and loss of nutrients.

Soil erosion can occur by the action of both water and wind, although water is the primary source of erosion.

Problems involving soil erosion were aggravated in the 1970s and early 1980s when high prices for farmland resulted in the intensive cultivation of high-income crops, particularly corn and soybeans. These crops grow in rows with bare soil in between, which tends to wash away with each rainfall. Furthermore, the practice of planting corn and soybeans year after year without intervening plantings of soilrestoring clover or grass became widespread. The problem of decreased productivity due to soil erosion has been masked somewhat by increased use of chemical fertilizers.

The preservation of soil from erosion is commonly termed **soil conservation**. There are a number of solutions to the soil erosion problem. Some are old, well-known agricultural practices such as terracing, contour plowing, and periodically planting fields with cover crops such as clover. For some crops **no-till agriculture**, now commonly called **conservation tillage**, greatly reduces erosion. This practice consists of planting a crop among the residue of the previous year's crop without plowing. Weeds are killed in the newly planted crop row by application of a herbicide prior to planting. The surface residue of plant material left on top of the soil prevents erosion.

Another, more experimental, solution to the soil erosion problem is the cultivation of perennial plants which develop large root systems and come up each spring after being harvested the previous fall. For example, a perennial corn plant has been developed by crossing corn with a distant, wild relative, teosinte, which grows in Central America.

Unfortunately, the resulting plant does not give outstanding grain yields. It should be noted that an annual plant's ability to propagate depends upon producing large quantities of seeds, whereas a perennial plant must develop a strong root system with bulbous growths called rhizomes which store food for the coming year. However, it is possible that the application of genetic engineering may result in the development of perennial crops with good seed yields. The cultivation of such a crop would cut down on a great deal of soil erosion.

The best known perennial plants are trees, which are very effective in stopping soil erosion. Wood from trees can be used as biomass fuel, as a source of raw materials, and as food (see below). There is a tremendous unrealized potential for an increase in the production of biomass from trees. For example, the production of biomass from natural forests of loblolly pine trees in South Carolina has been about three dry tons per hectare per year.

This has now been increased at least four-fold through selection of superior trees, and 30 tons may eventually be possible. In Brazil, experiments were conducted with a species of Eucalyptus, which has a seven-year growth cycle. With improved selection of trees, the annual yields for three successive cycles of these trees in dry tons per hectare per year were 23, 33, and 40.

The most important use for wood is, of course, as lumber for construction. This use will remain important as higher energy prices increase the costs of other construction materials such as steel, aluminum, and cement. Wood is about 50 percent cellulose, which can be hydrolyzed by rapidly improving enzyme processes to yield glucose sugar. The glucose can be used directly as food, fermented to ethyl alcohol for fuel (gasohol), or employed as a carbon and energy source for proteinproducing yeasts. Given these and other potential uses, the future of trees as an environmentally desirable and profitable crop is very bright.

Soil and Water Resources

The conservation of soil and the protection of water resources are strongly interrelated. Most fresh water falls initially on soil, and the condition of the soil largely determines the fate of the water and how much is retained in a usable condition. The land area upon which rainwater falls is called a **watershed**. In addition to collecting the water, the watershed determines the direction and rate of flow and the degree of water infiltration into groundwater aquifers.

Excessive rates of water flow prevent infiltration, lead to flash floods, and cause soil erosion. Measures taken to enhance the utility of land as a watershed also fortunately help prevent erosion. Some of these measures involve modification of the contour of the soil, particularly terracing, construction of waterways, and construction of waterretaining ponds. Waterways are planted with grass to prevent erosion, and waterretaining crops and bands of trees can be planted on the contour to achieve much the same goal. Reforestation and control of damaging grazing practices conserve both soil and water.

Agriculture and health

Some authorities hold that soil has an appreciable effect upon health. An obvious way in which such an effect might be manifested is the incorporation into food of micronutrient elements essential for human health. One such nutrient (which is toxic at overdose levels) is selenium. It is definitely known that the health of animals is adversely affected in selenium-deficient areas as it is in areas of selenium excess.

Human health might be similarly affected. There are some striking geographic correlations with the occurrence of cancer, some of which may be due to soil type. A high incidence of stomach cancer occurs in areas with certain types of soil in the Netherlands, the United States, France, Wales, and Scandinavia. These soils are high in organic matter content, are acidic, and are frequently waterlogged. A "stomach cancer-prone lifestyle" has been described,15 which includes consumption of home-grown food, consumption of water from one's own well, and reliance on native and uncommon foodstuffs.

One possible reason for the existence of "stomach cancer-producing soils" is the production of cancer-causing secondary metabolites by plants and microorganisms.

Secondary metabolites are biochemical compounds that are of no apparent use to the organism producing them. It is believed that they are formed from the precursors of primary metabolites when the primary metabolites accumulate to excessive levels.

The role of soil in environmental health is not well known, nor has it been extensively studied. The amount of research on the influence of soil in producing foods that are more nutritious and lower in content of naturally occurring toxic substances is quite small compared to research on higher soil productivity. It is to be hoped that the environmental health aspects of soil and its products will receive much greater emphasis in the future.

2.16 INDUSTRIAL ECOLOGY, RESOURCES AND ENERGY

Modern civilization depends upon a wide variety of resources consisting largely of minerals that are processed to recover the materials needed for industrial activities. The most common type of mineral material so used, and one that all people depend upon for their existence, is soil, used to grow plants for food. Also of crucial importance are metal ores. Some of these metal sources are common and abundant, such as iron ore; others, such as sources of chromium, are rare and will not last long at current rates of consumption. There are also some crucial sources of nonmetals.

Sulfur, for example, is abundant and extracted in large quantities as a by-product of sulfur-rich fuels. Phosphorus, a key fertilizer element, will last only for several generations at current rates of consumption. The materials needed for modern societies can be provided from either **extractive** (nonrenewable) or **renewable** sources. Extractive industries remove irreplaceable mineral resources from the earth's crust. The utilization of mineral resources is strongly tied with technology, energy, and the environment.

Perturbations in one usually cause perturbations in the others. For example, reductions in automotive exhaust pollutant levels to reduce air pollution have made use of catalytic devices that require platinum-group metals, a valuable and irreplaceable natural resource. Furthermore, automotive pollution control devices result in greater gasoline consumption than would be the case if exhaust emissions were not a consideration (a particularly pronounced effect in the earlier years of emissions control). The availability of many metals depends upon the quantity of energy used and the amount of environmental damage tolerated in the extraction of low-grade ores. Many other such examples could be cited. Because of these intimate interrelationships, technology, resources, and energy must all be considered together. The practice of industrial ecology has a significant potential to improve environmental quality with reduced consumption of nonrenewable resources and energy.

In discussing nonrenewable sources of minerals and energy, it is useful to define two terms related to available quantities. The first of these is **resources**, defined as quantities that are estimated to be *ultimately* available. The second term is **reserves**, which refers to well-identified resources that can be profitably utilized with existing technology.

Minerals in the geosphere

There are numerous kinds of mineral deposits that are used in various ways. These are, for the most part, sources of metals which occur in **batholiths** composed of masses of igneous rock that have been extruded in a solid or molten state into the surrounding rock strata. In addition to deposits formed directly from solidifying magma, associated deposits are produced by water interacting with magma. Hot aqueous solutions associated with magma can form rich **hydrothermal** deposits of minerals. Several important metals, including lead, zinc, and copper, are often associated with hydrothermal deposits.

Some useful mineral deposits are formed as **sedimentary deposits** along with the formation of sedimentary rocks. **Evaporites** are produced when seawater is evaporated. Common mineral evaporites are halite (NaCl), sodium carbonates, potassium chloride, gypsum ($CaSO_4{\bullet}2H_2O$), and magnesium salts. Many significant iron deposits consisting of hematite (Fe_2O_3) and magnetite (Fe_3O_4) were formed as sedimentary bands when earth's atmosphere was changed from reducing to oxidizing as photosynthetic organisms produced oxygen, precipitating the oxides from the oxidation of soluble Fe^{2+} ion.

Deposition of suspended rock solids by flowing water can cause segregation of the rocks according to differences in size and density. This can result in the formation of useful **placer** deposits that are enriched in desired minerals. Gravel, sand, and some other minerals, such as gold, often occur in placer deposits.

Some mineral deposits are formed by the enrichment of desired constituents when other fractions are weathered or leached away. The most common example of such a deposit is bauxite, Al_2O_3, remaining after silicates and other more soluble constituents have been dissolved by the weathering action of water under the severe conditions of hot tropical climates with very high levels of rainfall. This kind of material is called a **laterite.**

Evaluation of Mineral Resources

In order to make its extraction worthwhile, a mineral must be enriched at a particular location in earth's crust relative to the average crustal abundance. Normally applied to metals, such an enriched deposit is called an **ore**. The value of an ore is expressed in terms of a **concentration factor**.

Obviously, higher concentration factors are always desirable. Required concentration factors decrease with average crustal concentrations and with the value of the commodity extracted. A concentration factor of 4 might be adequate for iron, which makes up a relatively high percentage of earth's crust. Concentration factors must be several hundred or even several thousand for less expensive metals that are not present at very high percentages in earth's crust. However, for an extremely valuable metal, such as platinum, a relatively low concentration factor is acceptable because of the high financial return obtained from extracting the metal.

Acceptable concentration factors are a sensitive function of the price of a metal. Shifts in price can cause significant changes in which deposits are mined. If the price of a metal increases by, for example, 50%, and the increase appears to be long term, it becomes profitable to mine deposits that had not been mined previously. The opposite can happen, as is often the case when substitute materials are found or newly discovered, richer sources go into production.

In addition to large variations in the concentration factors of various ores, there are extremes in the geographic distribution of mineral resources. The United States is perhaps about average for all nations in terms of its mineral resources, possessing significant resources of copper, lead, iron, gold, and molybdenum, but virtually without resources of some important strategic metals, including chromium, tin, and platinum-group metals. For its size and population, South Africa is particularly blessed with some important metal mineral resources.

Extraction and mining

Minerals are usually extracted from Eearth's crust by various kinds of mining procedures, but other techniques may be employed as well. The raw materials so obtained include inorganic compounds such as phosphate rock, sources of metal such as lead sulfide ore, clay used for firebrick, and structural materials, such as sand and gravel.

Surface mining, which can consist of digging large holes in the ground, or strip mining, is used to extract minerals that occur near the surface. A common example of surface mining is quarrying of rock. Vast areas have been dug up to extract coal. Because of past mining practices, surface mining got a well-deserved bad name. With modern reclamation practices, however, topsoil is first removed and stored.

After the mining is complete, the topsoil is spread on top of overburden that has been replaced such that the soil surface has gentle slopes and proper drainage. Topsoil spread over the top of the replaced spoil, often carefully terraced to prevent erosion, is seeded with indigenous grass and other plants, fertilized, and watered, if necessary, to provide vegetation. The end result of carefully done **mine reclamation** projects is a wellvegetated area suitable for wildlife habitat, recreation, forestry, grazing, and other beneficial purposes.

Extraction of minerals from placer deposits formed by deposition from water has obvious environmental implications. Mining of placer deposits can be accomplished by dredging from a boom-equipped barge. Another means that can be used is hydraulic mining with large streams of water. One interesting approach for more coherent deposits is to cut the ore with intense water jets, then suck up the resulting small particles with a pumping system. These techniques have a high potential to pollute water and disrupt waterways.

For many minerals, underground mining is the only practical means of extraction. An underground mine can be very complex and sophisticated. The structure of the mine depends upon the nature of the deposit. It is of course necessary to have a shaft that reaches to the ore deposit. Horizontal tunnels extend out into the deposit, and provision must be made for sumps to remove water and for ventilation. Factors that must be considered in designing an underground mine include the depth, shape, and orientation of the ore body, as well as the nature and strength of the rock in and around it; thickness of overburden; and depth below the surface.

Usually, significant amounts of processing are required before a mined product is used or even moved from the mine site. Such processing, and the by-products of it, can have significant environmental effects. Even rock to be used for aggregate and for road construction must be crushed and sized, a process that has the potential to emit air-polluting dust particles into the atmosphere. Crushing is also a necessary first step for further processing of ore. Some minerals occur to an extent of a few percent or even less in the rock taken from the mine and must be concentrated on site so that the residue does not have to be hauled far. For metals mining, these processes—as well as roasting, extraction—and similar operations, are covered under the category of **extractive metallurgy**.

One of the more environmentally troublesome by-products of mineral refining consists of waste **tailings**. By the nature of the mineral processing operations employed, tailings are usually finely divided and, as a result, subject to chemical weathering processes. Heavy metals associated with metal ores can be leached from tailings, producing water runoff contaminated with cadmium, lead, and other pollutants. Adding to the problem are some of the processes used to refine ore. Large quantities of cyanide solution are used in some processes to extract low levels of gold from ore, posing obvious toxicological hazards.

Environmental problems resulting from exploitation of extractive resources—including disturbance of land, air pollution from dust and smelter emissions, and water pollution from disrupted aquifers are aggravated by the fact that the general trend in mining involves utilization of less rich ore. The average percentage of copper in ore mined in 1900 was about 4%, but by 1982 it was about 0.6% in domestic ore, and 1.4% in richer foreign ore. Ore as low as 0.1% copper may eventually be processed. Increased demand for a particular metal, coupled with the necessity to utilize lower grade ore, has a vicious multiplying effect upon the amount of ore that must be mined and processed and accompanying environmental consequences.

The proper practice of industrial ecology can be used to significantly reduce the effects of mining and mining by-products. One way in which this can be done is to entirely eliminate the need for mining, utilizing alternate sources of materials. An example of such utilization, widely hypothesized but not yet put into practice to a large extent, is the extraction of aluminum from coal ash. This would have the double advantage of reducing amounts of waste ash and reducing the need to mine scarce aluminum ore.

Metals

The majority of elements are metals, most of which are of crucial importance as resources. The availability and annual usage of metals vary widely with the kind of metal. Some metals are abundant and widely used in structural applications; iron and aluminum are prime examples. Other metals, especially those of the platinum group (platinum, palladium, iridium, rhodium) are very precious and their use is confined to applications such as catalysts, filaments, or electrodes for which only small quantities are required. Some metals are considered to be "crucial" because of their applications for which no substitutes are available and shortages or uneven distribution in supply that occur.

Such a metal is chromium, used to manufacture stainless steel (especially for parts exposed to high temperatures and corrosive gases), jet aircraft, automobiles, hospital equipment, and mining equipment. The platinum-group metals are used as catalysts in the chemical industry, in petroleum refining, and in automobile exhaust antipollution devices.

Metals exhibit a wide variety of properties and uses. They come from a number of different compounds; in some cases two or more compounds are significant mineral sources of the same metal. Usually these compounds are oxides or sulfides. However, other kinds of compounds and, in the cases of gold and platinum-group metals, the elemental (native) metals themselves serve as metal ore.

Metal resources and industrial and industrial ecology

Considerations of industrial ecology are very important in extending and efficiently utilizing metal resources.1 More than any other kind of resource, metals lend themselves to recycling and to the practice of industrial ecology. This section briefly addresses the industrial ecology of metals.

Aluminum metal has a remarkably wide range of uses resulting from its properties of low density, high strength, ready workability, corrosion resistance, and high electrical conductivity. Unlike some metals, such as toxic cadmium or lead, the use and disposal of aluminum presents no environmental problems. Further-ore, it is one of the most readily recycled of all metals.

The environmental problems associated with aluminum result from the mining and processing of aluminum ore. It occurs as a mineral called **bauxite**, which contains 40–60% alumina, Al_2O_3, associated with water molecules. Hydrated alumina is concentrated in bauxite, particularly in high-rainfall regions of the tropics, by the weathering away of more water-soluble constituents of soil. Bauxite ore is commonly strip mined from thin seams, so its mining causes significant disturbance to the geosphere.

The commonly used Bayer process for aluminum refining dissolves alumina, shown below as the hydroxide $Al(OH)_3$, from bauxite at high temperatures with sodium hydroxide as sodium aluminate,

$$Al(OH)_3 + NaOH \rightarrow NaAlO_2 + 2H_2O$$

leaving behind large quantities of caustic "red mud." This residue, which is rich in oxides of iron, silicon, and titanium, has virtually no uses and a high potential to produce pollution. Aluminum hydroxide is then precipitated in the pure form at lower temperatures and calcined at about 1200°C to produce pure anhydrous Al_2O_3. The anhydrous alumina is then electrolyzed in molten cyrolite, Na_3AlF_6, at carbon electrodes to produce aluminum metal.

All aspects of aluminum production from bauxite are energy intensive. Large amounts of heat energy are required to heat the bauxite treated with caustic to extract sodium aluminate, and heat is required to calcine the hydrated alumina before it can be electrolyzed. Very large amounts of electrical energy are required to reduce aluminum to the metal in the electrolytic process for aluminum production.

An interesting possibility that could avoid many of the environmental problems associated with aluminum production is the use of coal fly ash as a source of the metal. Fly ash is produced in large quantities as a by-product of electricity generation, so it is essentially a free resource. As a raw material, coal fly ash is very attractive because it is anhydrous, thus avoiding the expense of removing water; it is finely divided, and it is homogeneous. Aluminum, along with iron, manganese, and titanium, can be extracted from coal fly ash with acid. If aluminum is extracted as the chloride salt, $AlCl_3$, it can be electrolyzed as the chloride by the ALCOA process. Although this process has not yet been proven to be competitive with the Bayer process, it may become so in the future.

Gallium is a metal that commonly occurs with aluminum ore and may be produced as a byproduct of aluminum manufacture. Gallium combined with arsenic or with indium and arsenic is useful in semiconducter applications, including integrated circuits, photoelectric devices, and lasers. Although important, these applications require only miniscule amounts of gallium compared to major metals.

Chromium is of crucial importance because of its use in stainless steel and superalloys. These materials are vitally important to industrialized societies because of their applications in jet engines, nuclear power plants, chemical-resistant valves, and other applications in which a material that resists heat and chemical attack is required. Supplies of chromium are poorly distributed around the earth. It is important that chromium be handled according to good practices of industrial ecology.

Several measures may be taken in this respect. Chromium is almost impossible to recover from chrome-plated objects, and this use should be eliminated insofar as possible, as has been done with much of the decorative chromeplated adornments formerly put on automobiles. Chromium(VI) (chromate) is a toxic form of the metal and its uses should be eliminated wherever possible. The use of chromium in leather tanning and miscellaneous chemical applications should be curtailed. One important use of chromium is in the preparation of treated CCA lumber, which resists fungal decay and termites. The widespread use of this lumber has greatly extended the lives of wood products, which is in keeping with the practice of industrial ecology. However, its use of toxic arsenic, scarce copper, and even more scarce chromium are negatives, and alternative means of preserving lumber still need to be found.

Copper is a low-toxicity, corrosion-resistant metal widely used because of its workability (ductility and malleability), electrical conductivity, and ability to conduct heat. In addition to its use in electrical wire, where in some applications it is now challenged by aluminum, copper is also used in tubing, copper pipe, shims, gaskets and other applications.

There are at least two major environmental problems associated with the extraction and refining of copper. The first of these is the dilute form in which copper ore now occurs, such that in the U.S. 150–175 tons of inert material (not counting overburden removed in strip mining) must be processed and discarded to produce a ton of copper metal. The second problem is the occurrence of copper as the sulfide so that in the production of copper, large amounts of sulfur must be recovered as a by-product or, unfortunately in some less developed countries, released into the atmosphere as pollutant SO_2.

An advantage to copper for recycling is that it is used primarily as the metal, which represents "stored energy" in that it does not require energy for reduction to the metal. Recycling rates of scrap copper appear low in part because so much of the inventory of copper metal is tied up in long-lasting electrical wire, in structures, and other places where the lifetime of the metal is long. (This is in contrast to lead, where the main source of recycled metal is storage batteries, which last only 2–4 years.) An impediment to copper recycling is the difficulty of recovering copper components from circuits, plumbing, and other applications.

Cobalt is a "strategic" metal with very important applications in alloys, particularly in heat-resistant applications, such as jet engines. The major source of cobalt is as a byproduct of copper refining, although it can also be obtained as a byproduct of nickel and lead. As much as 50% of the cobalt in these sources is lost to tailings, slag, or other wastes, so there is a significant potential to improve the recovery of cobalt. Relatively low percentages of cobalt are recycled as scrap.

The industrial ecology of **lead** is very important because of the widespread use of this metal and its toxicity. Somewhat more than half of the lead processed by humans comes from the geosphere, mostly as lead mined for the metal, and with a very small fraction contained in coal that is burned. By far the greatest use of lead is in batteries, and the amount of battery lead recycled each year approaches that taken from the geosphere. A small fraction of lead is dissipated as wastes associated with the mining and refining of the metal and as lead in gasoline, an amount that is decreasing as use of unleaded gasoline becomes prevalent around the world. A significant quantity of lead goes into various uses other than batteries, including pigments, solder, cable sheathing, formed products, and shot in ammunition. Only a small fraction of lead from these uses is recycled, and this represents a potential improvement in the conservation of lead.

Another area in which improvements can be made is to eliminate or greatly reduce nonbattery uses of lead, as has been done in the case of lead shot and pigments. Although a large fraction of lead in batteries is recycled, about 1/3 of the lead used in batteries is lost; this represents another area of potential improvement in the utilization of lead.

Zinc is relatively abundant and not particularly toxic, so its industrial ecology is of less concern than that of toxic lead or scarce chromium. As with other metals, the mining and processing of zinc can pose some environmental concerns. Zinc occurs as ZnS (a mineral called sphalerite), and the sulfur must be reclaimed in the smelting of zinc. Zinc minerals often contain significant fractions of lead and copper, as well as significant amounts of toxic arsenic and cadmium.

Zinc is widely used as the metal, and lesser amounts are used to make zinc chemicals. One of the larger uses for zinc is as a corrosion-resistant coating on steel. This application, refined to a high degree in the automotive industry in recent years, has significantly lengthened the life span of automotive bodies and frames. It is difficult to reclaim zinc from zinc plating. However, zinc is a volatile element and it can be recovered in baghouse dust from electric arc furnaces used to reprocess scrap steel.

Zinc is used along with copper to make the alloy called brass. Brass is very well adapted to the production of various parts and objectives. It is recyclable, and significant quantities of brass are recycled as wastes from casting, machining, and as postconsumer waste.

Although a number of zinc compounds are synthesized and used, by far the most important of them is zinc oxide, ZnO. Formerly widely used as a paint pigment, this white substance is now employed as an accelerating and activating agent for hardening rubber products, particularly tires. Tire wear is a major vector for the transfer of zinc to the environment and, since it occurs with zinc, toxic cadmium is also dissipated to the environment by tire wear. The other two major compounds of zinc employed commercially are zinc chloride used in dry cells, as a disinfectant, and to vulcanize rubber, and zinc sulfide, used in zinc electroplating baths and to manufacture zinc-containing insecticides, particularly ***Zineb***.

Two aspects of zinc may be addressed in respect to its industrial ecology. The first of these is that, although it is not very toxic to animals, zinc is phytotoxic (toxic to plants) and soil can be "poisoned" by exposure to zinc from zinc smelting or from application of zinc-rich sewage sludge. The second of these is that the recycling of zinc is complicated by its dispersal as a plating on other metals. However, means do exist to reclaim significant fractions of such zinc, such as from electric arc furnaces as mentioned above.

Potassium deserves special mention as a metal because the potassium ion, K^+, is an essential element required for plant growth. It is mined as potassium minerals and applied to soil as plant fertilizer. Potassium minerals consist of potassium salts, generally KCl. Such salts are found as deposits in the ground or may be obtained from some brines. Very large deposits are found in Saskatchewan, Canada. These salts are all quite soluble in water.

Nonmetal mineral resources

A number of minerals other than those used to produce metals are important resources. As with metals, the environmental aspects of mining many of these minerals are quite important. Typically, even the extraction of ordinary rock and gravel can have important environmental effects.

Clays are secondary minerals formed by weathering processes on parent minerals). Clays have a variety of uses. About 70% of the clays used are miscellaneous clays of variable composition that have uses for a number of applications including filler (such as in paper), brick manufacture, tile manufacture, and portland cement production. Somewhat more than 10% of the clay used is fireclay, which has the characteristic of being able to withstand firing at high temperatures without warping.

This clay is used to make a variety of refractories, pottery, sewer pipe, tile, and brick. Somewhat less than 10% of the clay that is used is kaolin, which has the general formula $Al_2(OH)_4Si_2O_5$. Kaolin is a white mineral that can be fired without losing shape or color. It is employed to make paper filler, refractories, pottery, dinnerware, and as a petroleum-cracking catalyst.

About 7% of clay mined consists of bentonite and fuller's earth, a clay of variable composition used to make drilling muds, petroleum catalyst, carriers for pesticides, sealers, and clarifying oils. Very small quantities of a highly plastic clay called ball clay are used to make refractories, tile, and whiteware. U.S. production of clay is about 60 million metric tons per year, and global and domestic resources are abundant.

Fluorine compounds are widely used in industry. Large quantities of fluorspar, CaF2, are required as a flux in steel manufacture. Synthetic and natural cryolite, Na3AlF6, is used as a solvent for aluminum oxide in the electrolytic preparation of aluminum metal. Sodium fluoride is added to water to help prevent tooth decay, a measure commonly called water fluoridation. World reserves of high-grade fluorspar are around 190 million metric tons, about 13% of which is in the United States. This is sufficient for several decades at projected rates of use. A great deal of by-product fluorine is recovered from the processing of fluorapatite, $Ca_5(PO_4)_3F$, used as a source of phosphorus (see below).

Micas are complex aluminum silicate minerals that are transparent, tough, flexible, and elastic. Muscovite, $K_2O{\bullet}3Al_2O_3{\bullet}6\ SiO_2{\bullet}2H_2O$, is a major type of mica. Better grades of mica are cut into sheets and used in electronic apparatus, capacitors, generators, transformers, and motors. Finely divided mica is widely used in roofing, paint, welding rods, and many other applications. Sheet mica is imported into many countries, and finely divided "scrap" mica is recycled domestically. Shortages of this mineral are unlikely.

Pigments and **fillers** of various kinds are used in large quantities. The only naturally occurring pigments still in wide use are those containing iron. These minerals are colored by limonite, an amorphous brown-yellow compound with the formula $2Fe_2O_3{\bullet}3H_2O$, and hematite, composed of gray-black Fe_2O_3. Along with varying quantities of clay and manganese oxides, these compounds are found in ocher, sienna, and umber. Manufactured pigments include carbon black, titanium dioxide, and zinc pigments. About 1.5 million metric tons of carbon black, manufactured by the partial combustion of natural gas, are used in the U.S. each year, primarily as a reinforcing agent in tire rubber.

Over 7 million metric tons of minerals are used in the U.S. each year as fillers for paper, rubber, roofing, battery boxes, and many other products. Among the minerals used as fillers are carbon black, diatomite, barite, fuller's earth, kaolin (see clays, above), mica, limestone, pyrophyllite, and wollastonite ($CaSiO_3$).

Although **sand** and **gravel** are the cheapest of mineral commodities per ton, the average annual dollar value of these materials is greater than all but a few mineral products because of the huge quantities involved. In tonnage, sand and gravel production is by far the greatest of nonfuel minerals. Almost 1 billion tons of sand and gravel are employed in construction in the U.S. each year, largely to make concrete structures, road paving, and dams. Slightly more than that amount is used to manufacture portland cement and as construction fill. Although ordinary sand is predominantly silica, SiO2, about 30 million tons of a more pure grade of silica are consumed in the U.S. each year to make glass, high-purity silica, silicon semiconductors, and abrasives.

At present, old river channels and glacial deposits are used as sources of sand and gravel. Many valuable deposits of sand and gravel are covered by construction and lost to development. Transportation and distance from source to use are especially crucial for this resource. Environmental problems involved with defacing land can be severe, although bodies of water used for fishing and other recreational activities frequently are formed by removal of sand and gravel.

Phosphate minerals are of particular importance because of their essential use in the manufacture of fertilizers applied to land to increase crop productivity. In addition, phosphorus is used for supplementation of animal feeds, synthesis of detergent builders, and preparation of chemicals such as pesticides and medicines. The most common phosphate minerals are hydroxyapatite, $Ca_5(PO_4)_3(OH)$, and fluorapatite, $Ca_5(PO_4)_3F$. Ions of Na, Sr, Th, and U are found substituted for calcium in apatite minerals. Small amounts of PO_4^{3-} may be replaced by AsO43- and the arsenic must be removed for food applications. Approximately 17% of world phosphate production is from igneous minerals, primarily fluorapatites. About three-fourths of world phosphate production is from sedimentary deposits, generally of marine origin.

Vast deposits of phosphate, accounting for approximately 5% of world phosphate production, are derived from guano droppings of seabirds and bats. Current U.S. production of phosphate rock is around 40 million metric tons per year, most of it from Florida. Idaho, Montana, Utah, Wyoming, North Carolina, South Carolina, and Tennessee also have sources of phosphate. Reserves of phosphate minerals in the United States amount to 10.5 billion metric tons, containing approximately 1.4 billion metric tons of phosphorus.

Phosphate in the naturally occurring minerals is not sufficiently available to be used as fertilizer. For commercial phosphate fertilizer production, these minerals are treated with phosphoric or sulfuric acids to produce more soluble superphosphates.

$$2Ca_5(PO_4)_3F(s) + 14H_3PO_4 + 10H_2O \rightarrow 2HF(g) + 10Ca(H_2PO_4)_2{\bullet}H_2O$$

$$2Ca_5(PO_4)_3F(s) + 7H_2SO_4 + 3H_2O \rightarrow 2HF(g) + 3Ca(H_2PO_4)_2{\bullet}H_2O + 7CaSO_4$$

The HF produced as a byproduct of superphosphate production can create air pollution problems, and the recovery of fluorides is an important aspect of the industrial ecology of phosphate production.

Phosphate minerals are rich in trace elements required for plant growth, such as boron, copper, manganese, molybdenum, and zinc. Ironically, these elements are lost in processing phosphate for fertilizers and are sometimes added later.

Ammonium phosphates are excellent, highly soluble phosphate fertilizers. Liquid ammonium polyphosphate fertilizers consisting of ammonium salts of pyrophosphate, triphosphate, and small quantities of higher polymeric phosphate anions in aqueous solution can be used as phosphate fertilizers.

The polyphosphates are believed to have the additional advantage of chelating iron and other micronutrient metal ions, thus making the metals more available to plants.

There are at least two major reasons why the industrial ecology of phosphorus is particularly important. The first of these is that current rates of phosphate use would exhaust known reserves of phosphate within two or three generations. Although additional sources of phosphorus will be found and exploited, it is clear that this essential mineral is in distressingly short supply relative to human consumption; phosphate shortages, along with sharply higher prices, will eventually cause a crisis in food production. The second significant aspect of the industrial ecology of phosphorus is the pollution of waterways by waste phosphate, a plant and algal nutrient. This results in excessive growth of algae in the water, followed by decay of the plant biomass, consumption of dissolved oxygen, and an undesirable condition of eutrophication.

Excessive use of phosphate coupled with phosphate pollution suggests that phosphate wastes, such as from sewage treatment, should be substituted as sources of plant fertilizer. Several other partial solutions to the problem of phosphate shortages are the following:

- Development and implementation of methods of fertilizer application that maximize efficient utilization of phosphate
- Genetic engineering of plants that have minimal phosphate requirements and that utilize phosphorus with maximum efficiency
- Development of systems to maximize the utilization of phosphorus-rich animal wastes

Sulfur is an important nonmetal; its greatest single use is in the manufacture of sulfuric acid. However, the element is employed in a wide variety of other industrial and agricultural products. Current consumption of sulfur amounts to approximately 10 million metric tons per year in the United States. The four most important sources of sulfur are (in decreasing order) deposits of elemental sulfur, H2S recovered from sour natural gas, organic sulfur recovered from petroleum, and pyrite (FeS2). Recovery of sulfur from coal used as a fuel is a huge potential, largely untapped source of this important nonmetal. The resource situation for sulfur differs from that of phosphorus in several significant respects. Although sulfur is an essential nutrient like phosphorus, most soils contain sufficient amounts of nutrient sulfur, and the major uses of sulfur are in the industrial sector. The sources of sulfur are varied and abundant and supply is no problem either in the United States or worldwide; sulfur recovery from fossil fuels as a pollution control measure could even result in surpluses of this element.

About 90% of the use of sulfur in the world is for the manufacture of sulfuric acid. Almost 2/3 of the sulfuric acid consumed is used to make phosphate fertilizers, in which case the phosphorus ends up as waste "phosphogypsum," $CaSO_4$ x H_2O. Other uses of sulfur include lead storage batteries, steel pickling, petroleum refining, extraction of copper from copper ore, and the chemical industry.

The industrial ecology of sulfur needs to emphasize reduction of wastes and sulfur pollution, rather than supply of this element. Unlike many resources, such as most common metals, the uses of sulfur are for the most part dissipative, and the sulfur is "lost" to agricultural land, paper products, petroleum products, or other environmental sinks. There are two major environmental concerns with sulfur. One of these is the emission of sulfur into the atmosphere, which occurs mostly as pollutant sulfur dioxide and is largely manifested by production of acidic precipitation and dry deposition.

The second major environmental concern with sulfur is that it is used mostly as sulfuric acid and is not incorporated into products, thus posing the potential to pollute water and create acidic wastes. Acid purification units are available to remove significant amounts of sulfuric acid from waste acid solutions for recycling.

Gypsum. Calcium sulfate in the form of the dihydrate $CaSO_4.2H_2O$ is the mineral **gypsum**, one of the most common forms in which waste sulfur is produced. As noted, large quantities of this material are produced as a by-product of phosphate fertilizer manufacture. Another major source of gypsum is its production when lime is used to remove sulfur dioxide from power plant stack gas,
$Ca(OH)_2 + SO_2 \rightarrow CaSO_3 + H_2O$ to produce a calcium sulfite product that can be oxidized to calcium sulfate. About 100 million metric tons of gypsum are mined each year for a variety of uses, including production of portland cement, to produce wallboard, as a soil conditioner to loosen tight clay soils, and numerous other applications.

Calcium sulfate from industrial or natural (gypsum) sources can be calcined at a very low temperature of only 159°C to produce $CaSO_4.1/2H_2O$, a material known as plaster of Paris, which was once commonly used for the manual application of plaster to walls. Plaster of Paris mixed with water forms a plastic material which sets up as the solid dihydrate,

$$CaSO_4.1/2H_2O + 3/2H_2O \rightarrow CaSO_4.2H_2O$$

Cast into sheets coated with paper, this material produces plasterboard commonly used for the interior walls of homes and other buildings. Historically, plaster of Paris was used for mortar and other structural applications, and it has the potential for similar applications today.

The very large quantities of gypsum that are mined suggest that by-product calcium sulfate, especially that produced with phosphate fertilizers and from flue gas desulfurization, should be a good candidate for reclamation through the practice of industrial ecology. The low temperature (see above) required to convert hydrated calcium sulfate to $CaSO_4.1/2H_2O$, which can be set up as a solid by mixing with water, suggests that the energy requirements for a gypsum-based by-products industry should be modest. Low-density gypsum blown as a foam and used as a filler in composites along with sturdy reinforcing materials should have good insulating, fire-resistant, and structural properties for building construction.

World energy resources

At present, most of the energy consumed by humans is produced from fossil fuels. Estimates of the amounts of fossil fuels available differ; those of the quantities of recoverable fossil fuels in the world before 1800 are given in figure further. By far the greatest recoverable fossil fuel is in the form of coal and lignite. Furthermore, only a small percentage of this energy source has been utilized to date, whereas much of the recoverable petroleum and natural gas has already been consumed. Projected use of these latter resources indicates rapid depletion.

Although world coal resources are enormous and potentially can fill energy needs for a century or two, their utilization is limited by environmental disruption from mining and emissions of carbon dioxide and sulfur dioxide. These would become intolerable long before coal resources were exhausted. Assuming only uranium-235 as a fission fuel source, total recoverable reserves of nuclear fuel are roughly about the same as fossil fuel reserves. These are many orders of magnitude higher if the use of breeder reactors is assumed.

Extraction of only 2% of the deuterium present in the earth's oceans would yield about a billion times as much energy by controlled nuclear fusion as was originally present in fossil fuels! This prospect is tempered by the lack of success in developing a controlled nuclear fusion reactor. Geothermal power, currently utilized in northern California, Italy, and New Zealand, has the potential for providing a significant percentage of energy worldwide. The same limited potential is characteristic of several renewable energy resources, including hydroelectric energy, tidal energy, and especially wind power. All of these will continue to contribute significant, but relatively small, amounts of energy. Renewable, nonpolluting solar energy comes as close to being an ideal energy source as any available. It almost certainly has a bright future.

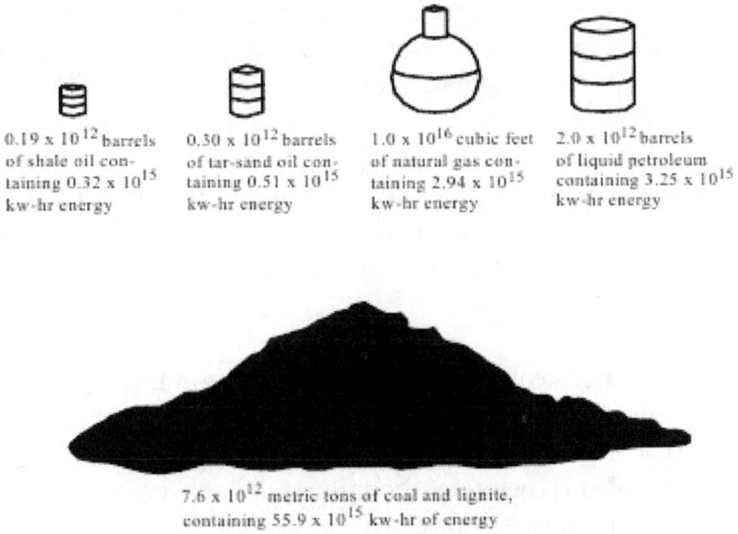

Fig. 144. *Original amounts of the world's recoverable fossil fuels (quantities in thermal kilowatt hours of energy based upon data taken from M. K. Hubbert, "The Energy Resources of the Earth," in Energy and Power, W. H. Freeman and Co., San Francisco, 1971).*

Energy conservation

Any consideration of energy needs and production must take energy conservation into consideration. This does not have to mean cold classrooms with thermostats set at 60°F in mid-winter, nor swelteringly hot homes with no air-conditioning, nor total reliance on the bicycle for transportation, although these and even more severe conditions are routine in many countries. The fact remains that the world has wasted energy at a deplorable rate. For example, U.S. energy consumption is higher per capita than that of some other countries that have equal, or significantly better, living standards. Obviously, a great deal of potential exists for energy conservation that will ease the energy problem.

Transportation is the economic sector with the greatest potential for increased efficiencies. The private auto and airplane are only about one-third as efficient as buses or trains for transportation. Transportation of freight by truck requires about 3800 Btu/ton-mile, compared to only 670 Btu/ton-mile for a train. It is terribly inefficient compared to rail transport (as well as dangerous, labor-intensive, and environmentally disruptive). Major shifts in current modes of transportation in the U.S. will not come without anguish, but energy conservation dictates that they be made.

Household and commercial uses of energy are relatively efficient. Here again, appreciable savings can be made. The all-electric home requires much more energy (considering the percentage wasted in generating electricity) than a home heated with fossil fuels. The sprawling ranch-house style home uses much more energy per person than does an apartment unit or row house. Improved insulation, sealing around the windows, and other measures can conserve a great deal of energy. Electric generating plants centrally located in cities can provide waste heat for commercial and residential heating and cooling and, with proper pollution control, can use municipal refuse for part of their fuel, thus reducing quantities of solid wastes requiring disposal.

As scientists and engineers undertake the crucial task of developing alternative energy sources to replace dwindling petroleum and natural gas supplies, energy conservation must receive proper emphasis. In fact, zero energy-use growth, at least on a per capita basis, is a worthwhile and achievable goal. Such a policy would go a long way toward solving many environmental problems. With ingenuity, planning, and proper management, it could be achieved while increasing the standard of living and quality of life.

Petroleum and natural gas

In 1994 world petroleum consumption was at a rate of about 65 million barrels per day. Liquid petroleum is found in rock formations ranging in porosity from 10 to 30%. Up to half of the pore space is occupied by water. The oil in these formations must flow over long distances to an approximately 15-cm diameter well from which it is pumped.

The rate of flow depends on the permeability of the rock formation, the viscosity of the oil, the driving pressure behind the oil, and other factors. Because of limitations in these factors, **primary recovery** of oil yields an average of about 30% of the oil in the formation, although it is sometimes as little as 15%. More oil can be obtained using **secondary recovery** techniques, which involve forcing water under pressure into the oil-bearing formation to drive the oil out.

Primary and secondary recovery together typically extract somewhat less than 50% of the oil from a formation. Finally, **tertiary recovery** can be used to extract even more oil, normally through the injection of pressurized carbon dioxide, which forms a mobile solution with the oil and allows it to flow more easily to the well. Other chemicals, such as detergents, may be used to aid in tertiary recovery. Currently, about 300 billion barrels of U.S. oil are not available through primary recovery alone.

A recovery efficiency of 60% through secondary or tertiary techniques could double the amount of available petroleum. Much of this would come from fields which have already been abandoned or essentially exhausted using primary recovery techniques.

Shale oil is a possible substitute for liquid petroleum. Shale oil is a pyrolysis product of oil shale, a rock containing organic carbon in a complex structure of biological origin from eons past called kerogen.

Shale oil may be recovered from the parent mineral by retorting the mined shale in a surface retort. A major environmental disadvantage is that this process requires the mining of enormous quantities of mineral and disposal of the spent shale, which has a volume greater than the original mineral. *In situ* retorting limits the control available over infiltration of underground water and resulting water pollution.

Water passing through spent shale becomes quite saline, so there is major potential for saltwater pollution. Large investments were made in these operations, and huge expenditures were projected for commercialization. Falling crude oil prices caused all these operations to be canceled. A large project for the recovery of oil from oil sands in Alberta, Canada, was also canceled in the 1980s.

Natural gas, consisting almost entirely of methane, has become more attractive as an energy source with recent discoveries and development of substantial new sources of this premium fuel. In addition to its use as a fuel, natural gas can be converted to many other hydrocarbon materials. It can be used as a raw material for the Fischer-Tropsch synthesis of gasoline. New unconventional sources of natural gas, such as may exist in geopressurized zones, could provide abundant energy reserves for the world, though at substantially increased prices.

Coal

From New Time until World War II, coal was the dominant energy source behind industrial expansion in most nations. However, after World War II, the greater convenience of lower-cost petroleum resulted in a decrease in the use of coal for energy in the world. Annual coal production in the world fell by about one third, reaching a low of approximately 400 million tons in 1958. Several statistics illustrate the importance of coal as a source of energy by earth's population.

Overall, about onethird of the energy used by humankind is provided from coal. The percentage of electricity generated by coal is even higher, around 45%. Almost three-fourths of the energy and coke used to make steel, the commodity commonly taken as a measure of industrial development, is provided by coal.

The general term *coal* describes a large range of solid fossil fuels derived from partial degradation of plants. Chemically, coal is a very complex material and is by no means pure carbon.

The extent to which coal can be used as a fuel depends upon solutions to several problems, including (1) minimizing the environmental impact of coal mining; (2) removing ash and sulfur from coal prior to combustion; (3) removing ash and sulfur dioxide from stack gas after combustion; (4) conversion of coal to liquid and gaseous fuels free of ash and sulfur; and, most important, (5) whether or not the impact of increased carbon dioxide emissions upon global climate can be tolerated. Progress is being made on minimizing the environmental impact of mining.

As more is learned about the processes by which acid mine water is formed, measures can be taken to minimize the production of this water pollutant. Particularly on flatter lands, stripmined areas can be reclaimed with relative success. Inevitably, some environmental damage will result from increased coal mining, but the environmental impact can be reduced by various control measures. Washing, flotation, and chemical processes can be used to remove some of the ash and sulfur prior to burning. Approximately half of the sulfur in the average coal occurs as pyrite, FeS2, and half as organic sulfur.

Although little can be done to remove the latter, much of the pyrite can be separated from most coals by physical and chemical processes. The maintenance of air pollution emission standards requires the removal of sulfur dioxide from stack gas in coal-fired power plants. Stack gas desulfurization presents some economic and technological problems.

Magnetohydrodynamic power combined with conventional steam generating units has the potential for a major breakthrough in the efficiency of coal utilization. A schematic diagram of magnetohydrodynamic

(MHD) generator is shown in **Figure 3**. This device uses a plasma of ionized gas at around 2400°C blasting through a very strong magnetic field of at least 50,000 gauss to generate direct current. The ionization of the gas is accomplished by injecting a "seed" of cesium or potassium salts.

In an MHD generator, the ultra-high-temperature gas issuing through a supersonic nozzle contains ash, sulfur dioxide, and nitrogen oxides, which severely erode and corrode the materials used.

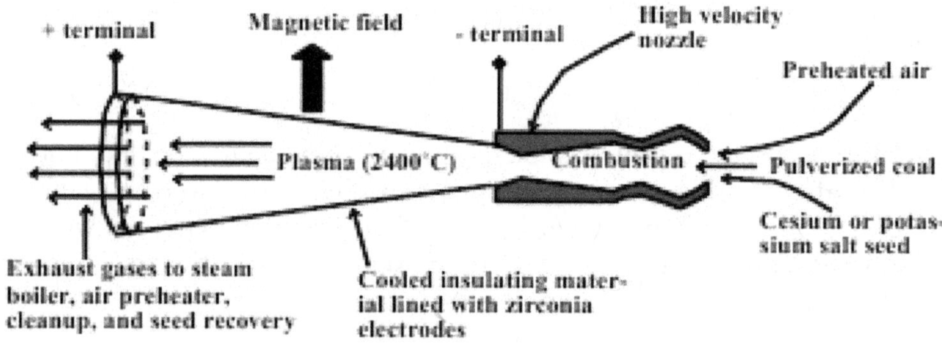

Fig. 145. *A magnetohydrodynamic power generator*

This hot gas is used to generate steam for a conventional steam power plant, thus increasing the overall efficiency of the process. The seed salts combine with sulfur dioxide and are recovered along with ash in the exhaust.

Pollutant emissions are low. The overall efficiency of combined MHD-steam power plants should reach 60%, one and one-half times the maximum of present steam-only plants. Despite some severe technological difficulties, there is a chance that MHD power could become feasible on a large scale, and an experimental MHD generator was tied to a working power grid in the former Soviet Union for several years. As of the early 1990s, the U.S. Department of Energy was conducting a proof-of-concept project to help determine the practicability of magnetohydrodynamics.

Coal Conversion

Coal can be converted to gaseous, liquid, or low-sulfur, low-ash solid fuels such as coal char (coke) or solvent-refined coal (SRC). Coal conversion is an old idea; a house belonging to William Murdock at Redruth, Cornwall, England, was illuminated with coal gas in 1792. The first municipal coal-gas system was employed to light Pall Mall in London in 1807.

The coal-gas industry began in the U.S. in 1816. The early coal-gas plants used coal pyrolysis (heating in the absence of air) to produce a hydrocarbon-rich product particularly useful for illumination. Later in the 1800s the water-gas process was developed, in which steam was added to hot coal to produce a mixture consisting primarily of H_2 and CO. It was necessary to add volatile hydrocarbons to this "carbureted" water-gas to bring its illuminating power up to that of gas prepared by coal pyrolysis. At the peak of its use in 1947, the water-gas method accounted for 57% of U.S.-manufactured gas.

The gas was made in lowpressure, low-capacity gasifiers that by today's standards would be inefficient and environmentally unacceptable (many sites of these old plants have been designated as hazardous waste sites because of residues of coal tar and other wastes). During World War II, Germany developed a major synthetic petroleum industry based on coal, which reached a peak capacity of 100,000 barrels per day in 1944. A synthetic petroleum plant operating in Sasol, South Africa, reached a capacity of several tens of thousands of tons of coal per day in the 1970s.

The two broadest categories of coal conversion are gasification and liquefaction. Arguably the most developed route for coal gasification is the **Texaco process**, which gasifies a water slurry of coal at temperatures of 1250°C to 1500°C and pressures of 350 to 1200 pounds per square inch. Chemical addition of hydrogen to coal can liquefy it and produce a synthetic petroleum product. This can be done with a hydrogen donor solvent, which is recycled and itself hydrogenated with H2 during part of the cycle. Such a process forms the basis of the successful **Exxon Donor Solvent process**, which has been used in a 250 ton/day pilot plant.

A number of environmental implications are involved in the widespread use of coal conversion. These include strip mining, water consumption in arid regions, lower overall energy conversion compared to direct coal combustion, and increased output of atmospheric carbon dioxide. These plus economic factors have prevented coal conversion from being practiced on a very large scale.

Nuclear fission power

The awesome power of the atom revealed at the end of World War II held out enormous promise for the production of abundant, cheap energy. This promise has never really come to full fruition, although nuclear energy currently provides a significant percentage of electric energy in many countries, and it may be the only source of electrical power that can meet world demand without unacceptable environmental degradation, particularly through the generation of greenhouse gases.

It has been characterized as a "misunderstood" source of electricity. Nuclear power reactors currently in use depend upon the fission of uranium-235 nuclei by reactions such as

$$^{235}_{92}U + ^{1}_{0}n \rightarrow ^{133}_{51}Sb + ^{99}_{41}Nb + 4^{1}_{0}n$$

to produce two radioactive fission products, an average of 2.5 neutrons, and an average of 200 MeV of energy per fission. The neutrons, initially released as fastmoving, highly energetic particles, are slowed to thermal energies in a moderator medium. For a reactor operating at a steady state, exactly one of the neutron products from each fission is used to induce another fission reaction in a chain reaction:

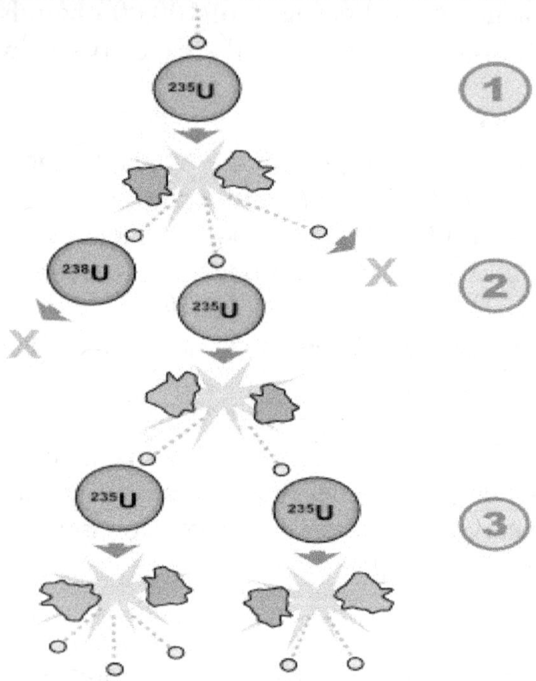

The energy from these nuclear reactions is used to heat water in the reactor core and produce steam to drive a steam turbine, as shown in Figure 7. Temperature limitations make nuclear power less efficient in converting heat to mechanical energy and, therefore, to electricity, than fossil energy conversion processes.

A limitation of fission reactors is the fact that only 0.71% of natural uranium is fissionable uranium-235. This situation could be improved by the development of **breeder reactors**, which convert uranium-238 (natural abundance 99.28%) to fissionable plutonium-239.

A major consideration in the widespread use of nuclear fission power is the production of large quantities of highly radioactive waste products. These remain lethal for thousands of years. They must either be stored in a safe place or disposed of permanently in a safe manner. At the present time, spent fuel elements are being stored under water at the reactor sites. Eventually, the wastes from this fuel will have to be buried.

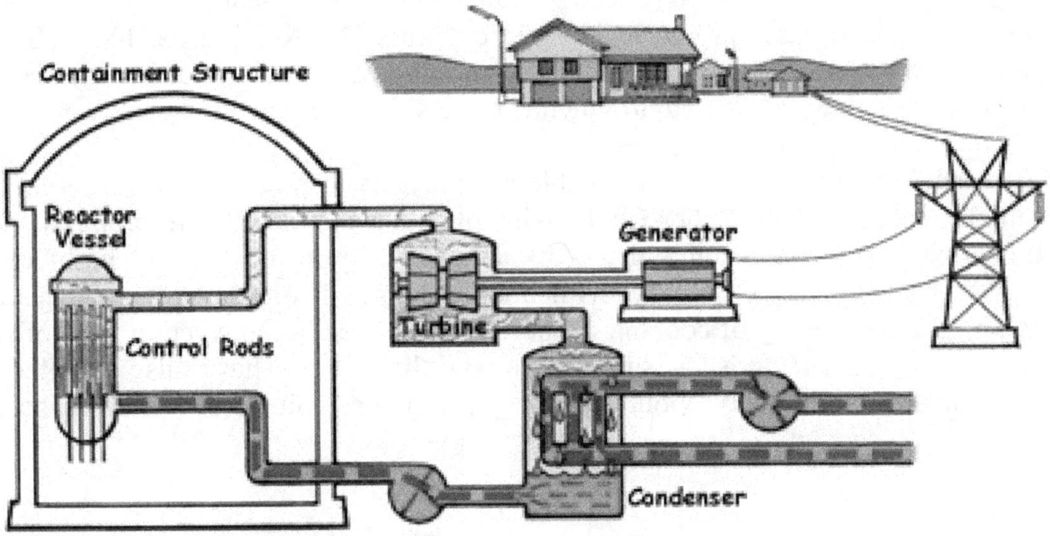

Fig. 146. *A typical nuclear fission power plant*

Another problem to be faced with nuclear fission reactors is their eventual ecommissioning. There are three possible solutions. One is dismantling soon after shutdown, in which the fuel elements are removed, various components are flushed with cleaning fluids, and the reactor is cut up by remote control and buried. "Safe storage" involves letting the reactor stand 30-100 years to allow for radioactive decay, followed by dismantling. The third alternative is entombment, encasing the reactor in a concrete structure.

The course of nuclear power development was altered drastically by two accidents. The first of these occurred on March 28, 1979, with a partial loss of coolant water from the Metropolitan Edison Company's nuclear reactor located on Three Mile Island in the Susquehanna River, 28 miles outside of Harrisburg, Pennsylvania. The result was a loss of control, overheating, and partial disintegration of the reactor core. Some radioactive xenon and krypton gases were released and some radioactive water was dumped into the Susquehanna River.

In August of 1993, cleanup workers finished evaporating the water from about 8 million liters of water solution contaminated by the reactor accident, enabling the reactor building to be sealed. A much worse accident occurred at Chernobyl in the Soviet Union in April of 1986 when a reactor blew up spreading radioactive debris over a wide area and killing a number of people (officially 31, but probably many more). Thousands of people were evacuated and the entire reactor structure had to be entombed in concrete. Food was seriously contaminated as far away as northern Scandinavia.

A much less serious, but still troublesome, nuclear accident occurred at the Tokaimura uranium processing plant in Japan on September 30, 1999, when a critical mass of enriched uranium was produced resulting in a chain reaction that exposed three workers to potentially lethal levels of radiation and contaminated 55 other workers and medical personnel.

In violation of procedures, the workers used buckets to mix uranyl nitrate in a mixing tank, leading to an accumulation of about 16 kg of solution, greatly exceeding the tank's safety limit of 2.4 kg. The uranium, prepared for use in Japan's Nuclear Cycle Development Instititute's experimental fast-breeder reactor, was enriched to 18.8% fissionable 235U, much higher than the 3–4% normally used in power reactors.

The single most important factor that may lead to renaissance of nuclear energy is the threat to the atmosphere from greenhouse gases produced in large quantities by fossil fuels. It can be argued that nuclear energy is the only proven alternative that can provide the amounts of energy required within acceptable limits of cost, reliability, and environmental effects.

New designs for nuclear power plants should enable construction of power reactors that are much safer and environmentally acceptable than those built with older technologies. The proposed new designs incorporate built-in passive safety features that work automatically in the event of problems that could lead to incidents such as TMI or Chernobyl with older reactors. These devices—which depend upon phenomena such as gravity feeding of coolant, evaporation of water, or convection flow of fluids—give the reactor the desirable characteristics of **passive stability**. They have also enabled significant simplification of hardware, with only about half as many pumps, pipes, and heat exchangers as are contained in older power reactors.

Nuclear fusion power

The two main reactions by which energy can be produced from the fusion of two light nuclei into a heavier nucleus are the deuterium-deuterium reaction,

$$^{2}_{1}H + ^{2}_{1}H \rightarrow ^{3}_{2}He + ^{1}_{0}n + 1 \text{ Mev (energy released per fusion)}$$

and the deuterium-tritium reaction:

$$^{2}_{1}H + ^{3}_{1}H \rightarrow ^{4}_{2}He + ^{1}_{0}n + 17.6 \text{ Mev (energy released per fusion)}$$

The second reaction is more feasible because less energy is required to fuse the two nuclei than to fuse two deuterium nuclei. However, the total energy from deuteriumtritium fusion is limited by the availability of tritium, which is made from nuclear reactions of lithium-6 (natural abundance, 7.4%). The supply of deuterium, however, is essentially unlimited; one out of every 6700 atoms of hydrogen is the deuterium isotope. The 3He byproduct of the fusion of two deuterium nuclei, reacts with neutrons, which are abundant in a nuclear fusion reactor, to produce tritium required for reaction 2.

The power of nuclear fusion has not yet been harnessed in a sustained, controlled reaction of appreciable duration that produces more power than it consumes. Most approaches have emphasized **magnetic confinement**, the "squeezing" of a plasma (ionized gas) of fusionable nuclei in a strong magnetic field. In 1994 a record power level pulse of 10.7 megawatts (MW) was achieved from the fission of deuterium with tritium by the Tokamak Fusion Test Reactor operated by Princeton University for the U.S. Department of Energy.

This level exceeds 20% of the power put into the reactor to achieve fusion, which of course must be boosted to well over 100% for a self-sustained fusion reactor. Within three years after this record power pulse was achieved the Princeton Tokamak Fusion Test Reactor was shut down for lack of funding, although experiments on controlled nuclear fusion have continued at the facility and others around the world. The United States also withdrew from a huge international Tokamak project after $1 billion had been spent on the undertaking, although a consortium of European countries, Russia, and Japan continue to support the project.

An alternative to magnetic confinement is **inertial confinement** in which a pellet composed of deuterium and tritium frozen on the inside of a plastic-coated pellet smaller than a pinhead is bombarded by laser beams or X-rays, heating the fuel pellet to a temperature of about 100 million °C and causing fission of the deuterium and tritium nuclei. Each such event, literally a miniature thermonuclear explosion, can release energy equivalent to the explosion of about 45 kg of TNT.

With both magnetic confinement and inertial confinement reactors, a central challenge is in harnessing the energy once it is released. Much of the energy is in the form of neutrons, which react with nuclei, such as those of iron and copper composing the reactor structure, making it radioactive and causing metal embrittlement, which would rapidly destroy the reactor. Most proposed power reactors now call for a replaceable lining, probably composed of lithium, which would absorb the neutrons and produce heat energy.

A great flurry of excitement over the possibility of a cheap, safe, simple fusion power source was generated by an announcement from the University of Utah in 1989 of the attainment of "cold fusion" in the electrolysis of deuterium oxide (heavy water). Funding was appropriated and laboratories around the world were thrown into frenetic activity in an effort to duplicate the reported results. Some investigators reported evidence, particularly the generation of anomalously large amounts of heat, to support the idea of cold fusion, whereas others scoffed at the idea. Since that time cold fusion has been disproven, and the whole saga of it, described in a detailed book about the topic,6 stands as a classic case of science gone astray.

Controlled nuclear fusion processes could be designed to produce almost no radioactive waste products. However, tritium used in the deuterium-tritium reaction is very difficult to contain, and some release of the isotope would occur. The deuteriumdeuterium reaction promises an unlimited source of energy. Either of these reactions would be preferable to fission in terms of environmental considerations. Therefore, despite the possibility of insurmountable technical problems involved in harnessing fusion energy, the promise of this abundant, relatively nonpolluting energy source makes its pursuit well worth a massive effort.

Geothermal energy

Underground heat in the form of steam, hot water, or hot rock used to produce steam has been used as an energy resource for about a century. This energy was first harnessed for the generation of electricity at Larderello, Italy, in 1904, and has since been developed in Japan, Russia, New Zealand, the Phillipines, and at the Geysers in northern California.

Underground dry steam is relatively rare, but is the most desirable from the standpoint of power generation. More commonly, energy reaches the surface as superheated water and steam. In some cases, the water is so pure that it can be used for irrigation and livestock; in other cases, it is loaded with corrosive, scale-forming salts. Utilization of the heat from contaminated geothermal water generally requires that the water be reinjected into the hot formation after heat removal to prevent contamination of surface water.

The utilization of hot rocks for energy requires fracturing of the hot formation, followed by injection of water and withdrawal of steam. This technology is still in the experimental state, but promises approximately ten times as much energy production as steam and hot-water sources. Land subsidence and seismic effects are environmental factors that may hinder the development of geothermal power. However, this energy source holds considerable promise, and its development continues.

The Sun: an ideal energy source

Solar power is an ideal source of energy that is unlimited in supply, widely available, and inexpensive. It does not add to the earth's total heat burden or produce chemical air and water pollutants. On a global basis, utilization of only a small fraction of solar energy reaching the earth could provide for all energy needs. This is still an enormous amount of land, and there are economic and environmental problems related to the use of even a fraction of this amount of land for solar energy collection.

Solar power cells (photovoltaic cells) for the direct conversion of sunlight to electricity have been developed and are widely used for energy in space vehicles. With present technology, however, they remain too expensive for large-scale generation of electricity, although the economic gap is narrowing. Most schemes for the utilization of solar power depend upon the collection of thermal energy followed by conversion to electrical energy. The simplest such approach involves focusing sunlight on a steam-generating boiler (see Illustration 8).

Parabolic reflectors can be used to focus sunlight on pipes containing heat-transporting fluids. Selective coatings on these pipes can be used so that most of the incident energy is absorbed. The direct conversion of energy in sunlight to electricity is accomplished by special solar voltaic cells. Such devices based on crystalline silicon have operated with a 15% efficiency for experimental cells and 11-12% for commercial units, at a cost of 25-50 cents per kilowatt-hour (kWh), about 5 times the cost of conventionally generated electricity. Part of the high cost results from the fact that the silicon used in the cells must be cut as small wafers from silicon crystals for mounting on the cell surfaces. Significant advances in costs and technology are being made with thin-film photovoltaics, which use an amorphous silicon alloy.

A new approach to the design and construction of amorphous silicon film photovoltaic devices uses three layers of amorphous silicon to absorb, successively, short wavelength ("blue"), intermediate wavelength ("green"), and long wavelength ("red") light, as shown in Figure 8. Thin-film solar panels constructed with this approach have achieved solar-to-electricity energy conversion efficiencies just over 10%, lower than those using crystalline silicon, but higher than other amorphous film devices.

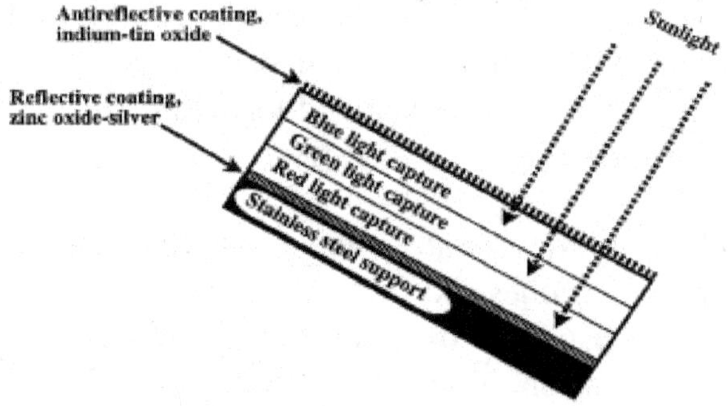

Fig. 147. *High-efficiency thin-film solar photovoltaic cell using amorphous silicon.*

The low cost and relatively high conversion efficiencies of these solar panels should enable production of electricity at only about twice the cost of conventional electrical power, which would be competitive in some situations A major disadvantage of solar energy is its intermittent nature. However, flexibility inherent in an electric power grid would enable it to accept up to 15% of its total power input from solar energy units without special provision for energy storage.

Existing hydroelectric facilities may be used for pumped-water energy storage in conjunction with solar electricity generation. Heat or cold can be stored in water, in a latent form in water (ice) or eutectic salts, or in beds of rock. Enormous amounts of heat can be stored in water as a supercritical fluid contained at high temperatures and very high pressures deep underground. Mechanical energy can be stored with compressed air or flywheels.

Hydrogen gas, H_2, is an ideal chemical fuel that may serve as a storage medium for solar energy. Solar-generated electricity can be used to electrolyze water:

$$2H_2O + \text{electrical energy} \rightarrow 2H_2(g) + O_2(g)$$

The hydrogen fuel product, and even oxygen, can be piped some distance and the hydrogen burned without pollution, or it may be used in a fuel cell. This may, in fact, make possible a "hydrogen economy." Disadvantages of using hydrogen as a fuel include its low heating value per unit volume and the wide range of explosive mixtures it forms with air. Although not yet economical, photochemical processes can be used to split water to H_2 and O_2 that can be used to power fuel cells.

No really insurmountable barriers exist to block the development of solar energy, such as might be the case with fusion power. In fact, the installation of solar space and water heaters became widespread in the late 1970s, and research on solar energy was well supported in the U.S. until after 1980, when it became fashionable to believe that free-market forces had solved the "energy crisis." With the installation of more heating devices and the probable development of some cheap, direct solar electrical generating capacity, it is likely that during the coming century solar energy will be providing an appreciable percentage of energy needs in areas receiving abundant sunlight.

The Surprising Success of Wind Power

Wind power is mentioned here because it is an indirect form of solar energy. During the 1990s, wind power emerged as a cost-competitive source of renewable energy with a remarkably high growth rate. Denmark has led other countries in establishing wind power as a significant fraction of its electrical generating capacity.

In October 1996, the largest wind farm established up to that time in Europe was opened in Carno, Wales, by National Wind Power, Ltd. This was the 32nd wind farm in Britain, which was already generating enough electricity from wind to power 150,000 homes. Producing 33.6 megawatts of power, the 3-bladed turbines used to generate power at the Welsh facility are 56 meters in diameter and are mounted on towers 64 m high.

Northern regions, including parts of Alaska, Canada, the Scandanavian countries, and Russia often have consistently strong wind conditions conducive to the generation of wind power. Isolation from other sources of energy makes wind power attractive for many of these regions.8 Severe climate conditions in these regions pose special challenges for wind generators. One problem can be the buildup of rime consisting of ice condensed directly on structures from supercooled fog in air.

Energy from biomass

All fossil fuels originally came from photosynthetic processes. Photosynthesis does hold some promise of producing combustible chemicals to be used for energy production and could certainly produce all needed organic raw materials. It suffers from the disadvantage of being a very inefficient means of solar energy collection (a collection efficiency of only several hundredths of a percent by photosynthesis is typical of most common plants). However, the overall energy conversion efficiency of several plants, such as sugarcane, is around 0.6%.

Furthermore, some plants, such as *Euphorbia lathyrus* (gopher plant), a small bush growing wild in California, produce hydrocarbon emulsions directly. The fruit of the Philippine plant, *Pittsosporum reiniferum*, can be burned for illumination due to its high content of hydrocarbon terpenes, primarily pinene and myrcene. Conversion of agricultural plant residues to energy could be employed to provide some of the energy required for agricultural production. Indeed, until about 80 years ago, virtually all of the energy required in agriculture—hay and oats for horses, home-grown food for laborers, and wood for home heating—originated from plant materials produced on the land. (An interesting exercise is to calculate the number of horses required to provide the energy used for transportation at the present time in the Los Angeles basin. It can be shown that such a large number of horses would fill the entire basin with manure at a rate of several feet per day.)

Annual world production of biomass is estimated at 146 billion metric tons, mostly from uncontrolled plant growth. Many farm crops and trees can produce 10-20 metric tons per acre per year of dry biomass, and some algae and grasses can produce as much as 50 metric tons per acre per year. The heating value of this biomass is 5000-8000 Btu/lb for a fuel having virtually no ash or sulfur. Current world demand for oil and gas could be met with about 6% of the global production of biomass. Meeting U.S. demands for oil and gas would require that about 6-8% of the land area of the contiguous 48 states be cultivated intensively for biomass production. Another advantage of this source of energy that is becoming increasingly important as more is learned about potential greenhouse warming is that use of biomass for fuel would not add any net carbon dioxide to the atmosphere.

As it has been throughout history, biomass is significant as heating fuel, and in some parts of the world is the fuel most widely used for cooking. For example, as of the early 1990s, about 15% of Finland's energy needs were provided by wood and wood products (including black liquor by-product from pulp and paper manufacture), about 1/3 of which was from solid wood. Despite the charm of a wood fire and the sometimes pleasant odor of wood smoke, air pollution from wood-burning stoves and furnaces is a significant problem in some areas. Currently, wood provides about 8% of world energy needs. This percentage could increase through the development of "energy plantations" consisting of trees grown solely for their energy content.

Seed oils show promise as fuels, particularly for use in diesel engines. The most common plants producing seed oils are sunflowers and peanuts. More exotic species include the buffalo gourd, cucurbits, and Chinese tallow tree.

Biomass could be used to replace much of the 100 million metric tons of petroleum and natural gas currently consumed in the manufacture of primary chemicals in the world each year. Among the sources of biomass that could be used for chemical production are grains and sugar crops (for ethanol manufacture), oilseeds, animal byproducts, manure, and sewage (the last two for methane generation).

The biggest potential source of chemicals is the lignocellulose making up the bulk of most plant material. For example, both phenol and benzene might be produced directly from lignin. Brazil has had a program for the production of chemicals from fermentation-produced ethanol.

Gasohol

A major option for converting photosynthetically produced biochemical energy to forms suitable for internal combustion engines is the production of either methanol or ethanol. Either can be used by itself as fuel in a suitably designed internal combustion engine. More commonly, these alcohols are blended in proportions of up to 20% with gasoline to give **gasohol**, a fuel that can be used in existing internal combustion engines with little or no adjustment.

Gasohol boosts octane rating and reduces emissions of carbon monoxide. From a resource viewpoint, because of its photosynthetic origin, alcohol may be considered a renewable resource rather than a depletable fossil fuel. The manufacture of alcohol can be accomplished by the fermentation of sugar obtained from the hydrolysis of cellulose in wood wastes and crop wastes. Fermentation of these waste products offers an excellent opportunity for recycling. Cellulose has significant potential for the production of renewable fuels.

Ethanol is most commonly manufactured by fermentation of carbohydrates. Brazil, a country rich in potential to produce biomass such as sugarcane, has been a leader in the manufacture of ethanol for fuel uses, with 4 billion liters produced in 1982. At one time Brazil had over 450,000 automobiles that could run on pure alcohol, although many of these were converted back to gasoline during the era of relatively low petroleum prices since about 1980. Significant amounts of gasoline in the United States are supplemented with ethanol, more as an octane-ratings booster than as a fuel supplement.

Methanol, which can be blended with gasoline, can also be produced from biomass by the destructive distillation of wood or by converting biomass, such as wood, to CO and H2, and synthesizing methanol from these gases.

Future energy sources

A tremendous potential exists for applying the practice of industrial ecology to lower the burden on virgin raw materials and sources of energy. These approaches include using less material (dematerialization), substitution of a relatively more abundant and safe material for one that is scarce and/or toxic, extracting useful materials from wastes (waste mining), and recycling materials and items. Properly applied, these measures can not only conserve increasingly scarce raw materials, but can increase wealth as it is conventionally defined. Corresponding measures can also be applied to energy resources. In recent decades energy conservation ("de-energization"); substitution of energy sources, such as inexhaustible wind power for coal in the generation of electricity; and burning of municipal refuse to raise steam for electricity generation have reduced the need to utilize diminishing fossil energy resources and to build new power plants.

The greatest potential for extending material resources is by recycling through the practice of industrial ecology. In a sense, too, energy resources can be recycled by using otherwise waste materials to generate energy and by using heat that might otherwise go to waste for beneficial purposes, such as heating buildings. Materials vary in their amenability to recycling. Arguably the most recyclable materials are metals in a relatively pure form. Such metals are readily melted and recast into other useful components.

Among the least recyclable materials are mixed polymers or composites, the individual constituents of which cannot be readily separated. The chemistry of some polymers is such that, once they are prepared from monomers, they are not readily broken down again and reformed to a useful form.

Possible Future Sources of Energy

Source	Principles
Coal conversion	Manufacture of gas, hydrocarbon liquids, alcohol, or solventrefined coal (SRC) from coal
Oil shale	Retorting petroleum-like fuel from oil shale
Geothermal	Utilization of underground heat
Gas-turbine electric	Utilization of hot combustion gases in a turbine, followed by a topping cycle involving steam generation
MHD	Electricity generated by passing a hot gas plasma through a magnetic field
Thermionics	Electricity generated across a thermal gradient
Fuel cells	Conversion of chemical to electrical energy
Solar heating and cooling	Direct use of solar energy for heating and cooling through the application of solar collectors
Solar cells	Use of silicon semiconductor sheets for the direct generation of electricity from sunlight
Solar thermal electric	Conversion of solar energy to heat followed by conversion to electricity
Wind	Conversion of wind energy to electricity
Ocean thermal electric	Use of ocean thermal gradients to convert heat energy to electricity
Nuclear fission	Conversion of energy released from fission of heavy nuclei to electricity
Breeder reactors	Nuclear fission combined with conversion of nonfissionable nuclei to fissionable nuclei
Nuclear fusion	Conversion of energy released by the fusion of light nuclei to electricity
Bottoming cycles	Utilization of waste heat from power generation for various purposes
Solid waste	Combustion of trash to produce heat and electricity
Photosynthesis	Use of plants for the conversion of solar energy to other forms by a biomass intermediate
Hydrogen	Generation of H2 by thermochemical or photochemical means for use as an energy-transporting medium

2.17 EXERCISE QUESTIONS

1. Discuss some of the advantages of physical-chemical treatment of sewage as opposed to biological wastewater treatment. What are some disadvantages?
2. Justify describing the sun as "an ideal energy source." What are two big disadvantages of solar energy?

3. What are some of the greater implications of the use of biomass for energy? How might such widespread use affect greenhouse warming? How might it affect agricultural production of food?
4. As it is now used, what is the principle or basis for the production of energy from uranium by nuclear fission? Is this process actually used for energy production? What are some of its environmental disadvantages? What is one major advantage?
5. Compare the most popular alternative sources of energy in the sense "price-effectiveness".
6. Characterize the hot problems related to waste generation in the whole African region and in Tanzania in particular.
7. Describe the generation of municipal solid waste in your country and compare these values with those typical for the whole African region.
8. Characterize the typical chemical composition of municipal wastewater and point out the possible sources of these contaminants.
9. Discuss the typical chemical composition of industrial wastewater and point out the possible sources of these contaminants.
10. Discuss the behavior of mercury in the modern environment. Why is this metal of special attention for environmental chemistry?
11. Discuss the chemistry of organic particle formation and explain the role of fuel combustion in these processes.
12. Characterize the chemical reactions that occur during dissolution of chlorine in treated water.
13. Discuss the advantages and disadvantages of chlorine application for treatment of raw water.
14. Compare nuclear energy to fossil fuel energy sources and defend or refute the statement, "Nuclear energy, with modern, safe, and efficient reactors, is gaining increasing attention as a reliable, environmentally friendly energy source."
15. Give the definition of air pollution. What are the primary pollutants, secondary pollutants, and precursors? Present examples.
16. Characterize the role of transport in urban air pollution in various African cities. Explain why two-wheeler engines emit more pollutants than car engines.
17. Explain the role of nitrogen dioxide and nitrate radicals in photochemical smog reaction during daytime and nighttime.
18. Show the profile of ozone concentrations in the stratosphere and present the explanation of such a distribution.
19. Explain the role of ozone in existence of live-beings on the Earth. Discuss the role of nitrous oxides in stratospheric chemistry of ozone. Discuss the relative importance of this species in ozone destruction.
20. Characterize water hardness and its origin
21. What are the main steps in treating raw water for drinking water distribution systems in Asia? Do all of them (or only some of them) need to be applied?
22. Explain the chemical reactions that occur at the stage of aeration during water treatment.
23. What are the reasons for wide application and sales of bottled drinking water in the African region?
24. Characterize the advantages of recycling and estimate the role of waste recycling in the waste disposal system. Explain the RR conception.
25. What are some characteristics of radionuclides that make them especially hazardous to humans?
26. What are the main features of the carbon cycle? Describe how the oxygen cycle is closely related to the carbon cycle.
27. Describe the role of organisms in the nitrogen cycle. Discuss the biogeochemical cycle of nitrogen in mixed ecosystems.

CURRICULUM VITAE OF THE AUTHORS

1. 0 Dr. H.C. Ananda Murthy

1.1Personal Information

Name : Dr. H.C. Ananda Murthy
Date of Birth : 06-07-1973 Age: 44 years Gender: Male
E-mail Id & Contact No. anandkps350@gmail.com, anandkps7@yahoo.com
& +255 768 406 138
Nationality : Indian
Marital Status : Married Number of Children: 2

1.2 A: Academic Information (Completed studies at Government Universities in India)

Qualification	Descipline/ subject	School/college/ University	Year of passing	Class with Percentage
B.Sc.	PCM	Bangalore University	1995	First Class 60.16%
M.Sc.	Chemistry (Inorganic Chemistry)	Bangalore University	1997	First Class 66.60%
M.Phil.	Chemistry (Composite materials)	Gharathidasan University, Tirchy	2996	First Class 76.40%
Ph.D.	Chemistry (Electrochemistry of composite materials) Physical Chemistry)	VT University, Kamataka	2012	

1. 2.B: Title of the Ph.D. Degree

A Study on the effect of Ceramic reinforcement on the corrosion behavior of Aluminium metal matrix composites.

1. 2.C: Major Courses Studied

SL.No.	Degree	Couses Leaned
1.	B.Sc	Physics, Chemistry and Mathematics
2.	M.Sc	Coordination Chemistry, bio-Inorganic Chemistry, Organometallic Chemistry & Catalysis, Molecular Spectroscopy, Solid State Chemistry, Industrial and Nuclear Chemistry, Analytical Chemistry.
3	M.Phil.	Apectroscopy, Inorganic Chemistry, Research Methodology, Corrosion Theory.
4	Ph.D	Corrosion Science, Mateial Science, Instrumental Methods of Analysis, Research Methodology.

1.3 A: Teaching Experience: Total 19 years

Period	Institution	Designation	Total Year of teaching
Before Ph.D (1997-2012)	Vivesvaraya Technological University, India	Lecturer (9 years) Senior Lecturer (3 years) Asst. Prf. & HOD (3 years)	15 years
After Ph.D (2012-2016)	Vivesvaraya Technological University, India The University of Dodoma, Tanzania	Asst. Prof. & HOD (1 year) Lecturer (3 Years)	4 years

1. 3. B: Major Courses Taught for U G Progam (Taught 19 years)

1. Coordination and Organometallic Chemistry
2. Transition Elements Chemistry
3. Organomitallic Chemistry
4. Electrochemistry and Corrosion Protection
5. General Chemistry
6. Engineering Chemistry

1. 3.C: Major Courses Taught for P G Progam (Taught for 1 year)

1. Advanced Coordination Chemistry
2. Surface Chemistry and Corrosion Science
3. Advanced Spectroscopy
4. Advanced Environmental Chemistry
5. Theoretical and Applied Electrochemistry

1.4 A: Scientific Research Publications: Total Number of Publications = 17

National and International Journal Papers = 13, Conference Papers = 4

SL.No.	Title of the Published Papers	Cooperated authors	Jornal (Volume, Issues, date)
1	Adsorption of Mercury (II) from Aqueous Solution Using Gum Acacia-Silica Nanocomposite: Kinetics, Equilibrium and Thermodynamics Studies	Somit Kumar Singh Vandana Singh	Advanced Material letters VBRI Press, Sweden -2016 (Accepted for publication in March, 2016)
2	Electoanalytical study on the corrosion behavior of $Ti0_2$ particulate reinforced Al 6061 composites	Single author	Material Science Research India, Vo. 12(2), 112-126 (2015). Doi.org/10.13005/msri/1205
3	Influence of TiC particulate reinforcement on the corrosion behavior of Al 6061 Metal matrix composites	Somit K. Singh	Advanced Materials letters VBRI Press, Sweden – 2015, 6 (7), 633-640.doi:

			10.5185/amlett.2015.5654
4	Effect of TiN particulate reinforcement on corrosion behavior of Aluminium 6061 composited in chloride medium	V Bheemaraju C Shiva Kumara	Bulletin of Material Science-Vol.36,No.6,Nov. 2013, pp. 1057. Indian Academy of Sciences (Springer)
5	Electrochemical investigation of corrosion inhibition of AA6063alloy in IM hydrochloric acid using Schiff base compounds	V Bheemaraju P F Sanaulla H B Lokesh	IOSR Journal of Applied Chemistry (IOSR-JAC) ISSN: 2278-5736. Vlume 2, Issue 5 (Nov.-Dec. 2012), Pp 37-47
6	Electrochemical Behaviour of AA6063 Alloy in Hydrochloric Acid using Schiff Base Compounds as Corrosion Inhibitors	V Bheemaraju P F Sanaulla H B Lokesh	International Journal of Engineering Research and Applications (IJERA) ISSN: 2248, Vol. 2, Issue 5, 2012.
7	Corrosion inhibition of AA 6061 and AA6063 alloys in hydrochloric acid media by Schiff base compounds	V Bheemaraju P F Sanaulla	Journal of Chilean Chemical Society, 57, N° 4 (2012)
8	Corrosion Characterization of TiO_2 Particulate Reinforced Al-6063 Composites in Chloride and Nitrate Media	V Bheemaraju P F Sanaulla	Asian Journal of Chemistry, ISSN 0970-7077, 2011, Volume 23, Issue 4, 1664 – 1668.
9	Influence of chlorides, nitrate and sulphate media on corrosion behaviour of TiO_2 particulate reinforced Al-6061 composites	V Bheemaraju P F Sanaulla	PortugaliaeElectrochimicaAc ta, ISSN 1647-1571, *2010, 28(5), 309-320*
10	Corrosion behaviour of TiO_2 particulate reinforced Al-6063 composites in sodium sulphate medium	V Bheemaraju P F Sanaulla	International Journal of Applied Chemistry, ISSN 0973-1792, Vol. 6, 2, (2010) p. 225
11	A Study on the effect of TiO_2 particulate on the corrosion behaviour of Al-6063 based composites	V Bheemaraju H B Lokesh	Material Science, An Indian Journal, MSAIJ, 4(4), June 2008 [328-334]
12	Estimation of copper in brass by iodo-potentiometric technique of analysis	O G Palanna	*Analytical Chemistry*, An Indian Journal, ACAIJ, 7(9),

			August 2008 [694-702]
13	Estimation of copper in copper salts at different pH conditions and buffer media by iodo-potentiometric technique of analysis	O G Palanna	*Analytical Chemistry*, An Indian Journal, ACAIJ, 7(9), July 2008 [684-693]

1.4. B: Paper Presentations in National/International Conferences

14 **Anand Murth H. C, Kusuma, ShobhaBhaskara, Bheema Raju V**
Corrosion behavior of 6061 Al and its TiB_2 particulate reinforced composites in chloride medium, International Conference on Electrochemical Science and Technology (ICONEST -2014), August 7-9, ECSI, Bengaluru, India.

15. **H.C. Ananda Murthy,** Suba Rameshand V. BheemaRaju Evaluation of corrosion behaviour of TiO_2 particulate reinforced Al-6061 composites by polarisation techniques**, National Conference on Recent Advances in Chemical and Environmental Sciences (NCRACES- 2011)** held on 28 & 29th of Dec 2011.

16. **H.C. Ananda Murthy**, O.G.Palanna, T. Jeevananda, Subaramesh, RamanathPrabhu, N. Suma A Novel iodo-Potentiometric method of estimation of percentage of MnO_2 in Pyrolusite, **National Conference on Recent Trends in Materials Chemistry and Engineering (RTMCE-2011)**, September 29-30, 2011, Department of Chemistry, RNS Institute of Technology, Bangalore 560098.

17. **H.C. Ananda Murthy** and V. BheemaRaju,
Corrosion behaviour of TiO_2 particulate reinforced Al-6061 in chloride, nitrate, sulphate and acidic media, International conference on Recent Advances in Industrial Electrochemical Science and Technology (ICRAIEST-2009), November 2009, Department of Chemistry, Mangalore University, Mangalagangothri, Mangalore.

2. Seminars and Training Programmes

Seminars/Course	**Duration**	**Organisers**
Workshop on "Research Methodologies and Report Writing	2 days	Visvesvaraya Technological University, Mysore on May 28th and 29th 2009
TEQUIP sponsored Short Term Training Programme	5 days	BMS college of Engineering And Dr. A I T, Bangalore from 26th 2006.
Workshop on "Industrial Corrosion Awareness & Prevention"	1 day	Mangalore University, Mangalagangothri, Mangalore on 25th February 2006.
AICTE-ISTE Sponsored short tern training programme	5 days	from 23rd to 27th August 2004 at Dr. Ambedkar Institute of Technology, Bangalore.

3. Written Books,Book Chaper and Compendia

Title (Written)	Publisher	Year
1. Corrosion Science for University Students	Lulu Press, Inc. 3101 Hillsborough St, Raleigh, NC 27607. ISBN: 978-1-329-97036-6	2016
2. Corrosion Behaviour of Reinforced Aluminium MMCs	LAMBERT Academic Publishing, Germany. ISBN: 978-3-659-56260-0	2014
3. Electrochemistry Chemistry and Corrosion protection (CH 314) for 3rd Year B.Sc. course.	The university of Dodoma, Tanzania.	2014
4. General Chemistry (CH 111) for 1st Year B.Sc. course.	The university of Dodoma, Tanzania.	2014
5. Coordination chemistry and Organo-metallic chemistry (CH 221) for 2nd Year B.Sc. course.	The university of Dodoma, Tanzania.	2014
6. Organometallic chemistry (CH 310) for 3rd year B.Sc. course.	The university of Dodoma,Tanzania.	2014
7. Engineering Chemistry	Wiley-India ISBN: 978-81-265-1988-0	2011

1.5 Details of Scholarship/Awards/Honours

i. Prathibha puraskar-2012 by Devanga Employees Association, Bangalore.

ii. NagarikaSanman- 2012 by Dr AIT employees welfare association, Bangalore.

1.6 Passport Details

Passport Number: K8778065
Date of Issue: 17/09/2012 Place of Issue: BANGALORE
Date of Expiry: 16/09/2022 Passport Validity: 10 years

1.7 DETAILS OF EXTRA-CURRICULAR ACTIVITIES

<u>The Innovation</u>: JAL-BHOOMI -THE AMPHIBIOUS CYCLE

I have successfully completed a project titled "Amphibious bicycle" sanctioned by National Innovation Foundation of India (NIF) with Three students of 7th semester Electronics and communications Engineering Branch in the year 2008 in RNS Institute of Technology. This Amphibious bicycle was displayed in VahanYatra held at Karnataka Trade Promotion Centre, Near ITPL, Bangalore in December 2008.

The amphibious bicycle comprises of a conventional bicycle modified with two extra attachments that enables it to run on both water and land.

I certify that the information given above is true, complete and correct to the best of my knowledge and belief.

Place: Dodoma, Tanzania.

2. Prof. Said Ali Hamad Vuai

1. BIODATA

1.1 Name : Said Ali Hamad Vuai

1.2 Date ofbirth : 09/11/1972

1.3 Sex : Male

1.4 Place of Birth : Kitambuu, Pemba

1.5 Nationality : Tanzanian

1.6 Marital status : Married

2.0 EDUCATIONBACKGROUND

(Start with the most recent awards)

2.1. June2013-April2014. University of Siena, Siena, Italy. Post-Doctoral on Environmental Sustainability Assessment

2.2.Oct. 2001-Sept.2004. University of the Ryukyus, Nishihara-cho, Okinawa, Japan. PhD in Marine and Environmental Sciences (Chemistry major).

2.3. Oct.1999-Sept. 2001,University of the Ryukyus, Nishihara-cho, Okinawa, Japan M Sc. in Chemistry, Biology and Marine Sciences (Chemistry major)

2.4. Oct.1993-June1997.University of Dar es Salaam, Tanzania B Sc. in Chemical and Processing Engineering.

a. Jan.1989-Nov.1992 Fidel Castro Secondary School, Vitongoji, Pemba, Zanzibar Certificate of Secondary Education & Advance certificate of Secondary Education

2.6. Jan. 1978- Nov. 1988 Mchangamdogo School, Mchangamdogo, Pemba Zanzibar

3.0 WORKEXPERIENCE/EMPLOYMENT

3.1 Record at the University of Dodoma

3.1.1. 10th April2013 to date Associate Professor, College of Natural and Mathematical Sciences

3.1.2. 6th Oct. 2008-9th April, 2013. Senior Lecture, College of Natural and Mathematical Sciences.

3.1.2. 4th Dec. 2008-30th June,2011. Head of Department of Physical Sciences

3.1.3 1st July. 2011 to date Dean, School of Mathematical Sciences

3.2 Work experience elsewhere

3.2.1 Dec. 2004– Oct. 2008 Lecturer Department of Sciences, The State University of Zanzibar (SUZA).

3.2.2.March. 1999-Sept. 1999 Quality Control Manager Fahari Beverages Ltd Bokoba Branch.

4.0 MEMBERSHIP IN COMMITTEES/LOCAL AND INTERNATIONAL BODIES/ORGANIZATIONS.

4.1 University Committee/Boards
4.1.1 University of Dodoma Senate since July2011
4.1.2 University of Dodoma Advisory Committee since July 2011
4.1.3 College of Natural and Mathematical Sciences since July 2011
4.1.4 Board of School of Mathematics Sciences since July 2011
4.1.4 Board of School of Physical Sciences since July 2011
4.1.6 University of Dodoma Human Resource Committee since July2011
4.1.5 Board of School of Natural Sciences and Mathematics 2009-2011
4.1.6 Senate of Research and Publication 2009-2010

4.2 National Committee/Boards:
4.2.1 National Stakeholder advisory Forum for Africa Stockpile Program
4.2.2 Public Health Research Group

4.3 International Committees/Board:
4.3.1. West Indian Ocean Marine Research Association (WIOMSA since2004
4.3.2. Society of Environmental Toxicology and Chemistry(SETAC) since2006

5.0 RESEARCH PROFILE

Date	Theme	Funding Agency
5.1. 2006-2008	Impact of Tourism on Child and Zanzibar	CODESRIA
5.2. 2006-2011	Socio-economic and ecological impact of Vicres periodic sedimentation of the livelihood of Flood plain Communities in the Lake Victoria Basin	COSTECH
5.3.2011	The potential of native Gracilaria seaweed an alternative cheap source of agar for tissue culture of vetetatively propagated crops in Tanzania	COSTECH

6.0 STUDENT'S SUPERVISION

6.1 Ali H.R. (2009). Determination of Chemical Composition of Cave water: A case study of Zanzibar Island, Tanzania. M. SC. Dissertation. Makerere University, Uganda.

6.2 Hamad B.A.(2010) Nutrients dynamic in Jozani Forest Reserve, Zanzibar.

M Sc. Dissertation. University of Dodoma6.3 Juma Mohammed (2011) Socio-economic and ecological importance of Bwawani Pond, Zanzibar. M Sc. Dissertation, University of Dodoma

6.4 Getachew R. Assessing domestic in water harvesting: Reliability, quality and point-of-use treatment systems in semi-arid city of Mekelle, Ethiopia. PhD Dissertation.

6.5 MorrisF.Hiji.AstudyonprocessofproductionofSulfuricacidfromSamena pyrites.

7.0 PUBLICATIONS

7.1. Publications in recognized/referred Journals

7.1.1 Abdallah H. Mtumwa , Edwin Paul and Said A.H. Vuai (2015) Determinants of under nutrition among Women in the Reproductive Age Group in Mainland-Tanzania. *South African Journal of Clinical Nutrition* (Accepted).

7.1.2 Fabrizio Saladini, Said A. Vuai, Benard Langat, Mathias Gustavsson, Richard Bayitse, Andrew Gidamis, Belmakki Mohammed, Amal Owis, Konanani Rashamuse, Daniel Sila, Simone Bastianoni (2015) Sustainability assessment of selected biowastes as feedstocks for biofuel and biomaterial production by emergy evaluation in five African countries. *Biomass & Bioenergy* (Accepted).

7.1.3 Getachew Redae Taffere, Yilma Seleshi, Abebe Beyene, Said A.H. Vuai, Janvier Gasana and Yilma Seleshi (2015)Location and time-specific investigation of roof rainwater quality is important to safeguard public health. Desalination and Water Treatment DOI: 10.1080/19443994.2015.1107757.

7.1.4Getachew Redae Taffere, Yilma Seleshi, Abebe Beyene, , Said A.H. Vuai, Janvier Gasana and Yilma Seleshi (2015) Reliability analysis of roof rainwater harvesting systems in a semi-arid region of sub-Saharan Africa: case study of Mekelle, Ethiopia. *Hydrological Sciences Journal* DOI:10. 1080/02626667.2015.1061195.

7.1.5 Said A. H. Vuai (2015) Aluminum, Silicon and Nutrients Characteristics in Precipitation of Semi-Arid Area in Dodoma Municipality, Tanzania. *International Journal of Environment and Toxicological Research* 92-100.

7.1.6 Morris Frank Hiji, Justin William Ntalikwa1, Said Ali Vuai (2014)Producing sulphuric acid in Tanzania and potential sources: A review. *American Journal of Chemistry and Applications*1(4): 40-44.

7.1.7 Said S. Bakari, Per Aagaard, Rolf D.Vogt, Fridtjov Ruden, Ingar Johansen and **Said Ali Vuai.** (*2013)* Strontium isotopes as tracers for quantifying mixing of groundwater in the alluvial plain of a coastal watershed, South-eastern Tanzania. *Applied Geochemistry,*(accepted).

7.1.8**Vuai, S. A. H.**, Mungai, N. W.and Ibembe, J. D.(2012) Effect of Land Use Activities on Spatial and Temporal Variation of Nutrients deposition in Mwanza Region: Implication to the Atmospheric loading to the Lake Victoria. *Atmospheric and Climate Sciences*(in press).

7.1.9 Hasan Rashid Ali, Nyakairu J. and **Said A. Vuai** *(2012).* Influence of tides to the cave water quality: A case study on Zanzibar Island, Tanzania. *International Journal of Current Research*Vol.4, Issue, 04,pp.327-334.

7.1.10 **Vuai, S. A. H.,** Mungai, N. W. and Ibembe, J. D.*(2012)*Spatial variation of Nutrients in Sondu-miriu and Simiyu-Duma Rivers: Implication on sources and factors influencing their transportation into the Lake Victoria. *Journal of Earth Sciences and Climate Change. 3:119.* doi:10.4172/2157-7617.1000119

7.1.11 **Vuai, S. A. H** *(2012).* Microbialand nutrient contamination of domestic well in Urban-West Region, Zanzibar, Tanzania. *Airand Water Born Diseases,*1:102. doi:10.4172/2167-7719.1000102.

7.1.12 Said Suleiman Bakari, Per Aagaard, Rolf D. Vogt, Fridtjov Ruden, Ingar Johansen and **Said Ali Vuai** (2012)**.** Delineation of groundwater proven an ceina coastal aquifer using statistical and isotopic methods, Southeast Tanzania. *Environmental Earth Sciences: 1-14.*

7.1.13 Ali, H.R., Nyakairu, G.W. and **Vuai, S. A. (2012**) Determination of Chemical Composition of Cave water: A case study of Zanzibar Island, Tanzania International Journal of Current Research. 4:327-334, 2012.

7.1.14 Nancy W. Mungai, Njue A. M., Abaya Samuel G.,**Vuai Said A. H**. and Ibembe John D. *(2011)* Periodic flooding and land use effects on soil properties in Lake Victoria basin *African Journal of Agricultural Research Vol. 6(16),*4613-4623.

7.1.15 JohnDanielBakibinga-Ibembe1,**VuaiA.Said**andNancyW.Mungai.(2011) Environmental laws and policies related to periodic flooding and sedimentation in the Lake Victoria Basin (LVB) of East Africa. *African Journal of Environmental Science and Technology Vol.5(5),*367-380.

7.1.16 **Said Ali Hamad Vuai** and Tokuyama,A. (2011) Trend of trace metals in precipitation around Okinawa Island, Japan. *Atmospheric Research*, 99, 80-84.

7.1.17**Said Ali Hamad Vuai**. (2010). Characterization of MS Wand related waste- derived composition Zanzibar municipality. *Waste Management and Research*, 28, 177-184.

7.1.18**Said Ali Vuai** and Akira(2007). Tokuyama. Solute generation and CO_2 consumption in the silicate rock area under subtropical climatic condition. *Chemical Geology* 236, 199–216.

7.1.19 Mohamed MkadamKombo, Tetsuya Yonaha, **Said Ali Vuai** and Akira Tokuyama (2005). Dissolved aluminum and silica release on the interaction of Okinawan subtropical red soil and seawater at different salinities: Experimental and fieldobservations. *Geochemical Journal,Vol.* 61pp. 591-601.

7.1.20 Mohamed Mkadam Kombo, **Said Ali Vuai**, Maki Ishiki and Akira Tokuyama (2005). Influence of salinity on pH and aluminum upon the interaction of acidic red soil with seawater.*JournalofOceanography,* Vol. 61, pp. 591-601.

7.1.21 **Said Ali Vuai**, Kazuyo Nakamura and Akira Tokuyama (2003). Geochemical characteristics of run of from acid sulfate soils in northern area of Okinawa Island, Japan, *Geochemical Journal*, Vol. 37, pp. 579-592.

7.1.22 **Said Ali Vuai,** Maki Ishiki and Akira Tokuyama. Acidification of fresh water by red soils in subtropical silicate rock area, Okinawa, Japan (2003). *Limnology,* Vol. 4. pp. 63-71.

7.1.23Said Suleiman Bakari, **Said Ali Vuai** and Akira Tokuyama (2004). Forest soil fertility in Okinawa Island, Subtropical Region in Japan. *Bulletin of Faculty of Science, University of the Ryukyus,* Vol. 78, pp.276-290.

7.1.24 Said Suleiman Bakari, **Said Ali Vuai**, F.L.Manuzi and Akira Tokuyama (2004). Assessment of Self-purification of Kizinga River-Dar es Salaam, Tanzania. *Bulletin of Faculty of Science, University of the Ryukyus,*Vol.78, pp.291-310.

7.1.25 Mohamed Mkadam Kombo, **Said Ali Vuai** and Akira Tokuyama(2003). Effect of ionic strength on aluminum and dissolve silicaunderred soil- seawater interaction. *Geochimicaet Cosmochimica Acta*, Vol.67, No18 (S1), pp. A230.

7.1.26 **Said Ali Vuai**, Kazuyo, Arakaki and Akira Tokuyama (2003). Dissolved Al, Fe, Mn, Cuand Znin surface and ground waters from the northern area of Okinawa Island, *Geochimicaet Cosmochimica Acta*, Vol.67, No18 (S1), pp. A 516.

7.1.27 Mohamed Mkadam Kombo, **Said AliVuai** and Akira Tokuyama(2003). Impact of acidic red soil on pH and aluminum under the soil-seawater interacting environment. *Bulletin of Faculty of Science, University of the Ryukyus*, No. 75 pp. 75-87.

7.1.28 **Said Ali Vuai**, Akira Tokuyama, Y. Tokashiki, and Moritaka Shimo, (2001). Particle size distribution, mineral sand chemical composition of red soil from silicate rock area of Central Okinawa. *Bulletin of Faculty of Science, University of the Ryukyus*, No. 71 pp. 71-77.7.1.25 **Said Ali Vuai**, Mariko Yonashiro and Akira Tokuyama (2000). Effect of Ionic strength on the absorbance in aluminum-Tiron complex. *Bulletin of Faculty of Science, University of the Ryukyus*, No. 70 pp. 97-102.

7.2 Publications in edited proceedings:

7.2.1 Bakibinga-Iembe, J.D., Mungai, N.W and Said A. Vuai (2011). Effect of Periodic flooding and sedimentation on Community Liveligoods in Sondu- Miriu Flood plain, Kenya, *Natural Resources Management and Land Use, Proceedings of the Cluster Workshop, Inter-University Council for East Africa, LakeVictoria Research Initiative. ISBN978-9970-452-02-6*

7.2.2 Ouma, K.O., Mungai, N.W., Said A. H. Vuai and Bembe, J.D.(2011) Spatial-Temporal Vairation of Nitrogen, Phosphorous and Sediments in surface run off and adjacent river system in Sondu-Miriu Bains, Kenya. *Natural Resources Management and Land Use, Proceedings of the Cluster Workshop, Inter- University Council for East Africa, Lake Victoria Research Initiative. ISBN978-9970-452-02-6*

7.3 Publication in Books

7.4 Thesis and dissertations:

7.4.1 Influence of acidic red soil and acid sulfate soil on the quality of fresh water located in the silicate rock area, northern Okinawa Island.
PhD Thesis, University of the Ryukyus, Okinawa, Japan, September, 2004.

7.4.2 Effect of red soil on chemical composition of natural waters.
Master's thesis, University of the Ryukyus ,Okinawa, Japan, September, 2001.

7.4.3 Design, construction and evaluation of non-convecting, saltgradient solar pond for storage of thermal energy.
Graduationthesis, University of Dar es Salaam ,Tanzania, June 1997

7.5 Student compendia

7.5.1 Coordination Chemistry: teaching Compendium for Structural Inorganic Chemistry (CH 202) and Inorganic Chemistry II (CH 330) 2010.

7.5.2 Basic principles of inorganic Chemistry: Teaching compendium for Basic inorganic Chemistry

Presentations in scientific conferences and workshops:

7.6.1 **Vuai, S. A. H.,** Mungai, N. W. and Ibembe, J. D. *Socio-economic and Ecological impacts of periodic sedimentation on the livelihoods of flood plain communities of Lake Victoria.* Lake Victoria Research (Vic Res) Initiative *B*iennial Forum from 9th to 13th October, 2011, Monyonyo, Kampala, Uganda.

7.6.2 **Vuai, S.A. H**. *Spatial and Temporal variation of nutrient along Simiyu-Duma and Sondu Miriu River*. Land Use and Natural Resource Cluster workshop, 30th October to 2nd November, 2011, Nairobi, Kenya.

7.6.3 **Vuai, S.A.H**. Spatial Variation of nutrient pollution in the Lake Victoria Basin. *Stakeholders Workshop for Vicres Projec*t December 7th 2011, Hotel Barbados, Kampala. 6.4 **Vuai S. A.**, Ali, H.R. and Bakari, S.S. *Geochemical Characteristics of Sapaleotherms and Water in Cave from Zanzibar Island, Tanzania.* Int. Conf. on Geosciences for Global Development (Geo Dev), Dhaka, Bangladesh, October 26-31, 2009. Abstract Volume, 84 pp.

7.6.5 Bakari, S.S., Aagaard, P., Vogt, R.D., Ruden, F. and **Vuai, S.A**. *Geochemical Characterization of the Upper and Lower Aquifer System in the Coastal Area, Tanzania.* Int. Conf. on Geosciences for

Global Development (Geo Dev), Dhaka, Bangladesh, October 26-31, 2009. Abstract Volume, 63 pp.

7.6.6 **Said A. H. Vuai**, Mohammed A. Sheikh and T. Oomori. Occurrence of tributylt in (TBT)compounds in the major ports of Tanzania. International conference on pesticide use in developing countries: environmental fate, effects and public health implications, Arusha,16-20 October 2006.

7.6.7 Kombo M., Tetsuya Y., **Vuai, S**. and Tokuyama A. Effect of ionic strength on dissolved silica in the interaction of acidic red soils from Okinawa Island and seawater solutions. 51st Annual Meeting of Geochemical Society of Japan, Shizuoka University, September 20-22, 2004.

7.6.8SaidSuleimanBakari,**SaidAliVuai**andAkira Tokuyama. Influence of underlain rock on the dissolved organic nutrients and metal release in the forest soil in Okinawa Island, Japan. 51st annual meeting of Geochemical Society of Japan, Shizuoko University, September 20-22, 2004.

7.6.9 **Said Ali Vuai**, Kazuyo Arakaki and Akira Tokuyama. Dissolved Al, Fe, Mn, Cu and Zn in surface and ground waters from the northern area of Okinawa Island. 13th V. M. Gold schimidt International Conference, Kurashiki, Japan September 7-12, 2003.

7.6.10MohamedMkadamKombo,**SaidAliVuai**,MakiIshikiandAkira Tokuyama.

Influence of salinity on pH and aluminum up on the interaction of acidic red soil with seawater solutions. 4th International workshop on the oceanography and fishery in the East China Sea. University of the Ryukyus, Okinawa, Japan. November, 8-9, 2003.

7.6.11 Mohamed Mkadam Kombo, **Said Ali Vuai** and AkiraTokuyama. Effect of ionic strength on aluminum and dissolved silica underred soil-seawater interaction.13th V.M. Goldschimidt International Conference, Kurashiki, Japan, September 7-12, 2003.

7.6.12 **Said Ali Vuai** and Akira Tokuyama. Characteristic sofrun off from acid sulfate soils in the northern area of Okinawa Island. 49th Annual Meeting of Geochemical Society of Japan, Kagoshima University, September 25-27, 2002.

7.6.13 Mohamed Mkadam Kombo, **Said Ali Vuai** and Akira Tokuyama. Effect of acidic red soil on pH and chemical composition of seawater. 49th Annual Meeting of Geochemical Society of Japan, Kagoshima University, September 25-27, 2002.

7.6.14 AkiraTokuyama, Mariko Yonashiro and **Said Ali Vuai**. 47 Annual Meeting of Geochemical Society of Japan,Yamagata University, Yamagata, Japan, September 25-27, 2002.

7.7 Other writings including consultancy reports and extension materials/Submitted Manuscripts:

7.7.1. Getachew Redaie Taffere, Abebe Beyene Hailu, Said Vuai and Janvier Gassana and Yilma Sileshi. Reliability analysis of roof rain water harvesting systems in semi-arid area of Mekelle City, northern Ethiopia. Journal of Hydrology (2015).

7.7.2. Fabrizio Saladini, Said A. Vuai, Benard Langat, Mathias Gustavsson, Richard Bayitse, Andrew Gidamis, Belmakki Mohammed, Amal Owis, Konanani Rashamuse, Daniel Sila, Simone Bastianoni. Sustainability assessment by means of emergy evaluation of selected biowaste as feed stocks for biofuels and biomaterials production in five African countries. Biomas Bioenergy (2015).

REFERENCES

1. Housecraft, C.E. and Sharpe, A.G. Inorganic Chemistry, Prentice Hall, Second Edition 2004.
2. Douglas, B.E., Mcdaniels, D.H. and Alexander, J.J. Concepts and Models in Inorganic Chemistry, Wiley 1994.
3. Purcell, K.F. and Kotz, J.C., Inorganic Chemistry, Prentice Hall 1980.
4. Elschenbroich, C. and Salzer, A. Organometallics, A Concise Introduction, Wiley VCH Verlag, 2nd edition 1992.
5. House, J. Inorganic Chemistry, Elsevier Science and Technology 2008.
6. Elschenbroich, C. and Salzer, A. Organometallics, A Concise Introduction, Wiley VCH Verlag, 2nd edition 1992.
7. M. Gerloch, E.G. Constable, Transition Metal Chemistry, VCH Verlagsgesellschaft mbH, D-69451 Weinheim (Federal Republic of Germany), 1994.
8. S.F.A.Kettle, Physical Inorganic Chemistry, Spectrum, 1996.
9. R Gopalan and V. Ramalingam, Concise coordination chemistry, Vikas publishing house Pvt Ltd, 2014.
10. Electronic (Absorption) Spectra of 3d Transition Metal Complexes
    S. Lakshmi Reddy, Tamio Endo and G. Siva Reddy
11. ChandraShekara Reddy, Advanced Coordination Chemistry, Campus Books International, India, 2015.
12. P.S.Sindhu, Elements of Molecular Spectroscopy, New Age International Publishers, India, 2010.
13. R.C. Mehrotra, Organometallic chemistry, New Age International publishers, 2015.
14. Robert H Crabtree, The organometallic chemistry of transition elements, Wiley-Interscience, 2005.
15. Perez, N. Electrochemistry and Corrosion Science. Kluwer Academic Publishers, 2004.
16. Mars G Fontana, Corrosion Engineering Tata McGraw Hill, 2005
17. Engineering Chemistry, John Wiley International, 2010.
18. Koryta, J. Principles of Electrochemistry, 2nd ed. J. Wiley & Sons, 1993.
19. Plieth, W. Electrochemistry for Material Science. Elsevier B.V. 2008.
20. Comninellis, C. and Chen, G. Electrochemistry for the Environment. Springer, 2010.
21. H C Ananda Murthy, Corrosion Behaviour of Reinforced Al MMCs, Lambert Academic Publishing, Germany, 2014.
22. H C Ananda Murthy, Corrosion Science, Lulu Press Inc, USA, 2016.
23. Vitousek, P. M., Mooney, H. A., Lubchenco, J., and Melilli, J. M. Human domination of Earth's ecosystems. *Science*, **277**, July, 1997, 494--99.)
24. Various internet sources.

www.ingramcontent.com/pod-product-compliance
Lightning Source LLC
LaVergne TN
LVHW060448061126
851857LV00016B/1759
9789987070558